Engineering Principles - Welding and Residual Stresses

Edited by Kavian Omar Cooke and Ronaldo Câmara Cozza

Published in London, United Kingdom

Engineering Principles - Welding and Residual Stresses
http://dx.doi.org/10.5772/intechopen.96831
Edited by Kavian Omar Cooke and Ronaldo Câmara Cozza

Contributors
Bernardo Zuccarello, Asma Manai, Santanu Saha, Ronaldo Câmara Cozza, Vikas Verma, Jorge Humberto Luna-Domínguez, Wilian Silva da Silva Labiapar, Marcelo de Matos Macedo, João Pedro de Castro Valente Lenz Ferreira, Das Tanmoy, Ramy A. Ahmed Fouad, Essam Ahmed Ali, Ahmed Ramadan Shaaban, Ahmed E. El-Nikhaily, Behrouz Bagheri, Mahmoud Abbasi, Farzaneh Sharif, Lino A. S. Rodrigues, Pedro P.G. Ribeiro, Tárcio dos S. Cabral, Ednelson Da S. Costa, Eduardo De M. Braga, Malayathodi Rijesh, Akash Biradar, K.T. Thilagham, S. Muthukumaran, Manuela Zinke, Stefan Burger, Sven Juettner, Morteza Tayebi, Ali Rajaee, Hedayat Mohammad Soltani, Paolo Tripicchio, Salvatore D'Avella, Janusz Hołowaty, Bernard Wichtowski, Shamsu Tukur Tukur Auwal, Singh Ramesh, Tan Caiwang, Murtala Sule Dambatta, Robert M. Merton Stwalley III, Tyler J. McPheron

First published in London, United Kingdom, 2022 by IntechOpen
IntechOpen is the global imprint of INTECHOPEN LIMITED, registered in England and Wales, registration number: 11086078, 5 Princes Gate Court, London, SW7 2QJ, United Kingdom

British Library Cataloguing-in-Publication Data
A catalogue record for this book is available from the British Library

Additional hard and PDF copies can be obtained from orders@intechopen.com

Engineering Principles - Welding and Residual Stresses
Edited by Kavian Omar Cooke and Ronaldo Câmara Cozza
p. cm.
Print ISBN 978-1-80355-384-9
Online ISBN 978-1-80355-385-6
eBook (PDF) ISBN 978-1-80355-386-3

For EU product safety concerns:
IN TECH d.o.o., Prolaz Marije Krucifikse Kozulić 3, 51000 Rijeka, Croatia,
info@intechopen.com or visit our website at intechopen.com.

Meet the editor

Dr. Kavian Omar Cooke joined the University of Bradford, England, in 2017, where he is an Associate Professor of Material Science and Engineering and Director of Post Graduate Research Studies. Before coming to the University of Bradford, Dr. Cooke served as Programme Director of Mechanical Engineering at the University of Technology, Jamaica. He was awarded a Ph.D. in Materials Engineering from the University of Calgary in 2011, which focused on transient liquid phase diffusion bonding of particle-reinforced aluminum alloys. His qualifications also include a BEng with honours and an MPhil from the University of Technology, Jamaica in Mechanical Engineering and an MBA in Management from Patten University, California. Dr. Cooke is a reviewer for several international journals. In 2013, he won the prestigious President Research Excellence Award from the University of Technology, Jamaica, for research excellence in surface and interface engineering. Dr. Cooke is also a chartered engineer (CEng), a member of the Institution of Engineering and Technology (IET), a fellow of the UK Higher Education Academy (FHEA), and a member of the Caribbean Academy of Science.

Ronaldo Câmara Cozza received a degree in Mechanical Engineering from the University Center of the Educational Ignatius Foundation "Padre Saboia de Medeiros" (FEI), Brazil, in 2002, and a degree in Mathematics from the Methodist University of São Paulo in 2005. He obtained master's and doctorate degrees from the Department of Mechanical Engineering, Polytechnic School, the University of São Paulo in 2006 and 2011, respectively. In 2015, he completed a post-doctorate in Metallurgical and Materials Engineering from the same university. Currently, Dr. Cozza is a full professor at the University Center FEI and the "Paula Souza" State Center of Technological Education (CEETEPS), both in São Paulo, Brazil. He teaches courses on Technical Drawing, Mechanical Processes of Manufacturing, Machine Kinematics and Dynamics, and Tribology. His research interests include the fundamentals and tribological behavior of materials under conditions of micro-abrasive wear.

Contents

Preface

Over the last decade, there has been substantial development of welding technologies for joining advanced alloys and composites demanded by an evolving global manufacturing sector.

The evolution of these welding technologies has been substantial and finds numerous applications in engineering industries. It is driven by our desire to reverse the impact of climate change and fuel consumption in several vital sectors.

In many engineering situations, residual stresses and dynamic external loads can cause the mechanical components to become unusable at an early stage of work. However, depending on their nature – if "tensile" or "compressive" – and their magnitude, residual stresses can also be beneficial, exemplifying compressive residual stresses acting on thin films, which increase their wear resistance. This book reviews the most recent developments in this field. It is organized into three sections: "Principles of Welding and Joining Technology", "Microstructural Evolution and Residual Stress" and "Applications of Welding and Joining." Chapters in these parts address such topics as stresses in welding, tribology, thin-film metallurgical manufacturing processes, and mechanical manufacturing processes, as well as recent advances in welding and novel applications of these technologies for joining different materials such as titanium, aluminum, and magnesium alloys, ceramics, and plastics.

This book is a practical reference for scholars, engineers, and professionals who require an understanding of welding technologies and residual stresses.

Dr. Kavian Omar Cooke
Associate Professor in Material Science and Engineering,
Director of Post Graduate Research Studies,
University of Bradford,
Bradford, UK

Dr. Ronaldo Câmara Cozza
Assistant Professor,
Mechanical Engineering,
Department of Mechanical Manufacturing,
CEETEPS—State Center of Technological Education "Paula Souza",
Technology Faculty—FATEC,
University Center FEI – Educational Foundation of Ignatius Padre Sabóia de Medeiros,
Mauá, SP, Brazil

Section 1

Principles of Welding and Joining Technology

Chapter 1

Laser Welding

Morteza Tayebi, Hedayat Mohammad Soltani and Ali Rajaee

Abstract

Among novel techniques, laser welding is considered an adaptable and rapidly evolving method, finding innumerable applications in engineering industries. It is capable of welding narrow and impassable joints precisely, which can be operated under computer control. This chapter of the welding Handbook reviews the most recent developments in the field of laser welding, which are used for different applications. The first section provides an overview of laser welding basics and then moves on to the developments such as high-power CO_2 laser welding, laser micro-welding, and solid-state laser welding technologies. The second section underlines laser welding instruments used for joining different materials such as titanium, aluminum, and magnesium alloys, ceramics, and plastics. The third section highlights the advances in innovative laser welding methods with discussions on the applications of laser welding robots to improve the modeling and simulation of this technique. Lastly, the fourth section focuses on the use of laser welding technology in various industries including aerospace, automotive, railway, etc. The present Handbook is a practical reference for scholars, engineers, and professionals using laser welding techniques or requiring an understanding of the field of laser welding technologies.

Keywords: laser welding, CO_2 laser welding, laser micro-welding, solid-state laser welding

1. Introduction

Since 1962, there have been numerous reports on the metallurgical applications of lasers, including welding. The first laser welding operation was reported in 1963 for the butt welding of steel sheets using a pulsed ruby laser. In 1965, laser systems were developed to be used for welding electronic circuits, vacuum tubes interior, and other special applications that conventional technologies at the time could not provide. Due to the limited power source, until 1970, laser welding was limited to welding low-thickness materials at low speed. Later, high-power continuous laser sources were employed for welding procedures. In 1971, the presence of high-penetration welding or keyhole welding via laser beam welding (LBW) and electron beam welding (EBW) was reported. In other words, by this technique at high laser intensities (MW/cm^3), it is possible to make keyholes in metals, which is not practical for pulsed laser welding because the formation of keyholes needs prolonged times to form and does not occur simply [1, 2]. Since 1972, the use of continuous CO_2 lasers changed this direction and high-thickness stainless steel joints with full penetration were welded similar to those welded by the EBW process using the keyhole mode. These investigations were conducted in Japan, Germany, and the United Kingdom. Subsequent advances in CO_2 laser welding were focused on further optimization of laser resources, laser beam quality, and understanding the interaction of joint design, welding speed, radiation concentration, and the plasma effects on weldability. Studies in this field have continued using sources with the power of up to 12-15 kW. Using the neodymium-doped yttrium aluminum garnet (Nd:YAG) sources can be more applicable than CO_2 lasers due to their short wavelength and reduced radiance from the metallic materials (**Figure 1**). Currently, laser sources have many applications in the field of material processing. LBW as a new technology in recent years has found wide applications in various industries such as automotive, military, aerospace, shipbuilding, electronics, etc. Based on the assessments, it has been estimated that in the near future, the practice of diode lasers will have made great strides in LBW [3, 4].

The laser welding technique differs from conventional fusion welding methods in terms of equipment and operation. In laser welding, a thin and deep weld pool is achieved and the applied heat input to the joint is very low compared to conventional methods. This property allows the LBW to be used in certain applications where the welding depth requires a high width. The penetration depth and welding width can be

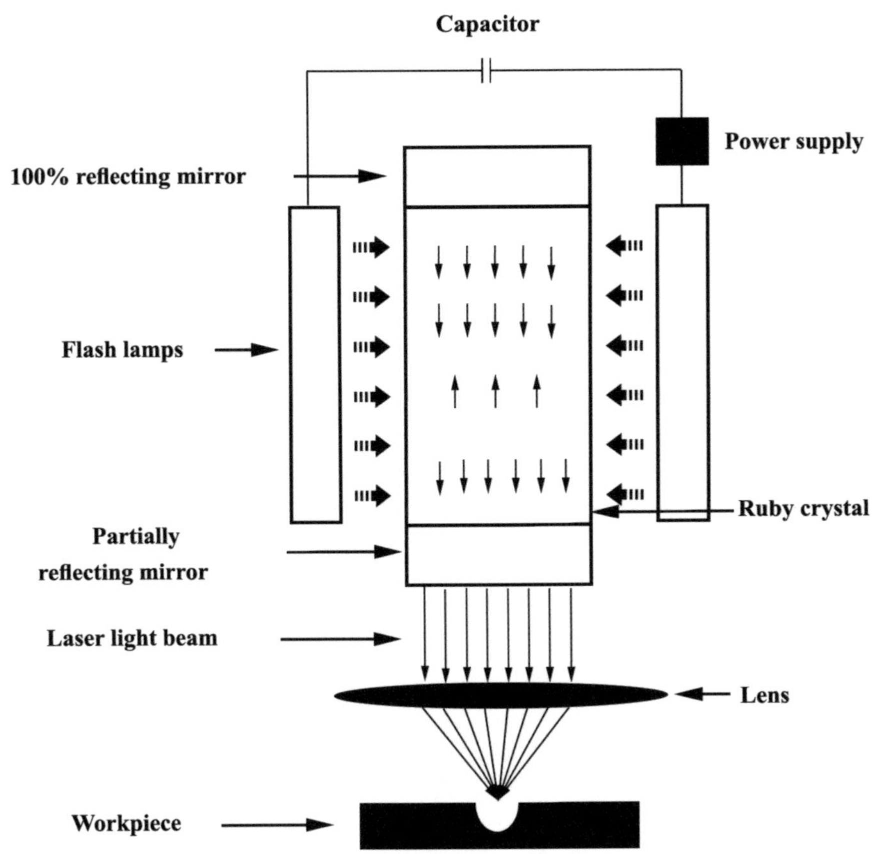

Figure 1.
Schematic of the laser beam production.

adjusted by controlling the laser power, changing the focal position of the beam, welding speed, distribution mode of energy transfer (pulsed or continuous mode), and shielding gas parameters. This enables the LBW to join and fabricate critical components with minimum risk. Features such as low welding width, high penetration depth, excellent joint strength, and low workpiece distortion in addition to fulfilling the need for low-weight joints with high corrosion resistance, proper weld appearance, low electrode consumption, eradicating machining process, and ability to weld unreachable areas have made industries interested in this technique.

There are two main methods for laser welding. The first is to move the workpiece rapidly underneath the beam to obtain continuous welding. The second route, which is more common, is to weld by irradiating a series of beams. Since during LBW both the melting and solidification processes occur in a few microseconds, almost no reactions take place between the melt and the surrounding atmosphere; therefore, generally using shielding gas is not necessary.

The most optimum joint design for LBW is the butt joint, however, due to the thickness limit, T-joints or corner joints are also desirable.

Hybrid processes that use a combination of laser and gas metal arc welding (GMAW) are also developed to be used in a fixed position. In addition, the equipment used to prepare the joint design is no longer needed. The alloys of the filler metals are specifically designed to make the joint physically uniform. In addition, the hybrid processes can significantly increase the production speed. Moreover, they also affect the penetration depth and sealing of the joint. Recent exclusive advances in the fabrication of laser diodes have provided a new opportunity to solve persistent industrial problems. These processes must be modified to be assimilated with the desired purposes [3, 4].

2. Laser welding process

LBW is one of the innovative methods of fusion welding. By irradiating a high-energy laser beam to the intended gap, the heat required to melt the edges and fill them is provided and the joining process is achieved. It should be noted that the equipment and operation of the LBW are dissimilar to the conventional fusion welding methods. In LBW, a thin and deep joint is achieved, and the heat input applied to the workpieces is so much lower than the conventional welding methods. This property allows LBW to be widely used in certain applications in which a high ratio of penetration depth to joint width is required [5–7].

LBW is a balance between heating and cooling in a certain volume of one or two solids, leading to the melting and solidification of a material. LBW is characterized by the creation of a molten region by the absorption of intensified radiation, which allows the molten pool to expand into the solid interface region, resulting in a continuous joint between the components. Failed joints occur when the molten area is too large, too small, or the material is excessively evaporated. Weld quality is analyzed according to the evaporation of alloying elements and the thermal gradient of the phenomena leading to the crack formation [8–11]. An imbalance in the volume and scale of the weld area leads to porosity. Achieving a balance between the heat input and output depends on the constant absorption of laser radiation and the uniform distribution of heat in the workpieces. The path of the laser beam to the weld pool is often disrupted by the accumulation of hot fumes at the beam focus point. In certain conditions, these hot fumes can turn into a plasma cloud that strongly affects the

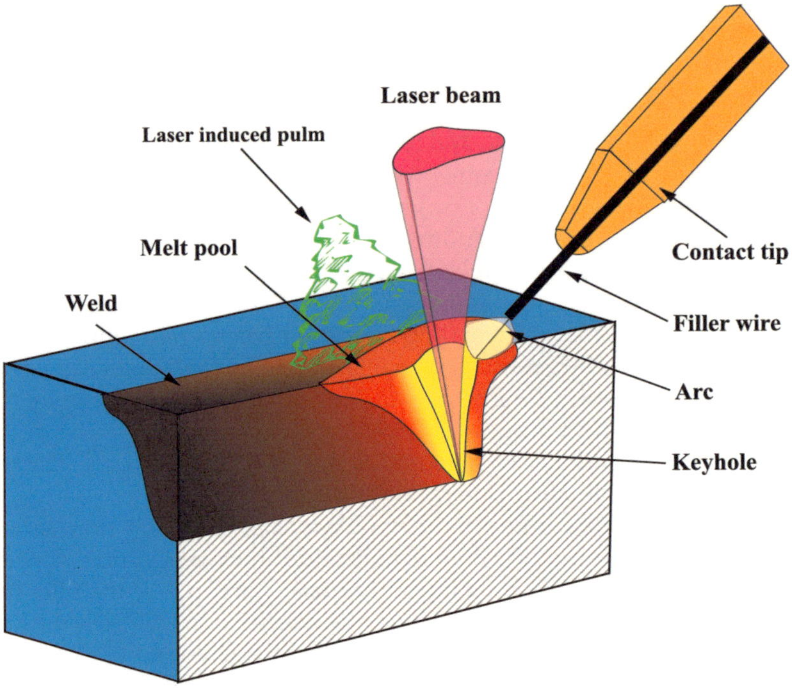

Figure 2.
The keyhole formation and laser welding using keyhole mode.

beam and absorbs and disperses it (**Figure 2**). The first stage in studying LBW is to determine the parameters affecting the heating/cooling balance, molten pool, welding area, reproducibility of the process, and the development of methods to control these parameters [5–7].

3. Advantages of laser beam welding

Proper joining rate, excellent welding quality, very high accuracy, high automation capability, and exceptional appearance of the welded joint are included as beneficial factors, leading to the application of LBW in various industries. Economically, the reduced production costs and low consumption of consumables have made this method one of the finest joining methods. To recognize why LBW is one of the best welding solutions, the top five advantages of this technique are listed below [5–7]:

3.1 Ability to join complex joints and high accuracy

LBW can weld complex joints successfully, especially it can join dissimilar materials or areas very difficult to reach using traditional welding techniques.

One of the main advantages of LBW is that it can offer a high level of accuracy and control, i.e. it can be used to join the smallest workpieces together without damaging them.

Strong potentials can be proposed for weight reduction and joint design opportunities. Typically edge welding is carried out by direct fusion of two base metals. Using

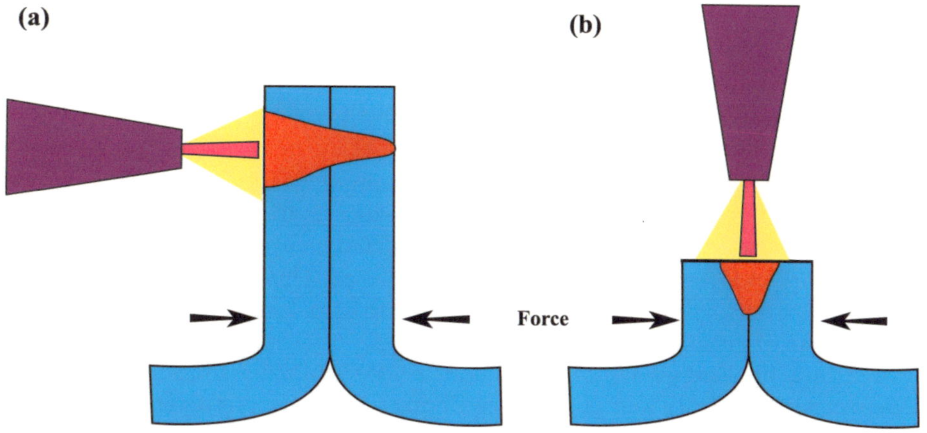

Figure 3.
Joining of edge seams by LBW: (a) flanged butt joint and (b) flare-V groove.

this tactic, it is vital to maintain an almost zero gap between the workpieces to ensure suitable joining. Using the LBW, high fusion depth can be gained while reducing flange length by more than 50% of current standards. This can be obtained by employing hybrid features of integrated clamping, optical seam tracking, and beam oscillation capabilities (known as laser welding optic).

Like other welding processes, during LBW it is challenging to guarantee the accurate positioning of the energy at the joint. But a combination of process robustness, workpiece tolerances, and robot accuracy results in obtaining proper welds. Finding the joint by optical seam tracking and laser triangulation provides accurate positioning for the laser spot during the process. This seam tracking data is then sent back to the optic controller, translating the required data for repositioning of the head galvo motors to point the laser beam to the desired coordinates. The system is capable of providing several inclination angles to accommodate the adjustments of joint position for two and three-layer joints as flange heights variation relative to one another. By adding an integrated clamping unit to the head not only the workpiece can be fixed at the desired position, but also provides the tooling costs to clamp the seam can be reduced (**Figure 3**).

The clamping unit design allows the reaching into flanges openings or structures and rapid open/close clamping mechanism (200 ms), providing a good foundation for high-volume applications. The innovative technologies offer extra advantages to meet the welding requirements for base metals such as ultra-high-strength steels, aluminum, and boron. By utilizing oscillation motors along with those directly tied to beam location two-axis oscillation can be obtained at frequencies up to 1 kHz, which eliminates the oxide layers, extra time for the gas to exist the weld pool, or post-weld annealing of brittle microstructures. An instance of the cleaning of the welding area can be seen during the zero-gap welding of galvanized materials. To this end, a gap (−0.1 mm) is characteristically required to provide a place for evaporation of the zinc at temperatures higher than 0.5 T_m of the base material. If not correctly set up, the gas expulsion can be trapped within the solidifying melt and form porosity in the final weld. The oscillation feature grants a remelting phenomenon for the weld pool and allows the zinc to escape to the surface and leave the weld. For structural applications, it is frequently necessary to join dissimilar materials for example boron steels to electrolytically or hot-dipped galvanized steels. According to the beam location control feature using oscillation, a melt pool is formed, which floats on the workpiece.

However, a distortion in the workpiece is not essentially attributed to the adaptive nature of the process.

3.2 Low heat input

LBW method uses a low heat input rate that minimizes the joint distortion. Hence, it is the preferred method for those who wish to make luxury products such as custom jewelry. Laser sources employ tremendously localized energy and allow non-contact use, which applies lower heat input on the workpieces. This method is ideal for non-contact applications, which protects other areas of the parent material from the adverse effects of heat.

The feature 'line energy' is commonly used as a denominator to compare welding processes carried out in 1F and 2F positions. Moreover, heat input determines the joint geometry, which can be controlled via the modification of the welding parameters. In other words, the heat input is directly correlated to the laser power (or arc power) and the welding speed. The heat input for the LBW process is calculated according to Eq. (1).

$$Q_{laser} = PL/vt \tag{1}$$

where Q_{laser} is the laser heat input (kJ/mm), PL is the output power of the laser source (kW), and vt (mm/s) is the travel speed. The heat input of the hybrid laser-arc welding (HLAW) considers the additional energy delivered by the arc and is determined by Eq. (2).

$$Q_T = Q_{arc} + Q_{laser} = \frac{UI60}{vt} + (PL60)/vQt \tag{2}$$

where Q_{laser} is the LBW heat input (kJ/mm), PL is the source output power (kW), vt (m/min) is the travel speed, QT is the HLAW heat input (kJ/mm), Q_{arc} is the arc welding heat input (kJ/mm), U is the arc voltage (V), and I is the arc current (A).

3.3 Compatibility and replicability

LBW can provide continuous and repeatable component fabrication. This helps industries to reduce their manufacturing costs significantly. LBW is far more quickly and much more versatile than the conventional methods. Laser welding can also be used for cutting and drilling.

When a lap fillet is the functional joint, which should be processed, similar issues are apparent as well as the joint location and overlap. To resolve the issues optical seam tracking and beam oscillation are employed. However, gap bridging technology can also be used instead of clamping equipment. In most LBW applications, zero-gap is a similar challenge as well as the joint location and ideal fusion between sheets. In lap edge configurations gaps should be seen; hence, options are currently developed to produce sound joints in this configuration.

If a filler wire is required for modification of the chemical composition or other gap-related conditions, options are developed to use the tactile seam tracking system and utilize the filler metal to bridge the gaps (**Figure 4**). But, using remote laser welding, it is not practicable to insert the wire into the melting pool when optical seam tracking is employed for beam placement. Additionally, it is possible to weld the gap automatically without filler metal. Remote laser welding – adaptive (RLWA) is a unit, utilizing a

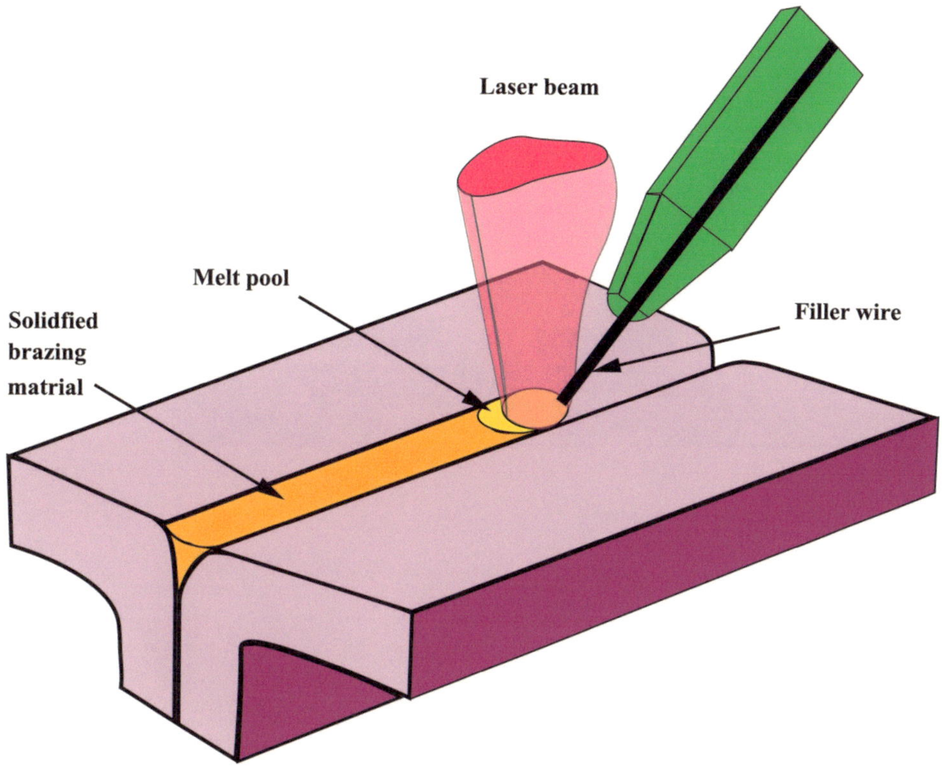

Figure 4.
Laser welding using brazing filler metal.

real-time seam finding and tracking by internal controlling of the head, which is called gap bridging. The final result is the dynamic control of the laser spot position relative to the seam, not only irradiating a predetermined point in coordinates. When the beam is accurately placed into the seam, joining a lap point with high reliability is possible. This issue is attributed to gaps in the material, which typical laser processes struggle to accommodate. With gap bridging algorithms, which are predefined in the system controls, the LBW optic can identify gaps in the joint via the seam tracking package and adjust various conditions automatically to process the joint.

By modulation of laser power, y offset of the beam relative to the joint edge, spot size, and using beam oscillation in the x and y directions, the melt can bridge the joint. Gaps with 50% or less of the upper sheet thickness can easily be addressed with both aluminum and steel materials, while recent studies show capabilities beyond that in certain situations (**Figure 5**).

3.4 High-strength joints

Since the heat input rate is significantly lower than the conventional methods, the heat affected zone (HAZ) of laser-welded joints is very small that allowing manufacturers to perform high-strength welds.

The laser beam machining (LBM) parameters such as laser intensity, beam distribution, scanning speed, spot size, and relative motion between the laser beam and workpiece can be adjusted according to different base materials. Currently, lasers are

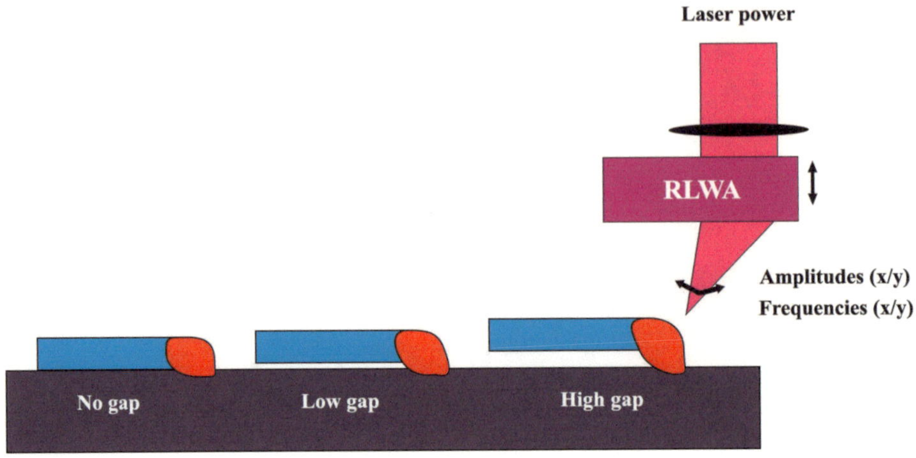

Figure 5.
Different LBW gaps and RLAW equipment.

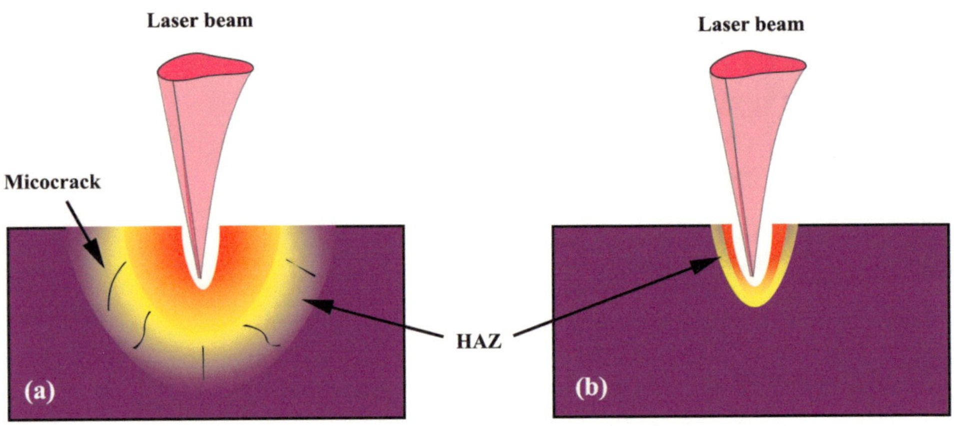

Figure 6.
Difference between the properties of (a) long-pulsed and (b) short-pulsed durations.

substituting conventional machining equipment because of their superior advantages. Major advances have been made in this area to shorten the pulse time for various machining processes. Prolonged pulse durations increase the HAZ and induce high thermal stresses, which result in the formation of cracks, voids, and surface debris. Short pulse times decline the thermal conduction, provide accurate machining operation, and proper surface finish. **Figure 6** reveals the difference between the properties of long and short pulse times [6].

3.5 Appearance, precision, and cleanliness

Since there is no need for filler metals in LBW, it provides an excellent welding quality and clean processing, so that it gains attention especially in the medical engineering industry where the quality of medical devices and components is very important.

The feature of weld optics focuses the beam down to a spot size range of 200-300 μm. On the contrary, larger spot sizes are rarely used, when low-shift welding is desired. However, large weld spots are not beneficial since they require higher overall energy and a larger heat input rate. The weld heads function with a collinear charge-coupled device (CCD) camera. Different focal planes of the beam are corrected in the visible and infrared spectra. During this modification process, a sharp CCD image indicates the proper adjustment of the weld head to its correct focal position. Moreover, a projected crosshair is centered to the position of the weld spot to be perfectly superimposed with the beam (**Figure 5**). Thus, machine vision algorithms are employed for automated precise adjustment of the welding optics to desired coordinates so that the beam is exactly irradiated to the gap between the workpieces to be joined.

To keep the power density constant, LBW should be always carried out using the waist of the beam with an accuracy of 10 μm. If the irradiated beam is not focused, the power density rapidly increases and causes an uneven weld pool. Since the strength of the contact between the joining surfaces is one of the key parameters of low-shift welding, the welding design should be consistent with this parameter. To adjust cylindrical parts to the most parallel state, a dome-shaped air bearing is used. A preset force moves the parts in contact with each other, which are self-adjusted as floating on an air bearing. When the surfaces are positioned parallel to each other, the angular position is fixed to activate the alignment. Predetermined offsets can compensate for the predicted weld shift in a planar setup. By optimizing the tolerance of the parts, the accuracy of the offset can be improved.

If a welded component shifts during the nano-welding process, additional employment of the weld energy in an opposite direction bends the workpiece back into the desired location. During this process, the effect of the unavoidable shrinkage is exploited. Careful experimenting should be carried out to determine the proper power and time of the corrective pulse. This procedure is known as laser-induced micro-adjustment (LIMA). Forming nanostructured weld joints requires a weld design, which is optimized for LBW. The LBW system has to consider the specific aspect ratio of the workpiece. To obtain an optimal result, it is suggested that the nanostructure designer and the manufacturer of the welding system work together from the beginning.

4. Laser welding limitations

Although LBW can be an amazing technology, it also has minor limitations. One of the most common disadvantages of LBW is the high cost of the process. LBW machines are very expensive and complicated so they are not accessible to many manufacturers. The other disadvantage is that it requires a highly-skilled workforce. Otherwise, LBW is almost perfect [5–7].

5. Laser sources and stimulated emission

Laser is a monochromatic light source (sometimes visible) and coherent with very high orientation and brightness. A comparison of the conventional light source and laser light source is given in **Figure 7** [6, 7].

The energy levels used in the laser radiation process are generally of two categories of atomic and molecular energy levels. Under normal conditions, atoms are generally in the ground state until they are excited by an external stimulus (pumping system). Then they return to the ground state in a very short time (nano/microseconds) by

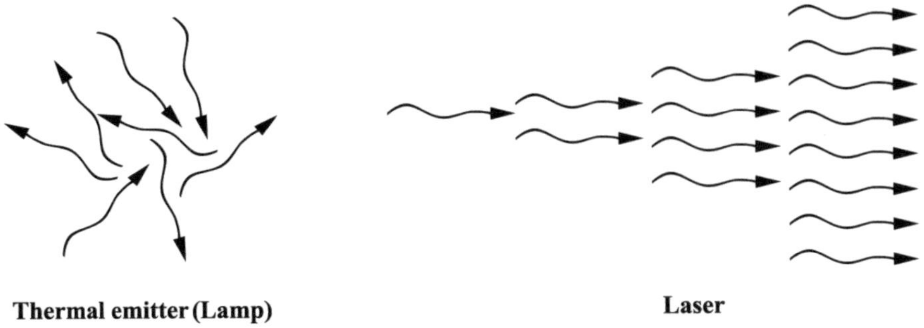

Figure 7.
The comparison between a plain light source and a laser source.

emitting a photon having an energy equal to the difference between the two levels. This return is possible in two ways [6, 7, 12], spontaneous emission and stimulated emission. Stimulated emission is considered in laser. When an atom is in the excited state, if a photon with an energy equal to its excitation energy passes by that atom, the photon induces it to be sunk. The photon is a laser photon if it is produced under the influence of another photon. Otherwise, it is the spontaneous emission [6, 7, 12].

6. Optical resonator

Each laser optical resonator consists of two mirrors and a pumping chamber, containing the active laser medium. Generally, one of the mirrors is a full reflector (input coupler), while the other is a partial reflector (output coupler). Photons, which are moving in the direction of the resonator axis are resonated to form the laser beam. Photons in other directions are destructed by adsorption on the walls of the active medium (**Figure 8**) [3–7].

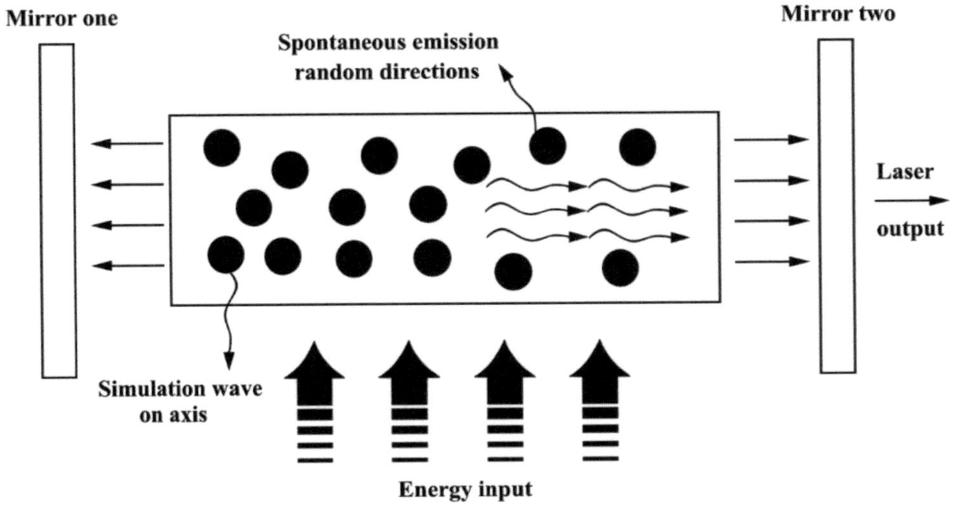

Figure 8.
The function of the optical resonator during laser beam production.

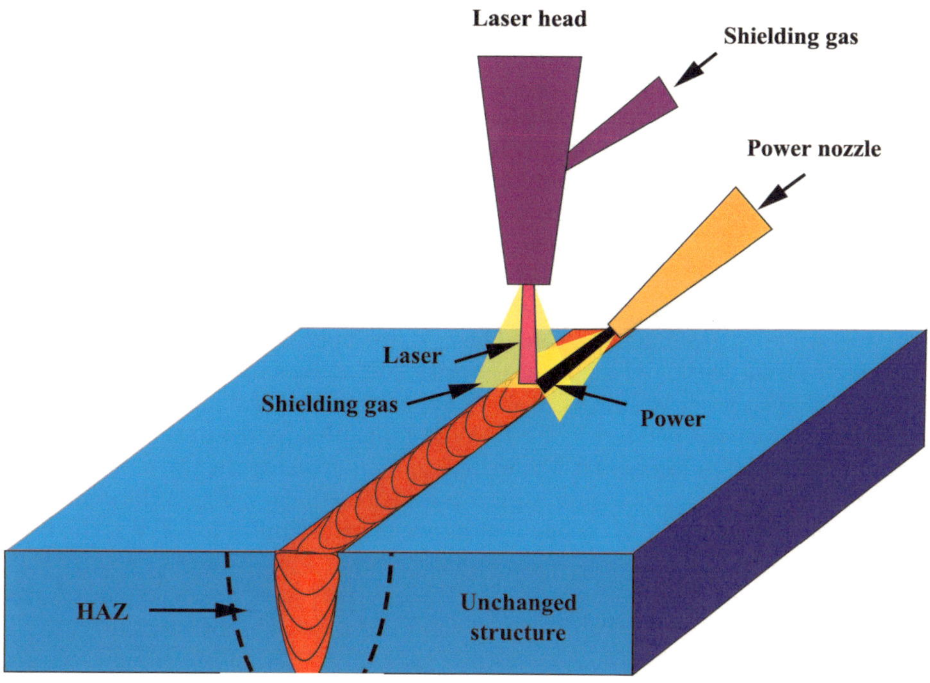

Figure 9.
The key components of laser welding procedure.

7. Laser welding procedure

The principles of the LBW process are not complicated. The procedure schematic is presented in **Figure 9**. (1) A pump, which is the energy source provides the energy required for the process. The pump stimulates the laser to such an extent that the electrons held by the atoms are moved to higher energy levels. (2) Electrons reduce their energy levels dramatically, releasing photons. The spontaneous emission of photons is what leads to the production of the laser beam. (3) Spontaneously emitted photons collide with the ones having higher energy levels. The collision reduces the energy levels of the electrons, leading to the emission of another group of photons. Both groups are now having the same wavelength and moving at the same speed. (4) Photons are emitted in all directions. However, they are all limited to travel in the same medium and hit the resonator before reflecting from the medium. The intensifying mirror then determines the level and direction of emission. To perform any type of amplification, the fraction of atoms must be higher than that of low-energy photons. (5) The laser beam is targeted and focused on the workpieces to be welded. Highly-focused light energy is converted to heat energy at the workpiece surface. (6) During a process known as surface conductivity, the generated heat melts the material surface. The generated heat is controlled to be below the boiling point of the parent material. This technique is an ideal solution when welding materials that have high thermal conductivity. Apart from welding, other procedures such as drilling, cutting, and stripping can be carried out using laser beam energy [3, 13–16].

By combining the LBW and GMAW techniques, the laser-GMAW hybrid welding is developed. This combination is an attractive tool with a great potential for welding lightweight structures, especially aluminum alloys. This hybrid welding technique is generally acknowledged for its efficiency, robustness, and flexibility. By combining a deep-penetrating laser beam with high filler feeding of GMAW the primary applications of LBW and GMAW can be improved significantly. The main benefits of this technique are high gap-bridging ability, deep and stable weld penetration, facile addition of the filler metal, and low distortion. This hybrid method allows much wider groove tolerance in comparison with LBW of specific alloys such as aluminum alloys. Furthermore, the distortion reduction decreases the required post-welding treatments and facilitates the assembling process because the hybrid-welded components are more dimensionally precise. Moreover, if very accurate metallurgical factors are needed, the hybrid process can be easily balanced with the filler metal, which declines the susceptibility to hot cracking, especially for specific aluminum alloys. This combined process can also enhance the weld bead shape appearance and quality (e.g. by elimination of undercut), reducing the porosity and increasing the welding speed.

Since hybrid laser arc welding (HLAW) apparatuses are influenced by each of the two processes, the weld geometry of HLAW is controlled by the heat input of each process as presented in **Figure 10**. For instance, by increasing the power of GMAW, the width to depth ratio of the weld is increased. Nevertheless, due to the contribution of both techniques, the HLAW-welded joints are usually similar to LBW at the bottom and similar to GMAW on the top of the joints.

In addition, due to the involvement of a high-density laser beam, keyhole formation is a characteristic of the HLAW process in most cases. On the other side, a

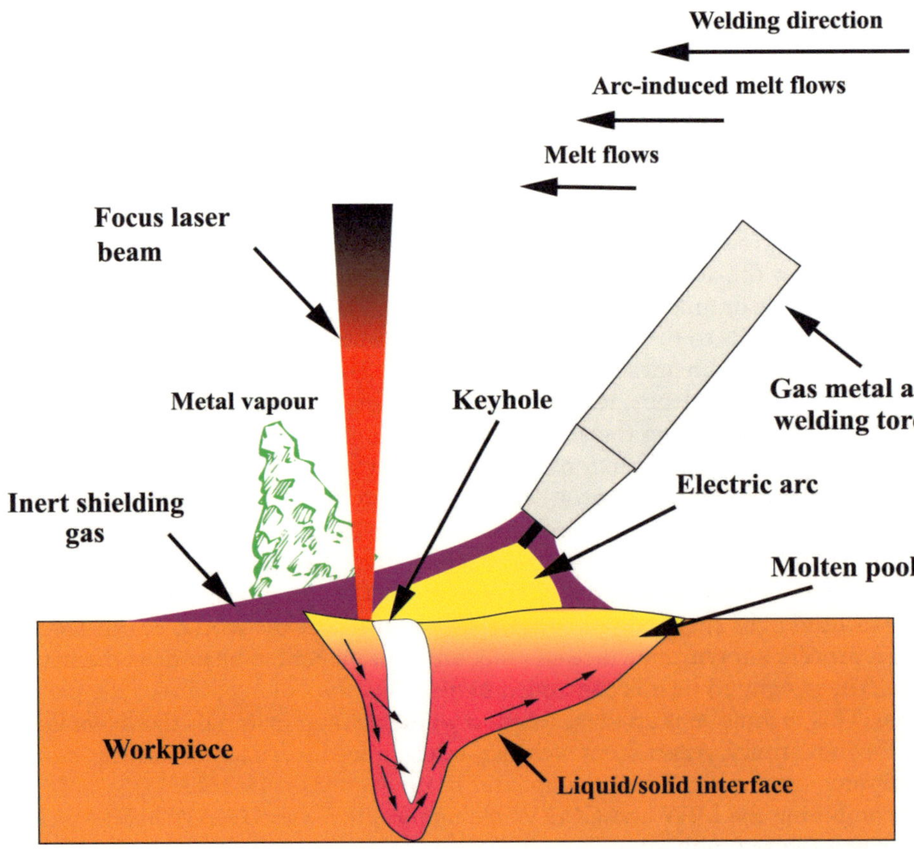

Figure 10.
Hybrid laser GMAW welding process.

conduction-like process without the formation of the keyhole is obtained if the beam is not focused or its power is insufficient. A previous study regarding the aluminum LBW showed that initially, the Nd:YAG laser beam absorption by the base metal surface could be as low as 10%. But, when the base metal was molten, the beam absorption greatly increased up to almost 100%, especially when the keyhole was formed. Interestingly, it has been reported that the arc stability of GMAW is increased when it is coupled with a laser beam. This enhancement is achieved when the arc is close enough to the beam and they share the same melting pool. For example, since the aluminum melt has a lower electrical resistance than that of the solid-state or oxide layer, the arc favors the path with the lower resistance. Besides, the interaction between the keyhole and arc plasmas increases the arc stability. The energy from the formed keyhole creates a metal plasma, ionizing the shielding gas of the GMAW process that facilitates the strike and stabilizes the arc. Furthermore, the HLAW arc has a finer geometry, a higher electrical conductivity, and a higher current density (up to 500% of the GMAW arc). On the other side, since in HLAW, the metal plasma is originated from both the base metal and the filler metal, more metal vapor is produced than that of the LBW. Consequently, the keyhole formation is much easier and process failure is prevented. The penetration of this technique is higher than the LBW due to the higher plasma pressure. Since the molten pool is larger during this process, the weld pool is in the liquid state for a longer time compared to the LBW. This is beneficial in the case of welding aluminum alloys due to the high hydrogen solubility in the molten aluminum. Hence, a larger melting pool gives more time to hydrogen bubbles to escape from the weld, resulting in the formation of fewer gas pores.

Because of the interaction between the two processes, the advantages are more than the drawbacks. For welding aluminum alloys, these advantages depend on the welding parameters, the alloy composition, and the joint type. A majority of the authors stated that the welding speed is increased by using the hybrid technique. Moreover, it improves the penetration of the weld seam, which is 10-20% and 20-50% higher than the LBW and GMAW, respectively. Additionally, many studies have expressed that the stability during aluminum welding is higher in comparison with LBW or GMAW processes. Additionally, the applied heat input is lower due to the elevated speed and high energy density of HLAW. By lowering the heat input the distortion of the welded components is directly decreased. Since the GMAW process generates a large welding seam, gap bridging is improved during HLAW compared to the LBW. It has been reported that the HLAW can increase the gap bridging from 1.05 to 1.19 mm compared to the GMAW, while the maximum gap tolerance of the autogenous LBW is around 0.3 mm. Another benefit of this process is its higher wire feed alignment compared to the LBW. Since in the HLAW process the feeding wire does not have to intersect with the laser beam and the weld pool, the addition of filler metal is more facile than that of the cold wire fed LBW. The reduction in the component distortion, high gap bridging, filler application, and wire misalignment tolerance are the main important aspects of automated HLAW that increase the robustness of this process for industrial applications compared to the primary original processes of LBW and GMAW.

8. Types of laser welding based on laser sources

Several types of lasers can be utilized for LBW. These include fiber lasers, Nd:YAG pulsed lasers, and Nd:YAG continuous-wave lasers. However, one should know that the type of employed laser source for LBW depends on the application [3, 6, 7, 13, 17].

Solid-sate lasers	Gas lasers	Semiconductor lasers	Liquid dye lasers
Ruby	ArF	AlGaInP	Stilbene
Alexandrite	KrF	AlGaAs	Coumarin 102
Ti-sapphire	XeCl	InGaAs	Rhodamine 6G
Nd-YLF	XeF	InGaAsP	
Nd:YAG	Argon		
Nd:glass	Krypton		
Er-YAG	HeCd		
	Copper vapor		
	Gold vapor		
	HeNe		
	CO_2		

Table 1.
Solid-sate, semiconductor, gas, and liquid dye lasers.

Fiber lasers can be used for a wide range of applications, from joining very small components used by medical engineering and electronics industries to welding thick components in the automotive and aerospace industries. Fiber lasers are versatile and inexpensive, which are suitable for achieving high-quality spot welds [3, 6, 7, 13, 17].

Laser sources are generally classified based on the state of matter of their active medium and their temporal modes. Hence, according to physics, lasers are categorized into solid-state, semiconductor, gas, and liquid dye lasers [4]. According to the temporal modes, they are classified into two modes of continuous-wave (CW) and pulsed mode. In the CW mode, the beam is continuously irradiated without interruption, whereas in the pulsed mode the beam is irradiated periodically. **Table 1** displays the laser types and their wavelengths.

These medium-power lasers use continuous-wave mode during the welding process. Therefore, the average power of the source must be higher than a certain limit. Continuous-wave lasers are ideal for high-speed and deep penetration welding. They produce joints with a very low heat input rate. Hence, they create a smaller HAZ [3, 6, 7, 13, 17]. Nd:YAG pulsed lasers generate discrete pulses of controllable energy that can be used for ideal welds. Although the average power of these lasers is often low, they employ high-power peaks for welding. They can be properly utilized for joining large spot welds as well as deep spot and seam welds.

The laser active medium is doped by a few numbers of impurity ions in solid-state layers. Among the solid-state lasers, Nd:YAG lasers are mostly employed for LBM applications. The solid-state sources (e.g. Nd:YAG, ruby, and Nd-glass lasers are vastly utilized for the machining of metals. However, Nd:YAG lasers can also be used for ceramic materials. Gas lasers are categorized into three types according to their composition (i.e. neutral atom, ion, and molecule). In general, gas lasers can be used in either CW or pulsed modes. They are also used with the transverse flow, axial flow, and folded axial flow for construction applications. Among them, CO_2 lasers are most commonly employed for machining ceramics, plastics, nonmetals, and even organic materials.

Although semiconductor lasers are made of solid materials, their functioning principles are different from solid-state sources. The function is based on the radiative recombination of charge carriers. These sources can produce wide beam divergence angles (around 40°).

In comparison with other lasers, liquid-state sources are easier to fabricate. Their main advantages are the simple cooling procedure and replacement of the laser cavities. Unique properties of liquid organic molecules allow the liquid dye lasers to be adjusted over a wide range of wavelengths (200-1000 nm).

9. Types of laser welding based on welding method

LBW is divided into three types of welding, conduction mode, conduction/penetration mode, and penetration or keyhole mode (**Figure 11**) [5].

9.1 Conduction welding

During the conduction mode of LBW, heat is transferred to the metal through thermal conductivity. In this mode, low energy is transferred per unit area (~0.5 MW/cm^2), which is often used for shallow joints to create a wide and shallow weld (**Figure 8**). Due to the low penetration depth in this mode, the gas absorption in the weld pool is low. It is also completely soundless. Typically, this mode can be utilized for applications, which require a high-quality weld e.g. battery sealing applications. The conduction mode is carried out in four steps [3, 5–7, 13, 17], (I) heating of the workpiece surface by the laser radiation, (II) formation of the molten pool, (III) no melt evaporation during the process, and (IV) determination of weld pool shape by the thermal conductivity (**Figure 12**).

9.2 Conduction/penetration welding

This mode occurs at medium energy density per unit area (~1 MW/cm^2) and leads to greater penetration than the conduction state (**Figure 11**) [3, 5–7, 13, 17]. In this case, the keyhole exists with shallow penetration that provides a characteristic aspect ratio (depth/width) of ~1. This mode is carried out almost exclusively via pulsed Nd:YAG laser for various spot and seam welding applications.

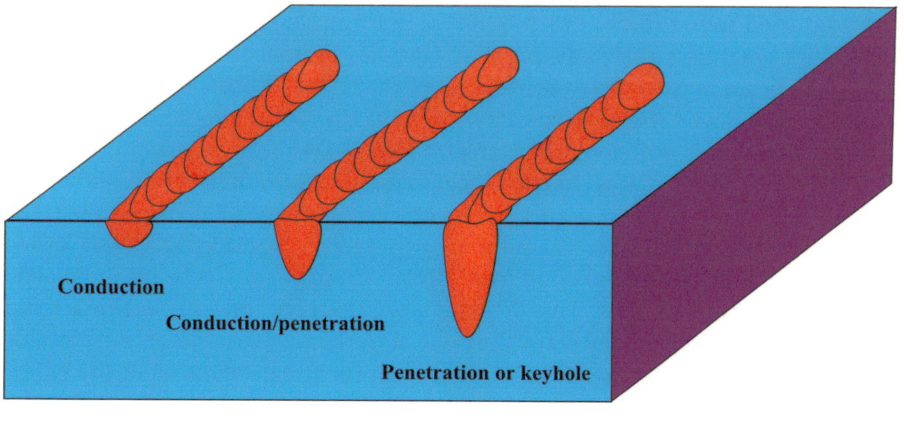

Figure 11.
Different modes of laser welding.

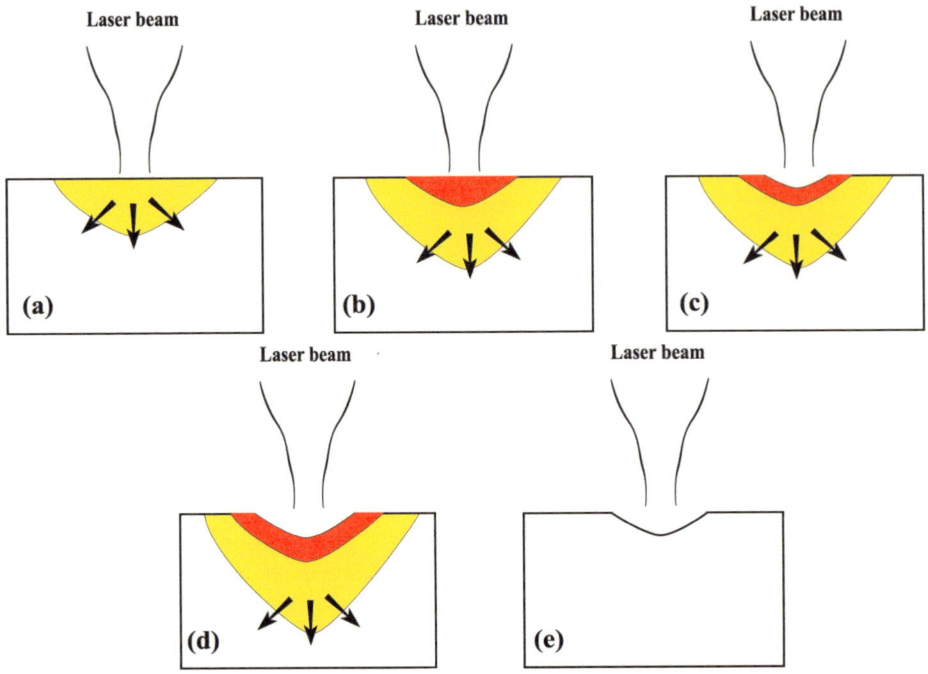

Figure 12.
The effect of thermal conductivity on the weld pool shape (a) heating, (b) surface melting, (c) surface vaporization, (d) plasma formation, and (e) ablation.

9.3 Keyhole welding

By increasing the peak power density (>1.5 MW/cm^2) the welding mode is shifted to the keyhole, i.e. deep narrow welds with an aspect ratio higher than 1.5. The laser beam heats up and melts the material quickly upon irradiation. If the intensity is high enough, a key-shaped cavity filled with the base metal vapor is formed, reflecting the generated heat into the material bulk that is sealed by the molten material behind the laser beam (**Figure 13**). In this case, welding is often performed by high-energy laser sources to join thick workpieces or fill cavities. This mode is called keyhole welding and it is accompanied by a muffled sound [3, 5–7, 13, 17]. Keyhole formation improves the laser heat absorption via two major mechanisms; first, through Fresnel absorption mechanism, absorbing the beam by successive reflections of the beam on the keyhole walls (**Figure 13**), and second, through the absorption of the laser energy into a vapor-filled cavity caused by a phenomenon called inverse Bremsstrahlung process [5, 17].

In the keyhole mode, the weld can be accomplished at either very high travel speeds (up to 20″/s with short depth welds, or very deep welds (i.e. up to 0.5″). The high-power density of the laser beam forms a thread of vaporized material, called a keyhole, extending into the bulk and providing a channel for the beam to be efficiently delivered into the joint. This direct energy delivery into the workpiece maximizes the weld depth and minimizes the heat input to the base metal, minimizing the HAZ and workpiece distortion. This mode is used for the production of many automotive and train components such as torque converters and gearboxes, which require up to 0.25″ penetration. The keyhole is surrounded by the melt, which tries to close it. Under steady-state conditions (optimized welding), the vapor pressure confined

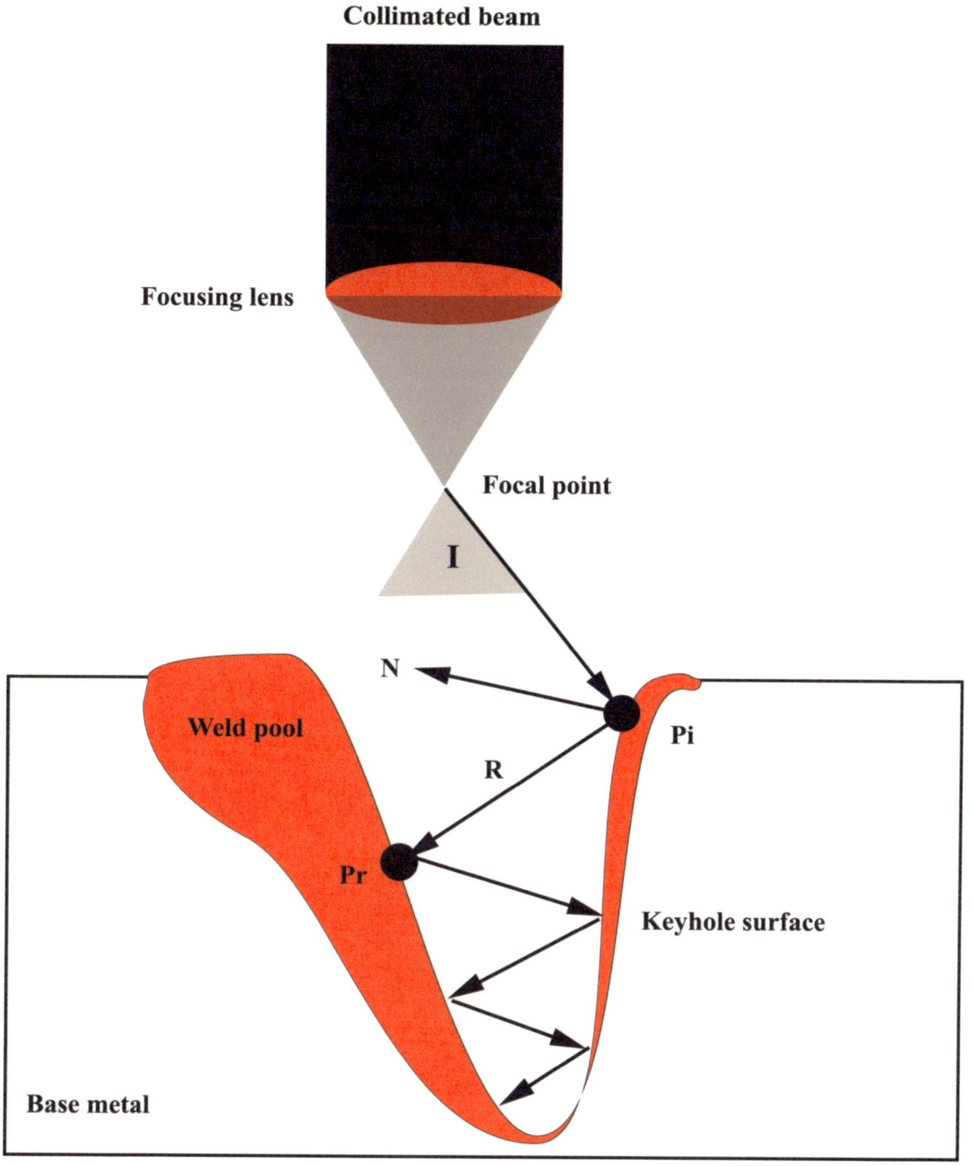

Figure 13.
Absorption of the laser beam due to successive radiations from the wall of the keyhole.

within the keyhole prevents the melt from collapsing in on itself permanently, which would interrupt the welding. But even during an optimized weld localized collapses of the keyhole may take place.

The main difference between these modes of LBW is that in the first mode the surface of the weld pool is not broken, but in the case of penetration welding, the surface of the weld is opened so that the laser beam penetrates the molten pool. Conduction welding is not prone to gas absorption during the process, which is due to the lack of penetration of the laser beam into the material bulk. In penetration welding, discontinuous closure of the keyhole increases the susceptibility to porosity formation in the weld pool (**Figure 14**) [3, 5–7, 13, 17].

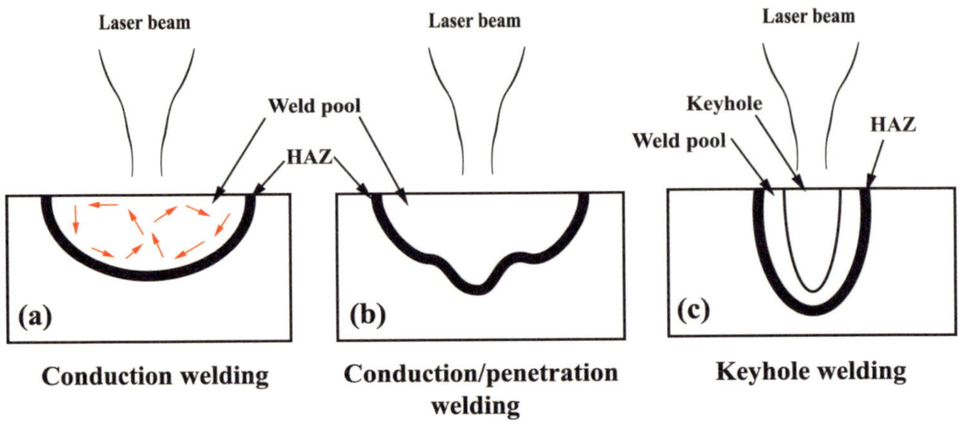

Figure 14.
The schematic of different modes of laser welding (a) conduction mode, (b) transition keyhole mode, and (c) keyhole/penetration mode.

Since LBW is a high energy density method, it does not require thermal conduction to achieve a deep penetration, which is in contrast to the conventional methods of arc welding and gas welding that obtain penetration via increasing the heat input. In conduction LBW, the weld width is often greater than the depth and the heat input is greater than the amount required for penetration. Moreover, in penetration or keyhole welding, the laser heat is transferred from the surface into the joint and creates a deep and thin weld pool (**Figure 14**) [4, 18].

10. Variables affecting laser welding

Achieving optimal conditions is a necessary and sufficient condition to achieve an ideal weld with a suitable appearance. Therefore, it is necessary to study the variables and parameters affecting LBW [3–7, 13, 17, 19–21].

10.1 Parameters

10.1.1 Parameters related to the laser source

- The type of laser source that determines the laser wavelength (CO_2, Nd:YAG, etc).
- Continuous (CW) or pulsed laser
- Medium laser power
- The energy of each laser pulse
- Pulse time width
- Pulse frequency

- Laser beam quality
- Optical specifications of the laser beam focusing system in the focus

10.1.2 Parameters related to the operation of the system and the welding process

- The parent material and its chemical composition
- The distance from the center of the beam to the surface of the workpiece
- Welding speed
- Conditions for starting and finishing welding
- Gas shielding type
- Gas discharge
- Nozzle geometry
- The geometry of the joint of the two sheets includes the distance between the two edges and the matching of the two edges
- Plasma plume or bubble formation when laser radiation on the metal surface

10.2 Effect of laser source type

The function basis of a laser source is depended on the state of matter of the source active medium (gas or solid-state). The most well-known gas lasers are CO_2 lasers, employing a combination of helium, nitrogen, and carbon dioxide gases with a ratio of 1:1:18. The sources operate based on the molecular energy levels of the gas. Hence, they are considered molecular lasers. The most important factor in optimizing the performance of a CO_2 laser is efficient cooling of the gas and prevention of decomposition and failure of the gas molecules. The new generation industrial CO_2 lasers are fabricated with low-power ranges (10-20 watts) to high-power ranges (about 6000 watts), using radio frequency (RF) waves as the pumping source called RF-excited lasers. The advantage of these lasers over other gas lasers is the possibility of operating in high-frequency pulse mode and the extended life span of the sealed-off tubes. Another type of gas laser is the gas dynamic laser. Since the function of these lasers is based on the sudden decrease of the gas pressure, the gas type of the source is very determinative during the pressure reduction and the sudden cooling mechanism of the active medium. This type of source is widely used for military purposes.

Nd:YAG lasers are the most widely used solid-state sources, having higher optical and physical properties, and greater efficiency compared to gas lasers. In Nd:YAG lasers, the active medium is a $Y_3Al_5O_{12}$ crystal, where some of the Y^{3+} ions have been replaced by Nd^{3+} ions, resulting in the formation of Nd:YAG crystal and providing a promising active medium with several high-intensity wavelengths in the infrared region. Another solid-state source is used in diode lasers (semiconductor lasers),

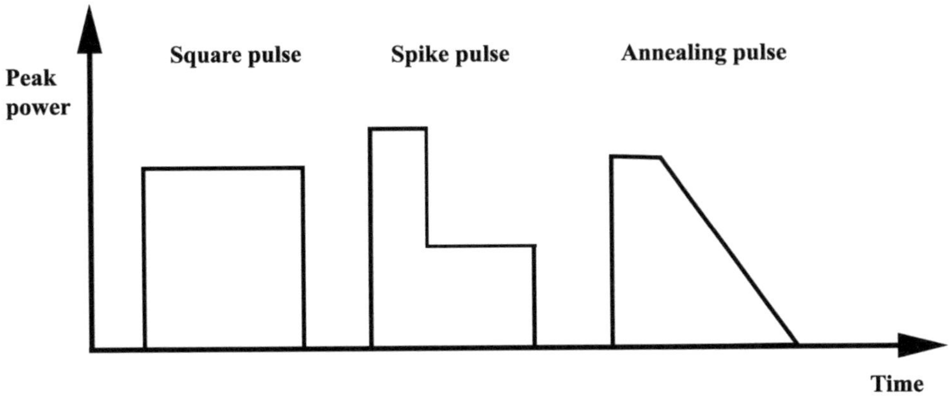

Figure 15.
Different pulse modes.

resulting in the fabrication of sources with ultrahigh efficiency and great tunability. However, the main limitation of these lasers is their high divergence.

Nd:YAG laser welding can be used more than CO_2 due to its shorter wavelength (1.06 mm) which allows displacement using optical fibers and also reduces the reflection of metal surfaces [3–7, 13, 17, 19, 20].

10.3 Pulse shape

In most cases, LBW uses a square pulse shape (**Figure 15**). But two other types of pulse shapes are also used in special welds. The first type (Spike pulse) is used for light-reflecting materials such as copper and aluminum (**Figure 15**). The second type (Annealing pulse) is used to minimize the radiant heat cycle during welding for crack-sensitive materials (**Figure 15**) [3–7, 13, 17, 19, 20].

It should be noted that the capability of pulse shaping in some lasers such as Nd: YAG sources can facilitate operations such as drilling and cutting. So that a chain of very short pulses with a higher peak power than the main pulse is descended immediately after the main pulse to form a quick coupling by the laser beam or numerous pre-pulses with a lower peak power before the main pulse (a pre-pulse should have lower power than the main pulse) are descended on thin foils to prevent the welding area from being punctured.

10.4 Peak power

The peak power of a laser source is the maximum power that the source can provide in either continuous welding (CW) or pulsed welding modes. It is measured in watts (W) or kilowatts (kW).

One of the important parameters of pulsed laser welding is pulse peak power. In fact, with this peak power, penetration welding can be created with a low-power laser. The peak power of a square pulse (P_p, (J/ms)) is equal to the pulse energy (J) divided by the pulse time or width (Pulse duration (ms)) (Eq. (3)) [3–7, 13, 17, 19, 20].

$$P_p = \frac{Pulse\ energy}{Pulse\ duration} \tag{3}$$

Obviously, by significantly reducing the pulse time in a low-power laser, a high pulse peak power can be achieved to create penetration welding.

An optimum peak power creates the deepest penetration in the given energy without the expulsion of materials. Welded joints, which are made with high peak power and short pulse widths are narrow and deep and require a high heat cycle.

10.5 Time width or pulse width

Pulse width is the duration of each laser pulse (ms). During the pulse on time, the workpiece senses the pulse power, and in the distance between the two laser pulses (pulse off), the parent material is cooling. The pulse width controls the heat input to the workpiece, the welding width, and the heat cycle. Increasing the pulse width expands the welding and HAZ dimensions due to the increased heat transfer time (**Figure 16**) [3–7, 13, 17, 19, 20].

In other words, the pulse width is a fine-tuning parameter, which is used to adjust the penetration and width of the weld and, if necessary, to stabilize the weld. By increasing the pulse width and prolonging the thermal transfer time, the weld dimensions (penetration, width, and HAZ). To increase the weld width, reduce the thermal cycle, and minimize depth variation, the pulse width must be increased.

It should be mentioned that the optimization of maximum power (peak power) and pulse width during LBW highly affect the joint quality. So that a very high maximum power causes spraying and improper joining. On the other hand, very small pulse width can cause spraying or lack of penetration.

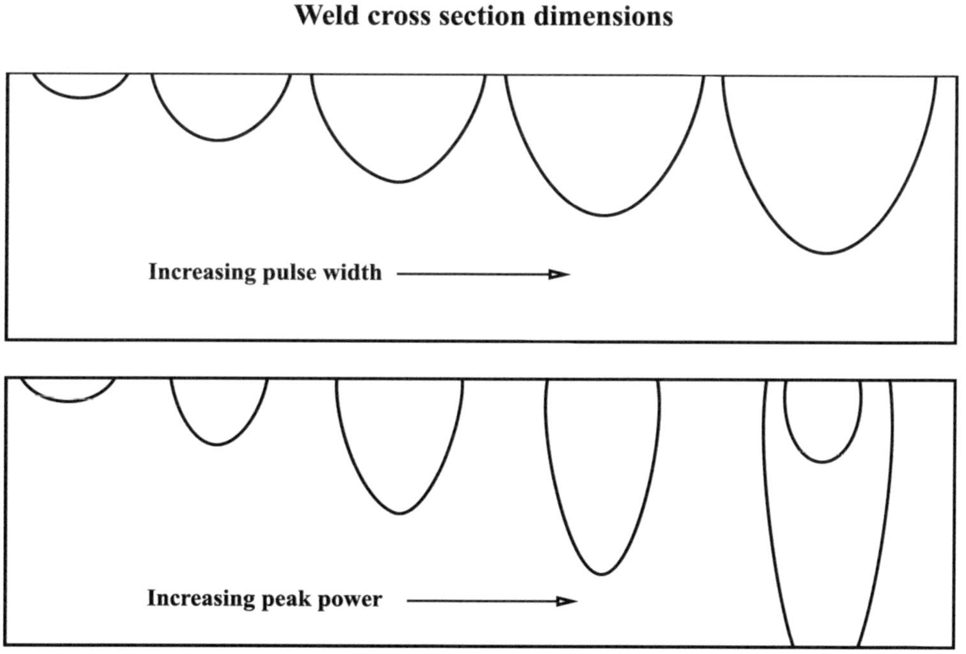

Figure 16.
Schematic of the pulse width and power peak effects on the weld shape.

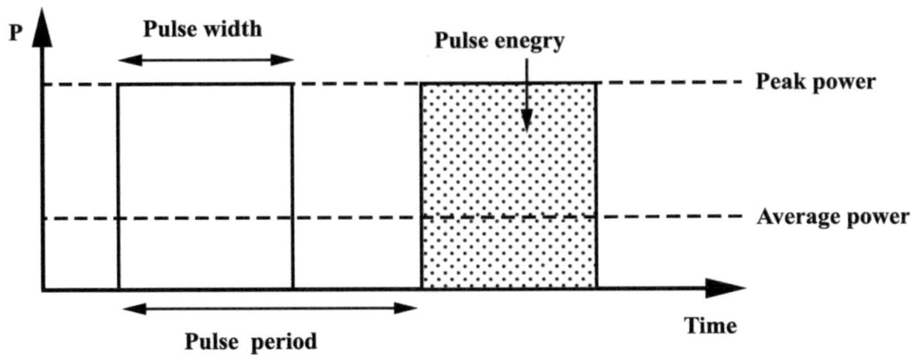

Figure 17.
Visual relationship between the frequency, pulse width, and energy level.

10.6 Pulse energy

From the point of view of the irradiated material, each laser pulse acts as a package of energy called pulse energy (pulse energy, (E, J)) and its relationship with the power peak (P_p, (J/ms), and pulse width (T, (ms)) (pulse width or pulse duration) (Eq. (4)) is in **Figure 17** [3–7, 13, 17, 19, 20].

$$E = P_p \times T \tag{4}$$

10.7 Frequency

Frequency (f) indicates the number of pulses of the flash lamp and therefore the number of laser pulses per second (Eq. (5)). Frequency is expressed in Hertz (Hz) or pulse per second (PPS) as given in Eq. (1). On the other hand, the frequency inverse (1/f) is equal to the interval between two consecutive pulses. By knowing the amplitude of the laser pulse (T), we can estimate the time between two pulses, i.e. the laser extinction time. It also controls the heat input to the workpiece and the heat cycle [3–7, 13, 17, 19, 20].

$$f = \frac{1}{Pulse\ period} \tag{5}$$

10.8 Average power

The importance of this parameter is for welds using more than one pulse. In fact, the average power (P_{ave}) of a laser source is obtained by multiplying the energy of each pulse by its frequency (Eq. (6)) [3–7, 13, 17, 19, 20].

$$P_{ave} = E \times F \tag{6}$$

Medium power is applied when more than one pulse is required for welding. As the average power increases, the heat input rate increases; hence, with increasing the heat input, the penetration depth and weld width increase. In general, at constant power, the smaller the beam diameter, the more concentrated the heat and the smaller the weld pool. The diameter of a laser beam output can be increased by increasing the power. For instance, lasers with 1, 5, 10, and 25 kW powers have diameters of 10, 25, 40, and 70 mm, respectively. The average power density of these diameters is between 6 and 13 W/cm^2.

10.9 Power intensity or density

The density or power intensity (I) at any given moment is equal to the amount of direct power equal to the cross-sectional area of the beam (D) at the parent material surface (Eq. (7)). The diameter of the laser spot in the focus depends on the type of laser and its beam quality and the beam focusing system. Power density is a function of the beam focusing tool and the maximum laser output power [3–7, 13, 17, 19, 20].

$$I = \frac{Power}{\pi \frac{D^2}{4}} \tag{7}$$

In short, the amount of beam intensity determines the state of the welding process and the formation or non-formation of the keyhole. On the power peak, the penetration rate of the weld, the pulse width usually controls the heat input to the workpiece, and the power density controls the penetration rate of the weld.

10.10 Optical specifications of the laser beam focusing system in the center

The choice of laser beam focusing system depends on the type of process, the type of laser, and the workpiece material. In fact, the cross-sectional area of the laser beam at the focus, which is one of the two main factors in determining the laser intensity at the focus, depends on the choice of the focal length of the laser beam focusing system. The relationship between the laser spot diameter at focus (D_F) and the focal depth or Rayleigh length (R_L) with the focal length of the beam focusing system is given in Eqs. (8) and (9).

$$D_F = M^2\left(\frac{4}{\pi}\right)\lambda\left(\frac{f}{D_L}\right) \tag{8}$$

$$R_L = D_F\left(\frac{f}{D_L}\right) \tag{9}$$

where f is the focal length of the beam focusing system, λ is the laser wavelength and M^2 is the quality factor of the laser beam. Rayleigh length is the distance at which the laser intensity reaches 70.7% of the maximum intensity at the focus and is considered as the focal depth or effective focal length. The larger the focal length of the focus system, the smaller the diameter of the laser spot in the focus (**Figure 18**) [3, 4, 6].

The intensity distribution due to the optical nature of lasers depends on the properties of the resonator, active medium, and pumping system. Although the intensity distribution at the cross-section of the beam is not uniform, it can be predicted according to the properties of the resonator, active medium, and pumping system and is considered as an intrinsic feature of each source. The best mode for intensity distribution is Hermite-Gaussian mode or TEM_{00}, having the highest intensity in the center of the beam with an M^2 factor of 1. High-power industrial lasers usually have a combination of TEM_{00} mode and higher modes in the beam. Therefore, the higher the share of higher modes, the larger the M^2 factor, the greater the divergence, and the

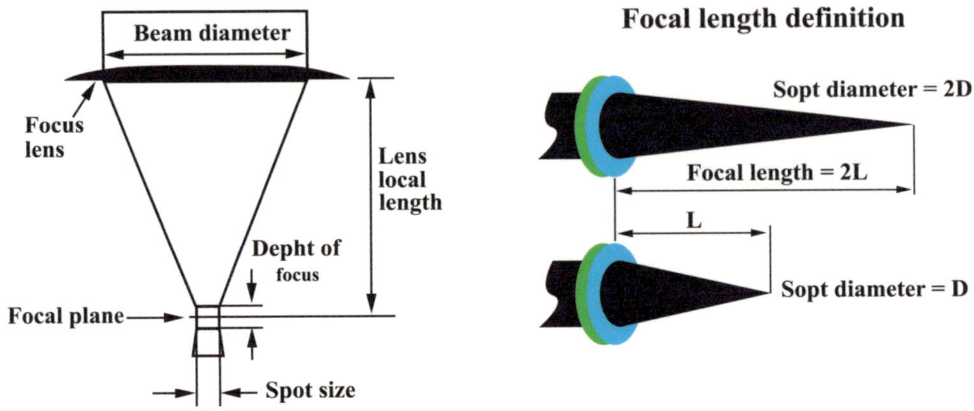

Figure 18.
Schematic of the laser beam focusing system.

lower the optical quality of the laser beam. To reduce the divergence and correct the beam of lasers with low optical beam quality, such as solid-state Nd:YAG sources, a special optical tool called a beam expander is used.

11. Joint and adaptation design (fit-up)

LBW is in most cases without metal fillers. Therefore, to have an ideal laser weld, the edges must fit perfectly at the joint. If the distance between the weld surfaces is too great, there will not be enough melt to fill the gap and an undercut will be created on the weld surface. To create the desired weld, the gap distance must be very small. Normally it should not be more than 10% of the material thickness [3, 4, 6].

12. Seam welding

Sequential irradiation of laser pulses on the workpiece, while the workpiece is moving at a slight velocity perpendicular to the axis of the radiation, creates a continuous chain of welding cones. Sometimes their overlapping extends to the lower sections and the depth of the workpiece (**Figure 19**). The percentage of overlap is a function of the speed, frequency, and diameter of the laser spot. This parameter is used to determine the most suitable laser conditions for work and to determine the total time of the welding cycle [3, 4, 6].

If a balance is struck between the parameters of pulse penetration, welding speed, and pulse frequency, seam welding has reached the desired state. Usually, at first, the laser parameters such as pulse width and power are selected and the welding speed is determined by considering the spot diameter and the overlap required (**Figure 20**). The overlap factor (Q_f) is given in Eq. (10) [3, 4, 6].

$$Q_F = [1 - (V/f)/(S + VT)] \times 100 \tag{10}$$

where V (mm/s) is the travel speed of the workpiece (Travel speed) below the welding head, f is the working frequency of the laser (Hz), φ is the beam spot size (mm) and T is the pulse width of the laser (ms)) [3, 4, 6].

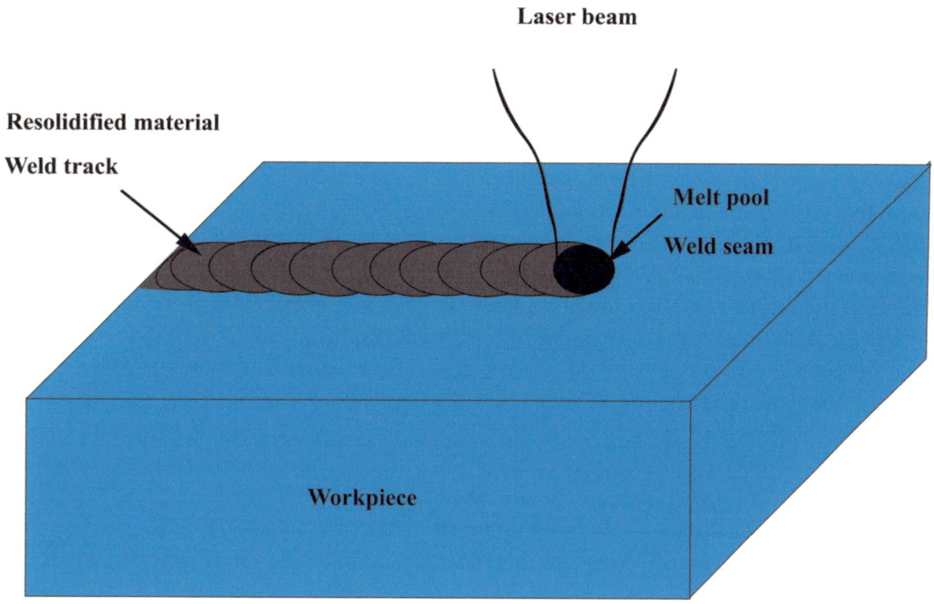

Figure 19.
Schematic of linear welding with a pulsed laser.

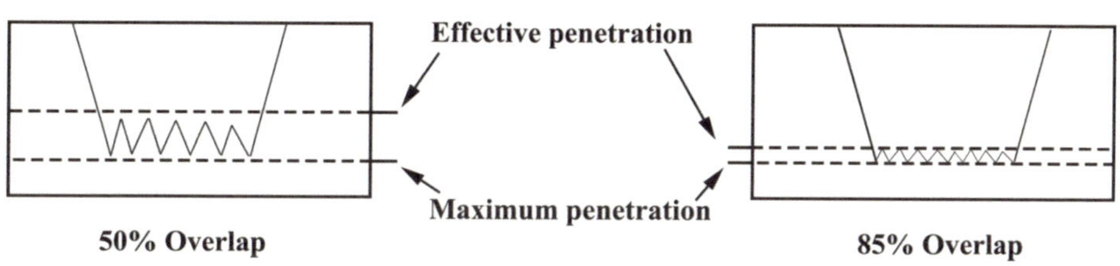

Figure 20.
The effect of overlap on penetration depth in seam laser welding.

The best conditions are obtained when complete and defectless welding is created with the least energy and number of pulses and with a reasonable speed [3, 4, 6].

13. Applications of laser welding

The application of laser beam welding (LBW) in the industry is increasingly expanding, from microelectronics to shipbuilding can use this welding process. This user potential can be attributed to the following factors [3–7, 13, 17, 19–21]:

- Limited heat input
- Small accepted heat area
- Slight unevenness
- High welding speed
- Potential of dissimilar joining

Some of these features make LBW the preferred option for some industries that have previously used the resistance welding process. The LBW process can also be combined with an electric arc welding process with a wire used in neutral shielding gas or the MIG welding process. These process combinations are designed to be placed on the surface to be welded. In addition, the special equipment used significantly reduces the tools needed to prepare the desired edge for welding. The existing filler wires with the appropriate chemical composition have provided the necessary conditions for the uniformity of the mechanical properties of the welding area. In addition, the combined processes can significantly increase the speed of work, are also effective in deep penetration and the overall sealing. Recent advances in laser diodes have provided a new opportunity to solve industry problems.

Powerful CO_2 lasers (2-10 kW) are currently used in automobile structure welding, heat exchangers, etc. For years, ruby lasers less than 500 W have been used to weld small workpieces such as small, delicate parts of medical instruments, electronic packs, and even razors. High-power ruby lasers use optical fibers to transfer the beam. This was done simply by robots and made possible a wide range of 3D applications such as cutting and laser welding automobile structures.

The laser beam is focused on a small point and creates at that point, it causes the metal to melt and even evaporate. To focus the power of powerful CO_2 lasers, water-cooled mirrors are used instead of lenses.

Fiber laser welding stands out as a robust technique when the joining of dissimilar materials is considered using the LBW in medical devices, electronics, automotive, and aerospace applications. It can simultaneously reduce manufacturing costs and offer design flexibility.

Theoretically, any material that can be joined by conventional methods, can be also welded by LBW. However, when welding dissimilar materials due to their different physical and chemical properties (e.g. melting point, boiling point, density, thermal conductivity, and coefficient of thermal expansion) various difficulties can occur, which makes the joint unacceptable. Moreover, good solid solubility is vital for the production of sound welds of dissimilar metals. This is only achieved when the materials have compatible melting points. If the vaporization temperature of a material is close to the melting point of the other one, the weldability is low, which results in low-quality joint and/or formation of brittle intermetallic phases.

Formerly, most dissimilar welds were performed by flashlamp pulsed Nd:YAG lasers. Lamp-pumped sources can produce multi-millisecond pulses, which have peak powers much higher than the average source power (with a low duty cycle). High peak power of the lamp-pumped Nd:YAG sources along with the pulse shaping capabilities results in an ideal option for welding dissimilar materials. Penetration depth is too deep during this technique, which may lead to defective joints. However, insufficient weld depths can be prevented by adjusting the initial and final powers based on the base metals and the joint geometry.

Researchers have developed various pulse shapes to improve weld quality and decrease cracks and porosity. These attempts have provided valuable solutions to join dissimilar materials in the absence of welding defects.

LBW is a versatile method that can be used for various types of metals (**Table 2**). Some of the metals commonly used in LBW are:

- Carbon steel [44, 55–57]

- Aluminum [26, 58, 59]

	Al	Cu	Mo	Ni	Fe	Ti	Mg
Al	[22]	[23, 24]		[25]	[26, 27]	[28–32]	[33, 34]
Cu			[35]		[36, 37]	[38]	[39]
Mo					[40]		
Ni				[41]	[42]	[43]	
Fe					[44–49]	[50]	[51, 52]
Ti						[53]	[54]
Mg							

Table 2.
Recent reports on dissimilar LBW joints.

- Titanium [60–64]
- Nickel [65–68]
- Stainless steel [18]
- Molybdenum [40]
- Kovar [69, 70]

Further advances have been made in this technology to expand its scope. It can now be used for many other types of metals and even dissimilar materials.

14. Conclusions

This chapter is focused on the fundamentals, parameters, and applications of laser beam welding. Currently, laser sources have many applications in the field of material processing. Laser beam welding as a new technology in recent years has found wide applications in various industries such as automotive, military, aerospace, shipbuilding, electronics, etc. An energy source provides the energy required for laser production. This source stimulates the electrons held by the atoms to move to higher energy levels. Electrons reduce their energy levels dramatically, releasing photons. The spontaneous emission of photons is what leads to the production of the laser beam. In LBW, a thin and deep joint is achieved, and the heat input applied to the workpieces is so much lower than the conventional welding methods. This property allows LBW to be widely used in certain applications in which a high ratio of penetration depth to joint width is required. LBW has a great power density (in the range of megawatts per cubic centimeter), which offers a very small HAZ due to its high heating/cooling rate. The weld pool size may vary between 0.2 and 13 mm. though only smaller sizes are used for welding. Different sources include fiber lasers, Nd:YAG pulsed lasers, and Nd:YAG continuous-wave lasers are used for LBW based on the application. LBW employs three types of modes including conduction mode, conduction/penetration mode, and penetration or keyhole mode to join the materials. The main difference between these modes is in the type of heating mode, weld pool filling, depth of penetration, and shape of the weld pool. Different parameters affect the LBW process such as the chemical composition of the parent material, welding gap, welding speed, gas shielding type, beam shape (geometry), joint type, etc. Many equations are suggested to determine the LBW parameters, pulse energy, frequency, power and power density, and beam focusing adjustments, which are described in detail in the chapter.

Conflict of interest

The authors declare no conflict of interest.

Author details

Morteza Tayebi[1*], Hedayat Mohammad Soltani[2] and Ali Rajaee[3]

1 Science and Research Branch, Islamic Azad University, Tehran, Iran

2 Mining and Metallurgical Engineering Department, Amirkabir University, Tehran, Iran

3 Department of Materials and Polymer Engineering, Hakim Sabzevari University, Sabzevar, Iran

*Address all correspondence to: mortezatayebi95@yahoo.com

References

[1] Svenungsson J, Choquet I, Kaplan AFH. Laser welding process – A review of keyhole welding modelling. Physics Procedia. 2015;**78**:182-191

[2] Stavridis J, Papacharalampopoulos A, Stavropoulos P. Quality assessment in laser welding: A critical review. International Journal of Advanced Manufacturing Technology. 2018;**94**:1-23

[3] Katayama S. Handbook of Laser Welding Technologies. Elsevier Science, Woodhead Publishing; 2013

[4] Milonni PW, Eberly JH. Laser Physics. Hoboken, New Jersey: John Wiley & Sons, Inc.; 2010

[5] Dawes CT. Laser Welding: A Practical Guide. Elsevier Science, Woodhead Publishing; 1992. ISBN: 9781855730342

[6] Katayama S. Fundamentals and Details of Laser Welding. Singapore: Springer; 2020

[7] Xinju L. Laser Technology. CRC Press; 2010. ISBN: 9781138372764

[8] Sadeghi B, Sharifi H, Rafiei M, Tayebi M. Effects of post weld heat treatment on residual stress and mechanical properties of GTAW: The case of joining A537CL1 pressure vessel steel and A321 austenitic stainless steel. Engineering Failure Analysis. 2018;**94**: 396-406

[9] Sharifi H, Raisi S, Tayebi M. The effect of stress relieving treatment on mechanical properties and microstructure of different welding areas of A517 steel. Materials Research Express. 2017;**4**(12):126508

[10] Bayati MS, Sharifi H, Tayebi M, Isfahani T. Effect of Al-B4C nanocomposite filler manufactured by accumulative roll bonding (ARB) method on the microstructure and mechanical properties of Weldings prepared by tungsten inert gas welding. Materials Research Express. 2019;**6**(10):106529

[11] Mohammad Soltani H, Tayebi M. Microstructural and mechanical investigation of brazing 304L stainless steel with corner joint using a Ni-based shim and wire. Materials Research Express. 2021;**8**(4):46532

[12] Cullity BD, Stock SR, Stock SR. Elements of X-Ray Diffraction. New York: Prentice Hall; 2001

[13] Pang S, Chen X, Shao X, Gong S, Xiao J. Dynamics of vapor plume in transient keyhole during laser welding of stainless steel: Local evaporation, plume swing and gas entrapment into porosity. Optics and Lasers in Engineering. 2016; **82**:28-40

[14] Society AW, Phillips AL. Welding Handbook. New York: American Welding Society; 1958

[15] Society AW, Weisman C, Kearns WH. Welding Handbook: Metals and their Weldability. University of Michigan: American Welding Society; 1976

[16] Department, U. S. W., and Office, U. S. W. Welding Theory and Application. United States: War Department, U.S. Government Printing Office; 1943

[17] Torkamany MJ, Hamedi MJ, Malek Ghaini F, Sabbaghzadeh J. The effect of process parameters on keyhole welding with a 400 W Nd : YAG pulsed laser. Journal of Physics D: Applied Physics. 2006;**39**:4563

[18] Soltani HM, Tayebi M, Mohammad Soltani H, Tayebi M. Comparative study

of AISI 304L to AISI 316L stainless steels joints by TIG and Nd:YAG laser welding. Journal of Alloys and Compounds. 2018; **767**:112-121

[19] Kou S. Welding Metallurgy. United States of America: Wiley; 2002

[20] Porter DA, Easterling E, Sherif M. Phase Transformations in Metals and Alloys. 3rd edition. Routledge; 2009

[21] Lippold JC, Kotecki DJ. Welding Metallurgy and Weldability of Stainless Steels. University of Michigan: Wiley; 2007

[22] Hu K, Muneer W, Zhang J, Zhan X. Effect of beam oscillating frequency on the microstructure and mechanical properties of dissimilar laser welding of AA2060 and AA6061 alloy. Materials Science and Engineering A. 2022;**832**: 142431

[23] Huang W, Wang H, Rinker T, Tan W. Investigation of metal mixing in laser keyhole welding of dissimilar metals. Materials and Design. 2020;**195**: 109056

[24] Mathivanan K, Plapper P. Laser welding of dissimilar copper and Aluminum sheets by shaping the laser pulses. Procedia Manufacturing. 2019; **36**:154-162

[25] Chen S, Yu G, Li S, Yang J, Huang J, Chen S. Interfacial microstructures and mechanical property of Ni/Al dissimilar butt joint made by laser welding. Journal of Manufacturing Processes. 2020;**50**: 17-23

[26] Sadeghian A, Iqbal N. A review on dissimilar laser welding of steel-copper, steel-Aluminum, Aluminum-copper, and steel-nickel for electric vehicle battery manufacturing. Optics and Laser Technology. 2022;**146**:107595

[27] Hu Y, Zhang Y, Mi G, Wang C, Zhang W, Zhang X. Effects of Si contents in filling wires on microstructure evolution and properties of Al-steel dissimilar joint by laser welding-brazing. Journal of Materials Research and Technology. 2021;**15**:1896-1904

[28] Behúlová M, Babalová E, Nagy M. Simulation model of Al-Ti dissimilar laser welding-brazing and its experimental verification. IOP Conference Series: Materials Science and Engineering. 2017;**179**:12007

[29] Chen S, Li LQ, Chen Y, Huang J. Joining mechanism of Ti/Al dissimilar alloys during laser welding-brazing process. Journal of Alloys and Compounds. 2011;**509**:891-898

[30] Alfieri V, Caiazzo F, Sergi V. Dissimilar joining of titanium alloy Ti-6Al-4V to Aluminum alloy 2024 via laser welding. International Congress on Applications of Lasers and Electro-Optics. 2013;**2013**(1):529-537

[31] Chen X, Jiang M, Chen Y, Lei Z, Zhao S, Lin S. Laser welding-brazing under temporal and spatial power modulation for dissimilar materials AA6061 to Ti6Al4V joints. Manufacturing Letters. 2021;**29**:70-73

[32] Xue X, Wu X, Liao J. Hot-cracking susceptibility and shear fracture behavior of dissimilar Ti6Al4V/AA6060 alloys in pulsed Nd:YAG laser welding. Chinese Journal of Aeronautics. 2021; **34**(4):375-386

[33] Meng Y, Lu Y, Li Z, Zhao S, Gao M. Effects of beam oscillation on Interface layer and mechanical properties of laser-arc hybrid lap welded Al/Mg dissimilar metals. Intermetallics. 2021;**133**:107175

[34] Halim SB, Bannour S, Abderrazak K, Kriaa W, Autric M. Numerical analysis

of intermetallic compounds formed during laser welding of Aluminum-magnesium dissimilar couple. Thermal Science and Engineering Progress. 2021;**22**:100838

[35] Zhou X-R, Ning J, Na S-J, Zhang L-J. Microstructures and properties of the dissimilar joint of pure molybdenum/T2 copper by single-mode laser welding. International Journal of Refractory Metals and Hard Materials. 2021;**101**:105667

[36] Xin J, Zhang H, Sun W, Huang C, Wang S, Wei J, et al. The microstructures and mechanical properties of dissimilar laser welding of copper and 316L stainless steel with Ni interlayer. Cryogenics (Guildf). 2021;**118**:103344

[37] Aghaee Attar M, Ghoreishi M, Malekshahi Beiranvand Z. Prediction of weld geometry, temperature contour and strain distribution in disk laser welding of dissimilar joining between copper & 304 stainless steel. Optik (Stuttg). 2020;**219**:165288

[38] Pugacheva NB, Makarov AV, Senaeva EI, Volkova EG. Crystallization of dissimilar Ti/Cu/steel laser welds. Journal of Crystal Growth. 2019;**526**:125212

[39] Tan C, He W, Gong X, Li L, Feng J. Influence of laser power on microstructure and mechanical properties of Fiber laser-tungsten inert gas hybrid welded Mg/Cu dissimilar joints. Materials and Design. 2015;**78**:51-62

[40] Gao X, Li L, Liu J, Wang X, Yu H. Analysis of Ni interlayer effects on laser beam welding of dissimilar pure Mo alloy to stainless steel. International Journal of Refractory Metals and Hard Materials. 2021;**100**:105654

[41] Shanthos Kumar G, Saravanan S, Raghukandan K. Effect of heat input on microstructure and mechanical properties of laser welded dissimilar grade nickel alloys. Optik (Stuttg). 2021;**248**:168106

[42] Ahmad GN, Raza MS, Singh NK, Muvvala G. Investigating the effect of process parameters on weld Pool thermal history and mechanical properties of laser welded Inconel 625 and duplex stainless steel 2205 dissimilar welds. Optik (Stuttg). 2021;**248**:168134

[43] Cai X, Li H, Ji B, Li M, Yao X, Wang Y, et al. Effect of single alloying element (Ti, Nb, V, Cu) on microstructure and mechanical properties of dissimilar TiAl/Ni-based Superalloy laser joints. Optics and Laser Technology. 2022;**146**:107575

[44] Hamada A, Ali M, Ghosh S, Jaskari M, Keskitalo M, Järvenpää A. Mechanical performance and formability of laser-welded dissimilar butt joints between medium-Mn stainless steel and high-strength carbon steel. Materials Science and Engineering A. 2022;**831**:142200

[45] Wu Y, Guo Y, Zhang W, Li L. Microstructure evolution and dynamic mechanical behavior of laser welded dissimilar joint between QP1180 and TRIP780. Journal of Materials Research and Technology. 2022;**16**:977-987

[46] Pramanick AK, Das H, Lee J-W, Jung Y, Cho H-H, Hong S-T, et al. Texture analysis and joint performance of laser-welded similar and dissimilar dual-phase and complex-phase ultra-high-strength steels. Materials Characterization. 2021;**174**:111035

[47] Chen L, Wang C, Zhang X, Mi G. Effect of parameters on microstructure and mechanical property of dissimilar

joints between 316L stainless steel and GH909 alloy by laser welding. Journal of Manufacturing Processes. 2021;**65**:60-69

[48] Zhou X, Zhao H, Liu F, Yang B, Chen B, Tan C. Influence of energy ratio on microstructure and mechanical properties in the transition zone of hybrid laser-MIG welded AH36/316L dissimilar joints. Journal of Materials Research and Technology. 2021;**15**: 4487-4501

[49] Akyel F, Gamerdinger M, Olschok S, Reisgen U, Schwedt A, Mayer J. Adjustment of chemical composition with dissimilar filler wire in 1.4301 austenitic stainless steel to influence residual stress in laser beam welds. Journal of Advanced Joining Processes. 2022;**5**:100081

[50] Zhang Y, Zhou J, Sun D, Gu X. Nd: YAG laser welding of dissimilar metals of titanium alloy to stainless steel without filler metal based on a hybrid connection mechanism. Journal of Materials Research and Technology. 2020;**9**(2):1662-1672

[51] Rong Y, Li H, Cheng D, Xiong Z, Chen Y, Liu Z, et al. Study on microstructure and properties of Mg/ steel laser welding-brazing joint assisted by alternating magnetic field with Ni interlayer. International Journal of Lightweight Materials and Manufacture. 2021;**4**(4):416-422

[52] Tan C, Li L, Chen Y, Guo W. Laser-tungsten inert gas hybrid welding of dissimilar metals AZ31B Mg alloys to Zn coated steel. Materials and Design. 2013; **49**:766-773

[53] Zhang K, Liu D, Cai J, Zhou Y, Niu H, Yan C, et al. Influence of laser offsets on microstructure and tensile properties of laser welded Ti-3Al-6Mo-2Fe-2Zr and TA15 dissimilar joints. Materials Characterization. 2021;**180**: 111441

[54] Zhang K, Liu J, Tan C, Wang G, Song X, Chen B, et al. Dissimilar joining of AZ31B Mg alloy to Ni-coated Ti-6Al-4V by laser heat-conduction welding process. Journal of Manufacturing Processes. 2018;**34**:148-157

[55] Prabakaran MP, Kannan GR. Effects of post-weld heat treatment on dissimilar laser welded joints of austenitic stainless steel to low carbon steel. International Journal of Pressure Vessels and Piping. 2021;**191**:104322

[56] Dong X, Wang G, Ghaderi M. Experimental investigation of the effect of laser parameters on the weld bead shape and temperature distribution during dissimilar laser welding of stainless steel 308 and carbon steel St 37. Infrared Physics & Technology. 2021; **116**:103774

[57] Xu T, Zhou S, Wu H, Ma X, Liu H, Li M. Dissimilar joining of low-carbon steel to Aluminum alloy with TiC particles added in a zero-gap lap joint configuration by laser welding. Materials Characterization. 2021;**182**:111574

[58] Gu H, Fu Q, Sheng B, Cai E, Tang G, Liu Q. Effect of pore-forming agent quantity on pore structure, phase composition, micro-hardness of gradient bioceramic coatings under optimal laser process parameters. Ceramics International. 2020;**46**(8, Part A): 11275-11281

[59] Cao X, Wang GG, Xing C, Tan C, Jiang J, Li T, et al. First-principles and experimental investigations on ductility/ brittleness of intermetallic compounds and joint properties in steel/Aluminum laser welding. Transactions of the Nonferrous Metals Society of China. 2021;**31**(10):2277-2286

[60] Zhang Y, Chen Y, Zhou J, Sun D, Li H, Xue R, et al. Use of pure vanadium and niobium/copper inserts for laser welding of titanium to stainless steel. Optics and Laser Technology. 2020; **244**(1):465-477

[61] Zhang Y, Chen Y, Zhou J, Xue R, Sun D, Li H. Characterization of laser beam offset welding of titanium to steel with 38Zn-61Cu alloy filler. Optics and Laser Technology. 2020;**127**:106195

[62] Fang Y, Jiang X, Song T, Mo D, Luo Z. Pulsed laser welding of Ti-6Al-4V titanium alloy to AISI 316L stainless steel using Cu/Nb bilayer. Materials Letters. 2019;**244**:163-166

[63] Zhang Y, Chen Y, Zhou J, Sun D, Li H. Experimental and numerical study on microstructure and mechanical properties for laser welding-brazing of TC4 titanium alloy and 304 stainless steel with Cu-Base filler metal. Journal of Materials Research and Technology. 2020;**9**(1):465-477

[64] Chattopadhyay A, Muvvala G, Sarkar S, Racherla V, Nath AK. Mitigation of cracks in laser welding of titanium and stainless steel by In-situ nickel interlayer deposition. Journal of Materials Processing Technology. 2022; **300**:117403

[65] Gook S, Krieger S, Gumenyuk A, El-Batahgy AM, Rethmeier M. Notch impact toughness of laser beam welded thick sheets of cryogenic nickel alloyed steel X8Ni9. Procedia CIRP. 2020;**94**: 627-631

[66] Wu D, Cheng B, Liu J, Liu D, Ma G, Yao Z. Water cooling assisted laser dissimilar welding with filler wire of nickel-based alloy/ austenitic stainless steel. Journal of Manufacturing Processes. 2019;**45**: 652-660

[67] Zhou S, Ma G, Dongjiang W, Chai D, Lei M. Ultrasonic vibration assisted laser welding of nickel-based alloy and austenite stainless steel. Journal of Manufacturing Processes. 2018;**31**: 759-767

[68] Baghjari SH, Malek Ghaini F, Shahverdi HR, Mapelli C, Barella S, Ripamonti D. Laser welding of niobium to 410 steel with a nickel interlayer produced by electro spark deposition. Materials and Design. 2016;**107**:108-116

[69] Mai TA, Spowage AC. Characterisation of dissimilar joints in laser welding of steel–Kovar, copper–steel and copper–aluminium. Materials Science and Engineering A. 2004;**374**(1): 224-233

[70] Baghjari SH, AkbariMousavi SAA. Experimental investigation on dissimilar pulsed Nd:YAG laser welding of AISI 420 stainless steel to Kovar alloy. Materials and Design. 2014;**57**:128-134

Chapter 2

Effect of Welding Variables on the Quality of Weldments

Ramy A. Fouad, Essam Ahmed Ali,
Ahmed Ramadan Shaaban and Ahmed E. El-Nikhaily

Abstract

The effect of nitrogen addition, heat input, and filler metals on weld metal microstructure and mechanical properties of alloy 316 ASS are studied. Autogenous gas tungsten arc welding (GTAW) is employed by adding up to 2vol. % N_2 in Ar. These variables affect a number of welding aspects, including arc characteristics and microstructure. The influence of shielding gas mixtures on microstructure and mechanical properties of GTAW of austenitic 316 stainless steel is studied. Mechanical properties of welds are determined through uniaxial tension, hardness measurements, impact, and bending tests. Weld defects, as porosity and inclusions are examined using radiographic testing. Weld specimens are free of porosity, inclusions, and hydrogen cracking. Mechanical properties and cooling rate are lower at higher heat input, but the cooling time, nugget area, and solidification time are higher. The addition of N_2 to Ar shielding gas leads to higher values of the ultimate tensile strength 'UTS', yield stress 'YS', and elongation percent. UTS, YS, and elongation of welds depend on heat input, filler metal, and N_2 content of shielding gas. Finally, a mathematical model is built depending upon the welding current, filler metals, and shielding gases.

Keywords: TIG welding, filler metals, nugget area, cooling rate, welding variables

1. Introduction

In this chapter, we will first introduce you to the field of welding processes using different welding variables examples. We will then provide an introduction to the classification of welding variables. Although most engineering programs or mechanical engineering programs require students to take welding technology courses, you should approach your study of welding technology as more than a mere requirement. Thorough knowledge of welding processes will make you a better engineer and designer. Welding science underlies all technological advances and an understanding of the basics of welding and its applications will not only make you a better engineer but will help you during the production process. In order to be a perfect designer, you must learn what welding will be appropriate to use in different applications. Also, in this chapter, previous works related to the current study are discussed. Various welding techniques will be presented. Finally, studies focused on welding and the effect of welding variables on stainless steel will be discussed. GTAW is considered one of the most productive welding methods since it is used in the welding of metals with high thickness. For this reason, it is used in the heavy industry and shipbuilding industry.

The shielding gas interacts with the base material and with the filler material, if any, to produce the basic strength, toughness, and corrosion resistance of the weld. It also affects weld bead shape and penetration pattern [1, 2]. Successful GTAW weldments of Monel 400 and AISI 304 were developed using ER304, ERNiCrMo-3, and ERNiCrMo-4 welding wires [3]. The tensile strength and yield strength of ERNiCrMo-3 weldments were comparable to those of parent metals. Tensile strength and yield stress of dissimilar ERNiCrMo-3 weld joints were better than ER304 and ERNiCrMo-4 weldments. Also, the effect of welding wires on the characteristics of dissimilar welding of SS316 L and carbon steel A516 GR 70 was studied [4] using three different filler materials ER80-Ni1, ER309L, and ER NiCrMO-3 (Inconel 625). Inconel 625 was found more suitable to weld dissimilar SS316 L and carbon steel A516 GR 70. Best results concerning UTS, and hardness were obtained using Inconel 625 as a welding electrode. The effect of welding electrodes on the characteristics of dissimilar AISI 420 and 304 L welds was studied [5] using three different filler rods ER312, ER316 L, and ER2209. The last rod produced welds with the highest impact toughness and lowest hardness. Kanigalpula et al. [6] developed mathematical models using central composite design methodology 'CCD' to determine the process variables that produce more stable weld bead geometry and microhardness in the electron beam welding process. Hackenhaar et al. [7] applied Box-Behnken design 'BBD' to investigate the effect of gas metal arc welding 'GMAW' parameters (wire feed speed, welding speed, and arc voltage) based on 3 responses (melting efficiency, bead on plate, and melting area of T-joint) in GTAW butt welding of 6.35 mm thick AISI 1010 steel plates. The melting efficiency showed a direct relationship with heat flow extraction in weld joint, thus, of joint geometry. Melting efficiency is lower for T-joint regardless used equation.

2. Basic welding parameters

The welding arc is formed by the arc plasma, which consists of ionized gas, molten metals, slags, vapors, and gaseous atoms and molecules. Arc welding variables are welding current, are voltage, welding speed, shielding gas and filler metal that cause much more effect on various weld joint properties such as strength, weld bead geometry, cooling rate and corrosion of stainless steel.

2.1 Welding current

Welding current is the most important variable affecting melting rate, metal deposition rate, depth of penetration, width of joint and the amount of molten base metal. The electrical energy can be calculated by Eq. (1) [8]:

$$Q = I * V\ J/S \text{ or } Q = I^2 * R_a\ J/S \tag{1}$$

where, Q = Consumed electrical energy I = Welding current.

V = Arc voltage R_a = Arc resistance.

The influence of welding current on AISI 316 welded by GTA was studied [9]. UTS and HV increased with increasing heat input. Best results for UTS and HV were obtained using 100A. Also, sigma phase and $Cr_{23}C_6$ in 316 SS welded samples increased with increasing heat input as shown in **Figures 1**–3.

2.1.1 Welding arc voltage

Welding arc voltage or arc voltage is the electrical potential difference between the welding wire tip and the molten weld zone surface. The welding arc voltage depends on arc length and type of electrode. Weld bead shape appearance depends on arc voltage. Increasing arc voltage causes porosity, welding reinforcement electrode, spatter flatten the weld bead, and increases the weld width [10].

2.1.2 Gas purity

Metals differ in their tolerance for foreign components of the shielding gas. Impurities of shielding gases affect weld quality and eventual fitness. Basyigit and Kurt [11] employed five different shielding gases such as, pure argon, 99%Ar + 1% N_2, 97%

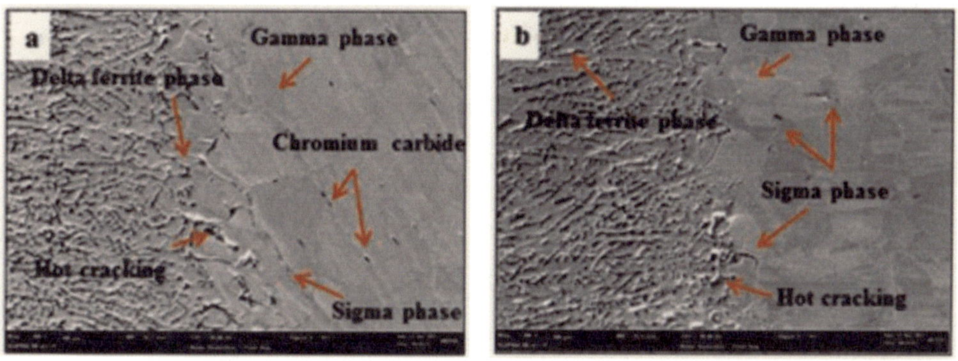

Figure 1.
Effects of welding current on the sigma and carbide phases distribution of 316 stainless steel, achieved by FESEM – EDX (a) current 90A and (b) current 110A.

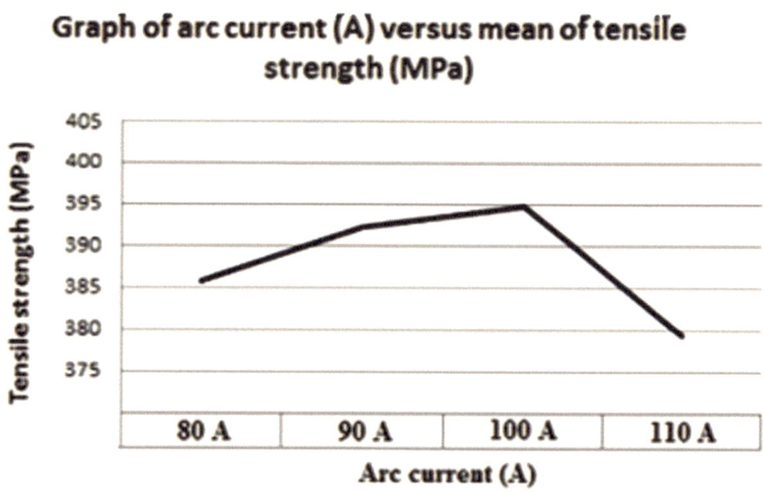

Figure 2.
Graph of arc current (A) versus mean of tensile strength (MPa) for316 stainless steel welded joint.

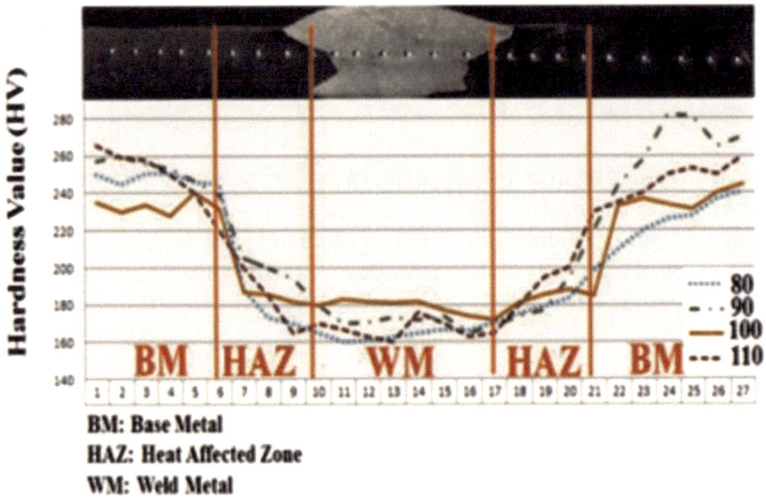

Figure 3.
Comparison of number of points versus hardness (HV) of 316 stainless steel welded joint.

Ar + 3% N_2, 94%Ar + 6%N_2, and 91%Ar + 9% N_2 for welding 2205 DSS using TIG welding. The austenitic structure increased with increasing N_2 content in shielding gas (**Figure 4**). Increasing N_2 content in argon shielding gas led to improve grain size, UTS and HV of DSS welds (**Figure 5**).

Besides that, Mosa et al. [12] studied the effect of shielding gas and heat input on mechanical properties of ASS 304 L welded by TIG welding. Tensile strength and hardness values decrease with increasing heat input, but the ferrite number, impact toughness, penetration depth and weld bead width increase. Addition of N_2 in Ar increases UTS and HV but reduces FN and impact toughness (**Figures 6–11**).

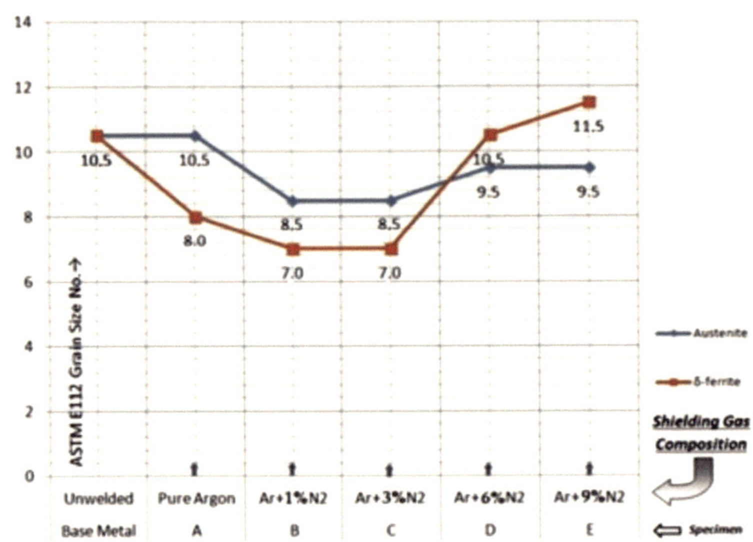

Figure 4.
The effects shielding gas composition on phases grain size.

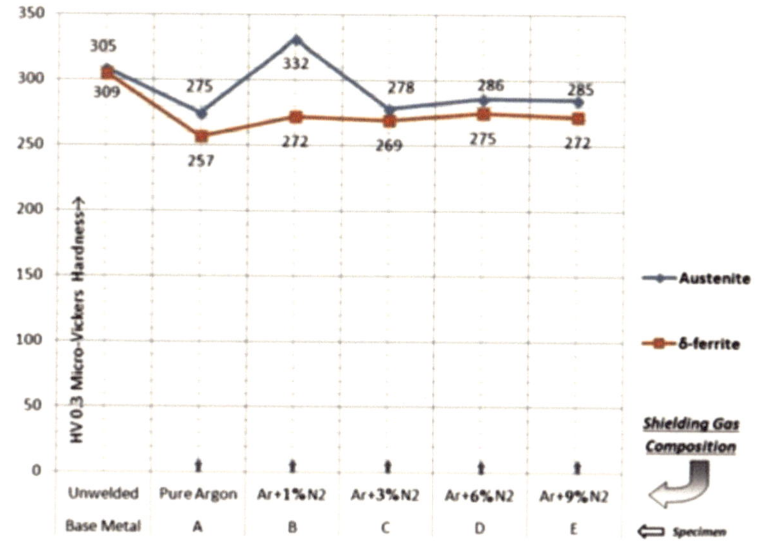

Figure 5.
The effects of shielding gas composition on micro-hardness values of base metal and weldments.

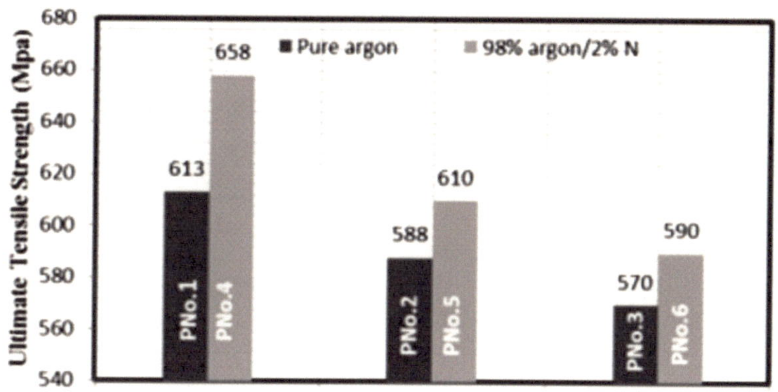

Figure 6.
Comparison between ultimate tensile strength of welding procedures using shielding gas pure Ar and 2%N/98%Ar.

2.1.3 Welding speed

Welding speed is the linear speed at which the arc moves with respect to the plate, along with the weld joint. The heat input and cooling rate increase with decreasing welding speed. Welding speed is calculated by Eq. (2) [13]:

$$\text{Welding Speed (mm/ min)} = \text{Electrode Travel/Arc Time} \tag{2}$$

Moreover, optimization of AISI 316 weld sample characteristics was studied [14] by Taguchi method (ANOVA). Besides, the effect of welding variables as welding speed and current, filler metal and root gap on UTS and bend strength was studied. Travel speed (46.51% contribution) has a greater influence on toughness (bend

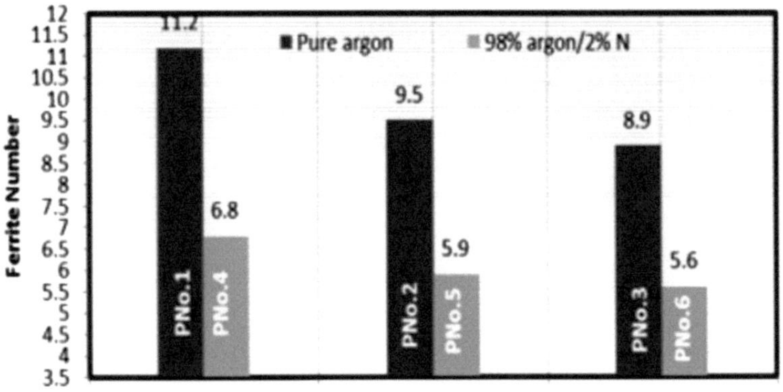

Figure 7.
Comparison between FN of welding procedures using shielding gas pure and 2% N/98%Ar.

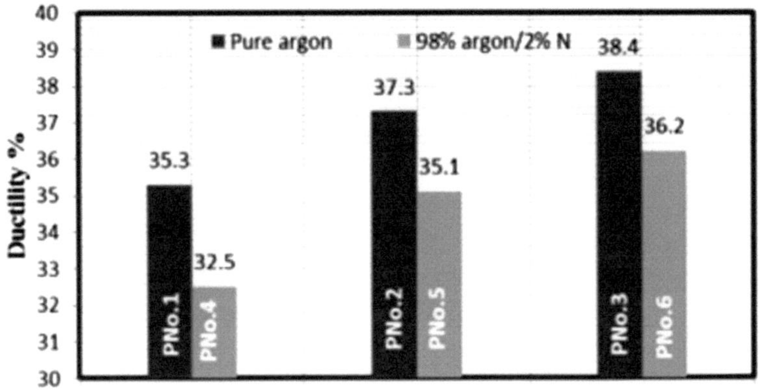

Figure 8.
Comparison between ductility of welding procedures using shielding gas pure Ar and 2%N/98%Ar.

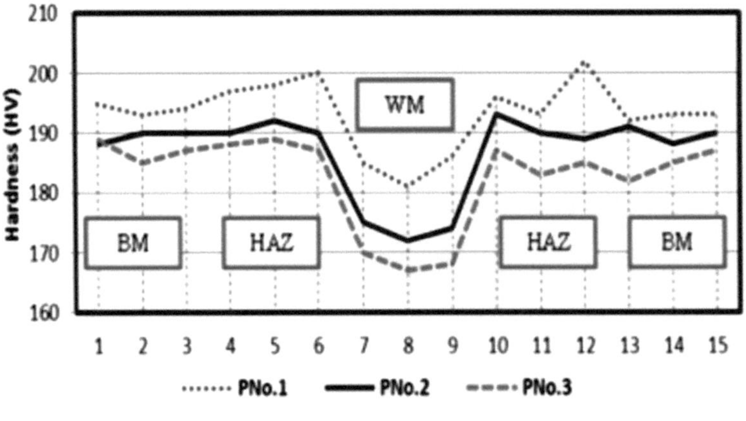

Figure 9.
Hardness values for PNo.1 (0.395 Kj/mm), PNo.2 (0.79 Kj/mm) and PNo.3 (0.998 Kj/mm) all with using shielding gas of pure argon.

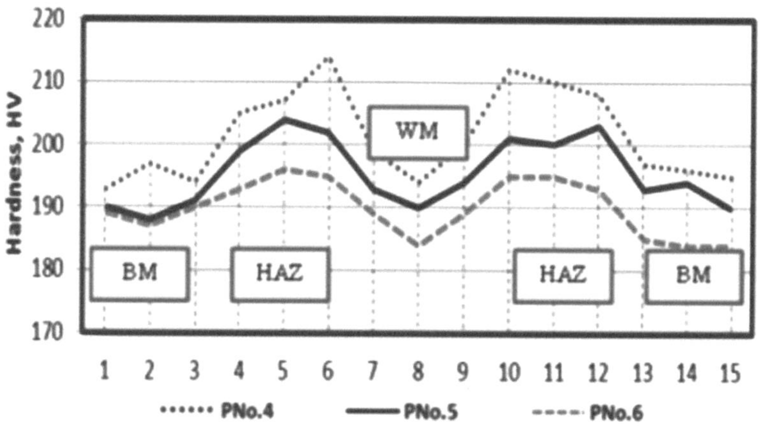

Figure 10.
Hardness values for PNo.4 (0.411 Kj/mm), PNo.5 (0.822 Kj/mm) and PNo.6 (1.053 Kj/mm) all with using shielding gas of (2%N/98% Ar).

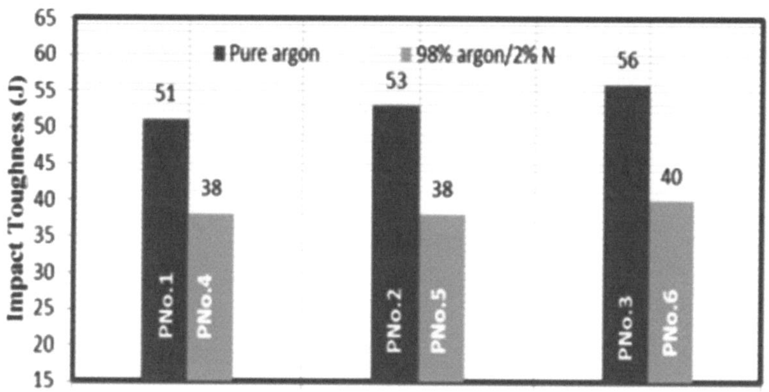

Figure 11.
Comparison between impact toughness of PNo.3 versus PNo.4.

strength) and welding current (96.75%) has maximum influences on UTS. The root gap has some effect on both tensile and bend strengths.

2.1.4 Effect of heat input

Heat input is a relative measure of the energy transferred per unit length of weld. It is an important characteristic because, like preheat and interpass temperature, it influences the cooling rate, which may affect the mechanical properties and metallurgical structure of weld region and HAZ [15]. It is calculated by Eq. (3).

$$HI\ (KJ/mm) = \eta * (V * I * 60)/(S * 1000) \quad (3)$$

where HI is the heat input (in KJ/mm), η is welding efficiency (η_{TIG} = 70%), V is arc voltage (in volt), I is welding current (in Amp) and S is welding speed (in mm/min).

The effect of heat input on microstructure and mechanical properties have been extensively investigated. For example, Movahedi and Ozlati [16] studied the influence of heat input on mechanical properties of dissimilar welds AISI 410 MSS and 2209 DSS rods. The heat input, UTS and %EL increase with increasing welding current. Best results of *UTS* and %EL are obtained at a welding current of 3.5 kA. Whereas the lowest results of UTS and %EL are obtained at a welding current of 2 kA (**Figure 12**). Moreover, Sergei Yu et al. [17] analyzed the influence of H_{net} on residual strain and phase content in AISI 304 stainless steel welds using different heat input values. The ferrite content increased with increasing heat input from 0.225 to 0.247 kJ/mm, while the austenite content decreased.

Singh and Kumar [18] investigated the characteristics of 304 stainless steel joints using SMAW-GTAW hybrid welding and different filler metals. The joint with 90A

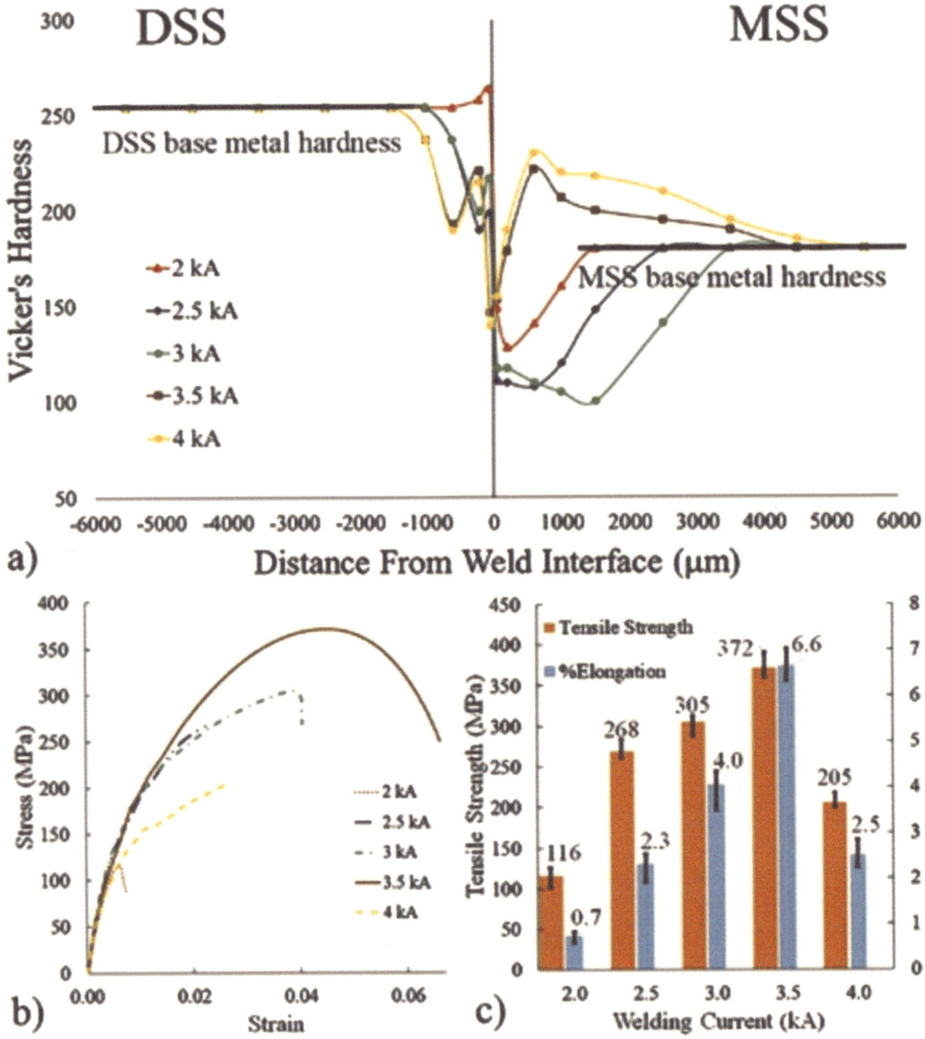

Figure 12.
(a) Hardness profile across the weld interface. (b) Stress-strain curves and (c) values of tensile strength and elongation for all samples [16].

has the highest hardness value and lowest toughness value. The toughness value of weld metal and HAZ increases with increasing heat input. Welding width and depth (penetration) increase with increasing heat input. It can be seen that root reinforcement deposited at 0.93 kJ/mm is wider than that deposited at 0.68 kJ/mm. Whereas, D. Bahar [19] studied the effect of welding parameters as; welding current, gas flow rate and welding speed on bending strength and weld geometry of dissimilar welds of SS 304 and mild steel 1018. The welding width, depth of penetration and bending strength increased with increasing welding current, gas flow rate and welding speed. Additionally, Bodude and Momohjimoh [20] explained the influence of welding variables on mechanical properties of low carbon steel welded by SMAW. Highest ultimate tensile strength and hardness were realized for samples welded at low current. Moreover, best results of weld toughness were obtained at welding current of 150A. HV and UTS increased with decreasing heat input, while the impact toughness increased with increasing heat input. Gupta et al. [21] studied the influence of H_{net} on the mechanical behavior of FSS 409 plate using two different filler metals (ER304L and ER308 L). Best *UTS* result, yield strength, hardness value and grain size is obtained using medium heat input '4 kj/mm', irrespective of the used filler metal. The mechanical behavior is also infl uenced by grain size of weld metal and heat input. Generally, the joints welded using 304 filler metal showed better results than using 308 filler wire. Moreover, the effect of heat input on the mechanical properties and fatigue life of AA6061 alloy welded by MIG welding was reported [22]. The weld penetration increased linearly with increasing the weld current and arc voltage, but with decreasing the welding speed. On the other hand, the fatigue life decreased with increasing the welding current and arc voltage, whereas, the fatigue life increased with increasing the welding speed. The impact toughness increased slightly with increasing heat input (**Table 1**). Swami et al. [23] studied the influence of MIG welding current, gas flow rate and shielding gas on UTS of 12 mm thick mild steel plates. MIG welding process variables affect UTS value of the weld metal. Best *UTS* result was recorded at 190 A, 15 (L/min) and 50% CO_2. Bansod et al. [24] investigated the change of mechanical properties of low-nickel ASS 304 welds using SMAW technique at various heat input values. Highest UTS and HV were obtained using low heat input values (**Table 2**). Hardness of the weld zone was lower than that of the heat affected zone and base metal (**Figures 13–15**). The ferrite number of weld region increased with decreasing the heat input (**Table 3**). Furthermore, Bansod et al. [25] studied the influence of heat input on physical metallurgy, mechanical behavior and corrosion rate of Cr-Mn ASS and low nickel ASS specimens. The width of HAZ increased with increasing heat input, while the FN and volume fraction of delta ferrite in weld region decreased. On the other hand, the hardness and tensile strength increased with decreasing heat input. Besides, the pitting resistance was improved with increasing delta ferrite. Ahmed et al. [26] studied the change of weld strength of ASS 316 welds using GTAW technique at various heat input values and filler metals. Using ERNiCrMo-3 as filler rod produced weldments with higher ultimate tensile strength and yield stress than using ER309L or ER316 L. The ultimate tensile strength, yield stress and elongation percent decrease with increasing heat input. Highest values are obtained using ERNiCrMo-3 filler rod at comparatively low welding current (80 A). The hardness is lower in weld zone than that of in heat affected zone and base metal. In general, it decreases with increasing heat input (welding current). Highest values are obtained using ERNiCrMo-3 filler with low heat input (80 A) (**Figures 16–18**).

Sample number	Heat Input (J/mm)	No. of cycles to failure	Impact energy (J)	Penetration (mm)
1	264	221	83.91	2.411
2	219.96	360	67.37	2.464
3	188.58	476	47.71	2.310
4	303.6	198	91.27	2.571
5	252.96	331	73.75	2.612
6	216.84	463	55	2.511
7	343.2	161	91.25	2.599
8	285.96	317	83.93	2.634
9	245.1	429	71.6	2.541
10	312	190	94.18	2.781
11	259.98	325	83.01	2.872
12	222.84	439	67.63	2.741
13	358.8	157	81.41	2.860
14	298.98	296	90.55	2.932
15	256.26	408	77.21	2.802
16	405.6	101	63.25	2.941
17	337.98	264	94.32	2.970
18	289.68	372	84.62	2.872
19	360	136	77.84	2.983
20	300	272	91.01	3.020
21	257.1	390	78.68	2.911
22	414	98	61.17	3.078
23	345	258	85.65	3.101
24	295.68	366	88.61	3.001
25	468	85	58.64	3.150
26	390	246	71.91	3.202
27	334.26	353	93.65	3.120

Table 1.
Life, impact energy and penetration of weld with respect to the weld parameters [22].

2.1.5 Cooling rate and solidification time

Cooling rate '*CR*' is the heat loss during welding per unit time. *CR* plays an important role in determining the final solidification microstructure and its properties. Merchant Samir [27] studied the influence of welding current, arc voltage and welding speed on cooling rate, solidification time and hardness value of mild steel welded by MMAW process. It was found that the cooling rate decreased with increasing the welding current, while the solidification time increased for samples welded using different current and voltage values. The cooling rate increased with increasing

	Base metals		Dissimilar joint welded by different fillers		
	Cr-Mn SS	**304 SS**	**316 L**	**310**	**308 L**
Yield strength (MPa)	222.6	335.2	336.7	330.4	333.3
Ultimate tensile strength (MPa)	808.7	670	667.50	660.0	667.9
% elongation	60.2	52.5	50.34	49.5	55.2
Fracture zone	—	—	304	304	304

Table 2.
Tensile test results [24].

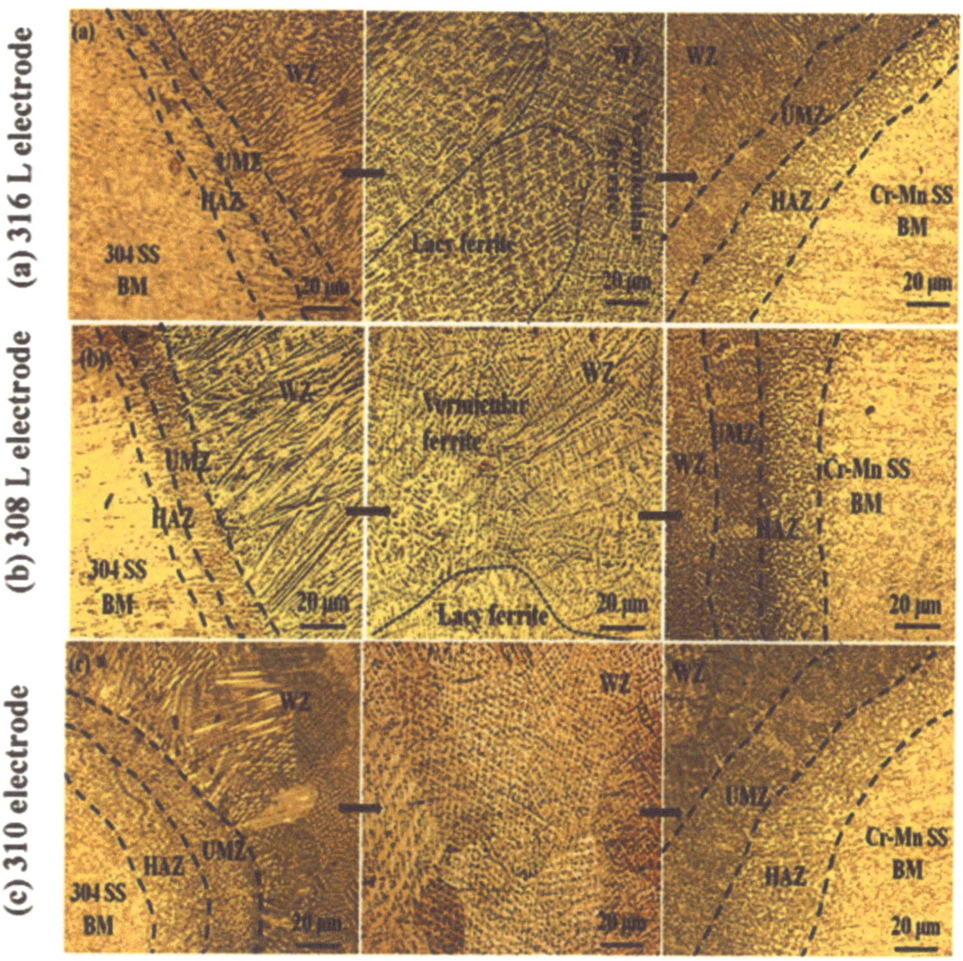

Figure 13.
Optical micrographs of weld samples (a) 316 L electrode, (b) 308 L electrode and (c) 310 electrode [24].

welding speed, while the solidification time decreased. The best result for *HRN* is obtained in HAZ of all samples. Besides, Rahul Kumar et al. [28] examined the effect of welding variables and cooling rate on the mechanical behavior of mild steel welded by SAW. They found that the cooling rate and hardness increased with reducing the

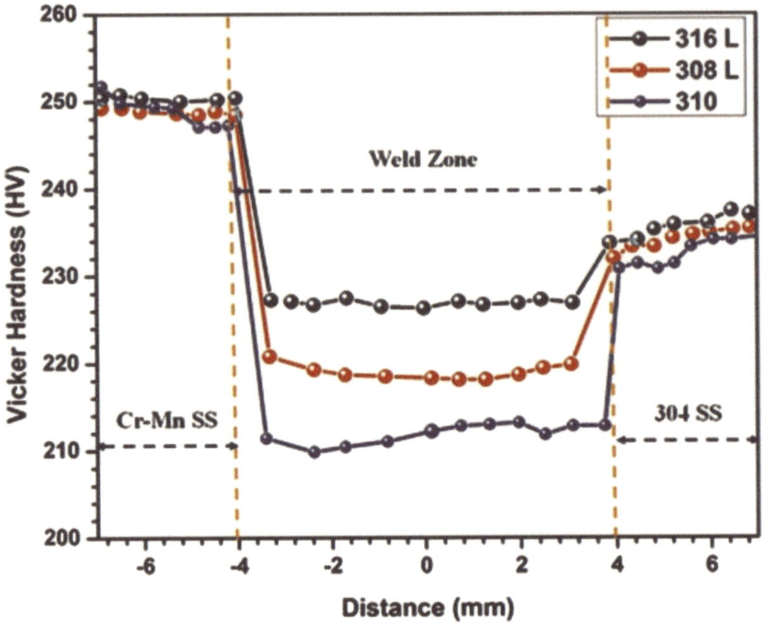

Figure 14.
Microhardness profile across the weld specimens [24].

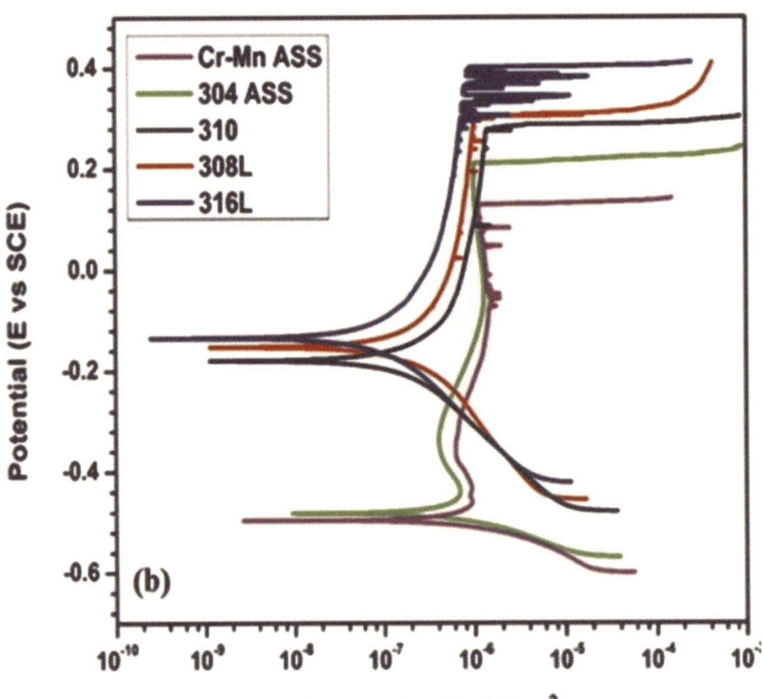

Figure 15.
Potentiodynamic polarization plots of various sample [24].

Welding Electrodes	Unmixed zone length (μm) 304 SS	HAZ width (μm)	Unmixed zone length (μm) Cr-Mn SS	HAZ width (μm)	Average level of dilution (%)	δ-ferriteby Ferritoscope (FN)		
						Weld zone	Base metal (304 SS)	Base metal (Cr-Mn SS)
316 L SS	180	245	257	498	24 ± 3	6.0	0.15	0.16
308 L SS	181	240	250	452	25 ± 3	5.1	0.15	0.16
310 SS	185	256	255	468	24 ± 2	0.14	0.15	0.16

Table 3.
Microstructural details of welding [24].

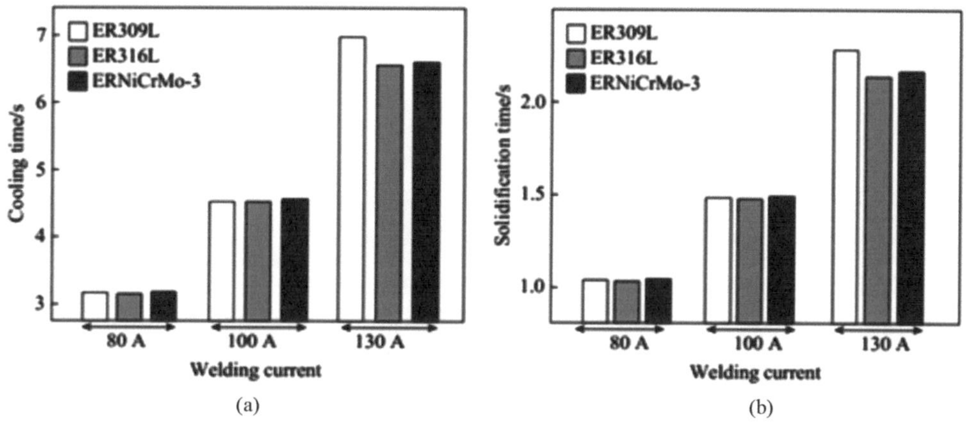

Figure 16.
Effect of welding current on (a) cooling time (b) solidification time of 316SS welding [26].

heat input. Whereas a finer grain size was formed at higher cooling rate and lower heat input.

Effect of cooling rate on solidification and segregation characteristics of SASS was studied [9]. The grain size was refined more with increasing cooling rate. Dendrite arm spacing decreased at welding begin, then decreased slowly with increasing cooling rate. Transition cooling rate was 20°C/sec. Also, the effect of heat input on cooling rate and PREN in SDSS welds was studied [14]. Grain size and cooling rate increased with increasing heat input. Best results for PREN were obtained at an intermediate heat input value of 1.4 kJ/mm. Besides, Ahmed et al. [29] examined the effect of heat input and shielding gas on the Performance of 316SS welded by GTAW. They found that the heat input, cooling time, solidification time, grain size and nugget area increase with increasing the welding current. Besides, the cooling rate decreases with increasing the welding current. Whereas the UTS, YS and EL% decrease with increasing heat input, and the addition of 2%N_2 to Ar shielding gas increases the mechanical properties of 316 stainless steel weld joints. The best mechanical properties are obtained at welding current 80 amp with Ar-2%N_2. The hardness is lower in the weld zone than in the heat affected zone and base metal, and the addition of 2% N_2 to shielding gas increases it. Moreover, the hardness decreases with increasing heat input (**Figures 19–26**).

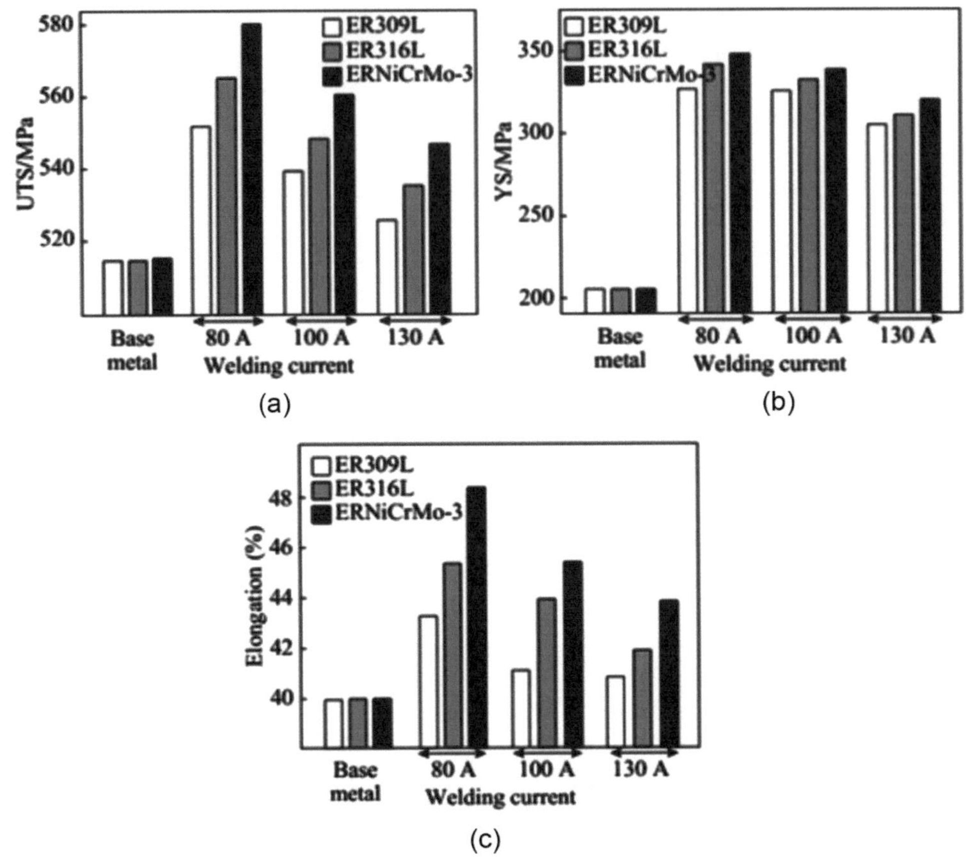

Figure 17.
Mechanical test results of TIG welded joints (a) tensile strength, (b) yield strength, and (c) percentage elongation [26].

The cooling rate in the temperature range 800–500°C is important for phase transformation of stainless steel. It determines the final solidification mode or microstructure of the weld metal and its properties [30]. The cooling rate and cooling time [26, 31] can be calculated using Eq. (4) and Eq. (5), respectively.

$$\left(\frac{\partial T}{\partial t}\right)_x = \left(\frac{\partial T}{\partial x}\right)_t * \left(\frac{\partial x}{\partial t}\right)_{xT} = -2\pi K * \left(\frac{(T - T_o)^2}{H_{net}}\right) \tag{4}$$

$$t_{8/5} = \frac{HI}{2\pi\lambda} * \left(\frac{1}{500 - T_o} - \frac{1}{800 - T_o}\right) \tag{5}$$

where, $(\partial T/\partial t)_x$ is the cooling rate '°C/sec', K or λ is the thermal conductivity (W/mmK), and T is the temperature near the pearlite nose on TTT diagram '550°C' and T_o is the initial temperature of the plate '20°C'.

The solidification time 'S_t' of welding joint depends on the cooling rate and heat input. The S_t time is important as it affects the microstructure and properties, and can be calculated using Eq. (6) [26, 31]:

$$St\,(sec) = L\,Hnet/2\pi K\rho c\,(T_m - T_o)^2 \tag{6}$$

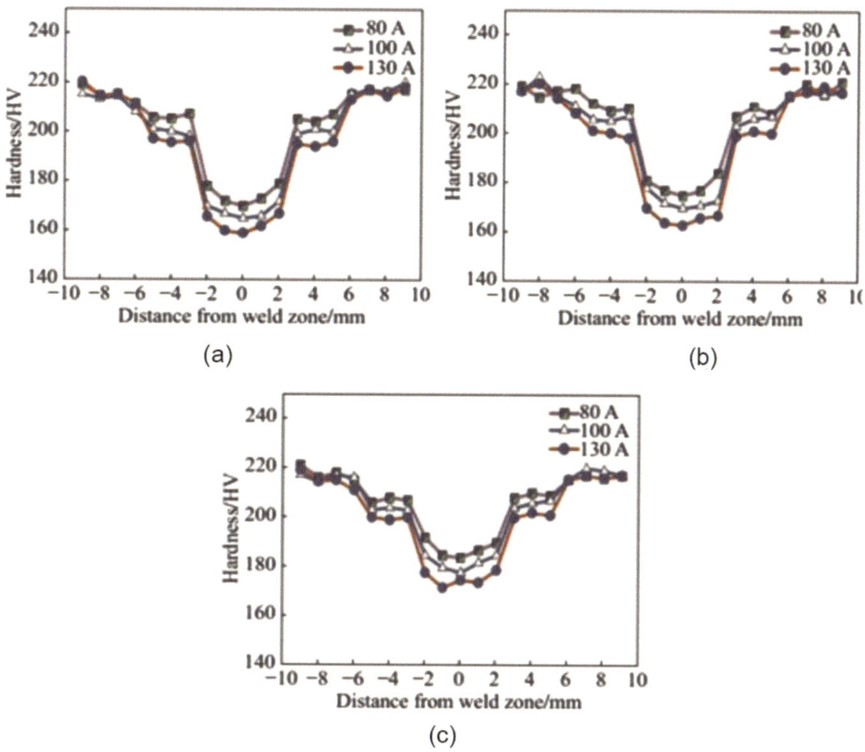

Figure 18.
Vickers hardness profiles of 316SS TIG joint cross sections for different values of welding current using (a) ER309L, (b) ER316 L and (c) ERNiCrMo-3 as filler rods [26].

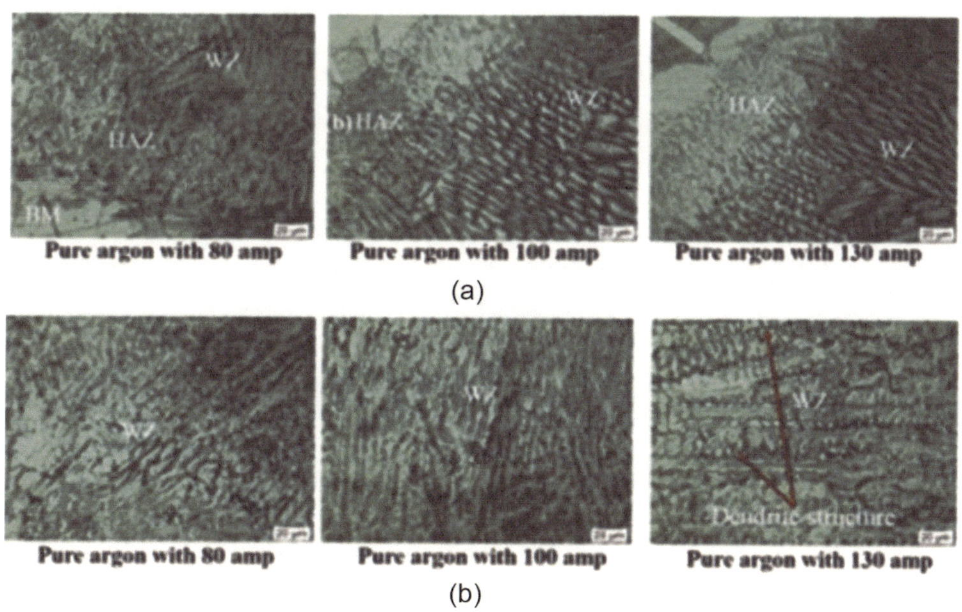

Figure 19.
Microstructure of weldments using various welding currents 80, 100 and 130 amp and pure argon as shielding gas, at different locations (a) and (b) [31].

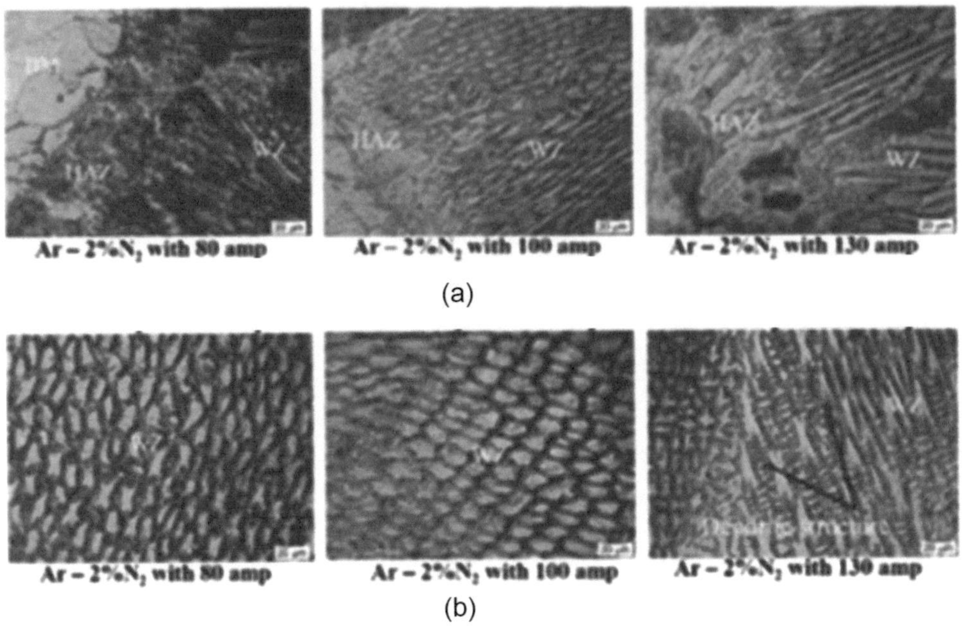

Figure 20.
Microstructure of weldments using various welding currents 80, 100 and 130 Amp and Ar-2%N_2 as shielding gas, at different locations (a) and (b) [31].

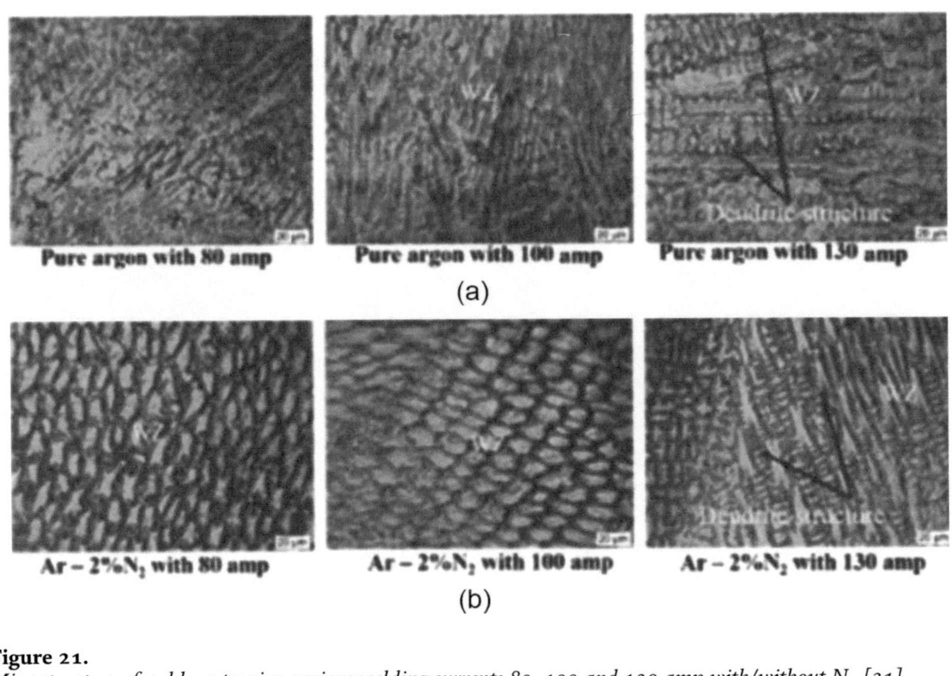

Figure 21.
Microstructure of weldments using various welding currents 80, 100 and 130 amp with/without N_2 [31].

2.1.6 Weld bead geometry

The effect of weld bead area on mechanical properties was investigated [26, 31], and it was found that the nugget area 'Na' increases with increasing weld current and

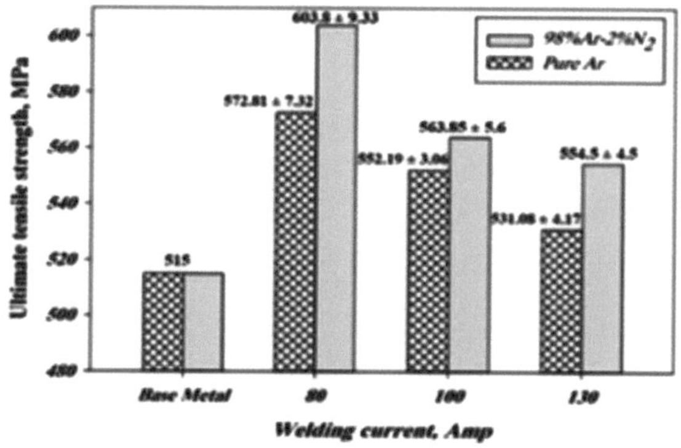

Figure 22.
Ultimate tensile strength of GTAW welded AISI 316SS using various welding currents 80, 100 and 130 Amp and Ar-2%N_2 as shielding gas [31].

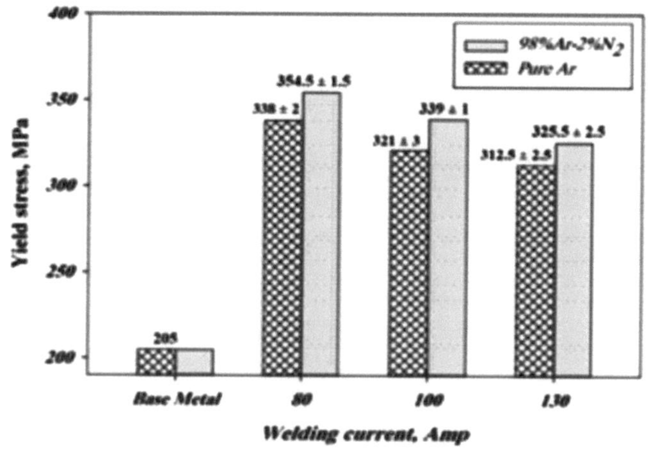

Figure 23.
Yield stress of GTAW welded AISI 316SS using various welding currents 80, 100 and 130 Amp and Ar-2%N_2 as shielding gas [31].

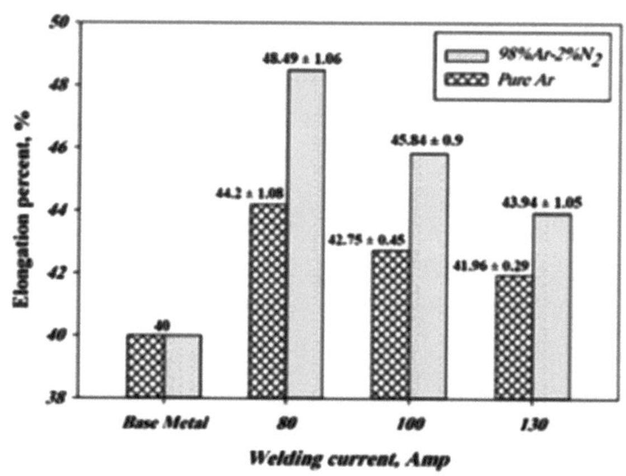

Figure 24.
Percentage elongation of 316SS welding specimens [31].

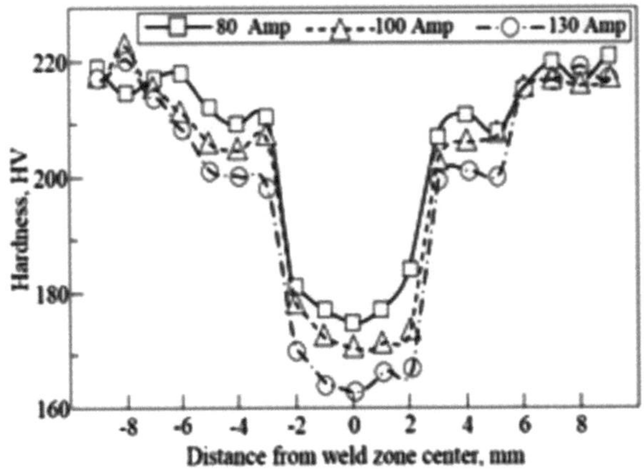

Figure 25.
Hardness profiles of 316SS GTAW joint cross sections of weld joints using pure argon as shielding gas [31].

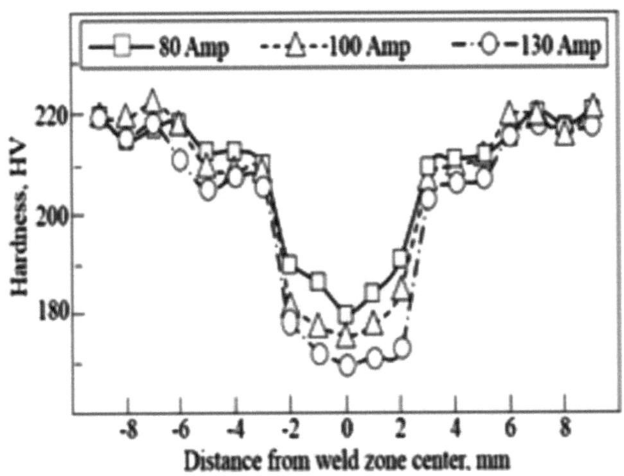

Figure 26.
Hardness profiles of 316SS GTAW joint cross sections of weld joints using Ar-2% N_2 *as shielding gas [31].*

arc voltage, but decreases with increasing welding speed, and can be calculated with Eq. (7) (**Table 4**).

$$\mathrm{Na}\left(\mathrm{mm}^2\right) = 33312 * 10^{-6} * [\mathrm{A1.55/S0.903}] \tag{7}$$

where, Na is nugget area 'mm^2', A is the welding current in 'amp', and S is the welding speed 'mm/sec'.

3. Conclusions

1. Increasing weld current, increases heat input, time of cooling and of solidification, metal deposition rate and nugget area, but decreases cooling rate.

Weld current	Electrodes	Nugget area
80 amp	ER309L	17.3
100 amp		26
130 amp		42.4
80 amp	ER316 L	17.2
100 amp		26
130 amp		40
80 amp	ERNiCrMo-3	17.4
100 amp		26.2
130 amp		40.3

Table 4.
Nugget weld area of 316SS specimens welded using different welding current values and filler metals and pure Ar shielding gas [26].

2. Ultimate tensile strength, yield stress, elongation percent and fatigue life decrease with increasing weld current. Using ERNiCrMo-3 filler rod, 80amp weld current and 2%N_2 in shielding gas leads to highest mechanical properties.

3. The hardness is lower in weld zone than in heat affected zone and base metal. The addition of 2% N_2 to shielding gas increases it, but the increase of weld current decreases it. Using ERNiCrMo-3 filler rod, 80 amp weld current and 2% N_2 in shielding gas leads to highest hardness.

4. Weld metal grain size and crystallization temperatur increase with increasing heat input and N_2 content in the shielding gas. Adding 2% N_2 to Ar shielding gas leads to $\alpha \rightarrow \gamma$ phase transformation.

5. Average size of secondary dendrite arm spacing and eutectic volume fraction increase with decreasing cooling rate. At lowest cooling rate "31.5°C/s" the volume fraction is ~1.96%.

6. Using ER309L filler rod leads to higher toughness, and adding 2%N_2 to Ar shielding gas leads to lower toughness, that increases with increasing weld current, and decreasing N_2 in shielding gas.

7. From the mathematical modeling, by adding 2%N_2 to shielding gas UTS, HV and spattering increase. While UTS, HV and spattering decrease with increasing the welding current.

Acknowledgements

The success of this research has been achieved due to the invaluable contributions of various individuals. I would like to take this opportunity to acknowledge their efforts:

I would also like to express gratitude to the headmaster of Petrojet Company (Suez) for providing welding process and collaboration.

Furthermore, I want to thank all my friends who gave me a wonderful time to support me to finish this chapter, especially, Dr. Eng. Ahmed M. Fouad, Dr. Asmaa Fouad, Dr. Fadel Shaaban, Dr. Waheed S. A. Barakat and Dr. Mohamed I. Habba. I would like to wish them all good luck in their further carrier, and of course. I hope we remain friends.

Author details

Ramy A. Fouad[1]*, Essam Ahmed Ali[2]*, Ahmed Ramadan Shaaban[1]* and Ahmed E. El-Nikhaily[3]*

1 Mechanical Department, Faculty of Technology and Education, Suez University, Egypt

2 Metallurgical and Materials Engineering, Faculty of Petroleum and Mining Engineering, Suez University, Egypt

3 Metal Forming Engineering, Faculty of Technology and Education, Suez University, Egypt

*Address all correspondence to: ramy_fouad12@suezuniv.edu.eg; essam.ahmed@suezuni.edu.eg; ahmed.essa@suezuni.edu.eg; ahmed.eassa@ind.suezuni.edu.eg

References

[1] Nabavi B, Goodarzi M, Amani V. Nitrogen effect on the microstructure and mechanical properties of nickel alloys. Welding Journal. 2015;**94**:53-60

[2] Lin YC, Chen PY. Effect of nitrogen content and retained ferrite on the residual stress in austenitic stainless steel weldments. Materials Science and Engineering. 2001;**A307**(307):165-171. DOI: 10.1016/s0921-5093(00)01821-9

[3] Yelamasetti B, Kumar S, Sridhar Babu B, Vishu Vardhan T, Gunda VR. Effect of filler wires on weld strength of dissimilar pulse GTA Monel 400 and AISI 304 weldments. Materials Today: Proceedings. 2019;**19**:1-5. DOI: 10.1016/j.matpr.2019.06.759

[4] Bahador A, Hamzah E, Mamat MF. Effect of filler metals on the mechanical properties of dissimilar welding of stainless steel 316l and carbon steel A516 GR70. Journal Teknologi (Sciences & Engineering). 2015:61-65. DOI: 10.11113/jt.v75.5174

[5] Moslemi N, Redzuan N, Ahmad N, Hor TN. Effect of current on characteristic for 316 stainless steel welded joint including microstructure and mechanical properties. 12th Global Conference on Sustainable Manufacturing. 2015;**26**:560-564. DOI: 10.1016/j.procir.2015.01.010

[6] Bharatha P, Sridharb VG, Senthil Kumar BM. Optimization of 316 stainless steel weld joint characteristics using taguchi technique. Procedia Engineering. 2014;**97**:881-891. DOI: 10.1016/j.proeng.2014.12.363

[7] Basyigit AB, Murat MG. The effects of TIG welding rod compositions on microstructural and mechanical properties of dissimilar AISI 304L and 420 stainless steel welds. Metals. 2018;**8**:1-14. DOI: 10.3390/met8110972

[8] Dhobale AL, Mishra HK. Review on effect of heat input on tensile strength of butt weld joint using MIG welding. International Journal of Innovations in Engineering Research and Technology [IJIERT]. 2015;**2**:1-13

[9] Hao Y, Li J, Li X, Liu W, Cao G, Li C, et al. Influences of cooling rates on solidification and segregation characteristics of Fe-Cr-Ni-Mo-N super austenitic stainless steel. Journal of Materials Processing Tech. 2020;**275**:1-9. DOI: 10.1016/j.jmatprotec.2019.116326

[10] Sameer SK. Study of influence of welding parameters on mild steel. International Advanced Research Journal in Science, Engineering and Technology. 2015;**2**:49-50. DOI: 10.17148/IARJSET.2015.2509

[11] Basyigit AB, Kurt A. The effects of nitrogen gas on microstructural and mechanical properties of TIG welded S32205 duplex stainless steel. Metals. 2018;**8**:1-13. DOI: 10.3390/met8040226

[12] Mosa ES, Morsy MA, Atlam A. Effect of heat input and shielding gas on microstructure and mechanical properties of austenitic stainless steel 304L. International Research Journal of Engineering and Technology (IRJET). 2017;**4**:370-377

[13] Suhail M, Hasan MF, Bharti PK. Effect of welding speed, current and voltage on mechanical properties of underwater welded mild steel specimen (C, Mn, Si) with insulated electrode E6013. International Journal of Mechanical Engineering. 2014;**4**:120-124

[14] Wang H-S. Effect of welding variables on cooling rate and pitting corrosion resistance in super duplex stainless weldments. Materials Transactions, The Japan Institute of Metals. 2005;**46**:593-601

[15] Singh DK, Sahoo G, Basu R, Sharma V, Mohtadi-Bonab MA. Investigation on the microstructure-mechanical property correlation in dissimilar steel welds of stainless steel SS 304 and medium carbon steel EN 8. Journal of Manufacturing Processes. 2018;**36**:281-292. DOI: 10.1016/j.jmapro.2018.10.018

[16] Movahedi M, Ozlati A. Effect of welding heat-input on tensile strength and fracture location in upset resistance weld of martensitic stainless steel to duplex stainless steel rods. Journal of Manufacturing Processes. 2018;**35**: 517-525. DOI: 10.1016/j.jmapro.2018.08.039

[17] Sergei YT, Filippov AV, Savchenko NL, Fortuna SV, Rubtsov VE, Kolubaev EA, et al. Effect of heat input on phase content, crystalline lattice parameter, and residual strain in wire-feed electron beam additive manufactured 304 stainless steel. The International Journal of Advanced Manufacturing Technology. 2018;**99**: 2353–2363. DOI: 10.1007/s00170-018-2643-0

[18] Randhawa HS, Kumar S. Effect of heat input on the characteristics of 304l austenitic stainless steel weld deposited by GTAW for root pass and SMAW for filler passes. International Journal of Research in Advent Technology. 2018;**6**: 1616-1621

[19] Bahar D. Optimization of process parameters for tungsten inert gas (Tig) welding to join a butt weld between stainless steel (SS304) and mild steel (MS1018). International Journal of Engineering Sciences & Emerging Technologies. 2017;**10**:1-8

[20] Bodude MA, Momohjimoh I. Studies on effects of welding parameters on the mechanical properties of welded low-carbon steel. Journal of Minerals and Materials Characterization and Engineering. 2015;**3**:142-153. DOI: 10.4236/jmmce.2015.33017

[21] Gupta SK, Raja AR, Vashista M, Khan MZ, Yusufzai MZK. Effect of heat input on microstructure and mechanical properties in gas metal arc welding of ferritic stainless steel. Materials Research Express. 2018;**6**:1-17. DOI: 10.1088/2053-1591/aaf492

[22] Ghazvinloo HR, Honarbakhsh-Raouf A, Shadfar N. Effect of arc voltage, welding current and welding speed on fatigue life, impact energy and bead penetration of AA6061 joints produced by robotic MIG welding. Indian Journal of Science and Technology. 2010;**3**:1-7

[23] Swami SA, Jadhav SM, Deshpande A. Influence of Mig welding process parameters on tensile properties of mild steel. European Journal of Engineering Research and Science. 2016;**1**:1-5

[24] Bansod AV, Patil AP, Moon AP, Shukla S. Microstructural and electrochemical evaluation of fusion welded low nickel and 304 ASS at different heat input. Journal of Materials Engineering and Performance. 2017;**26**:5847–5863. DOI: 10.1007/s11665-017-3054-3

[25] Bansod AV, Patil AP, Shukla S. Effect of heat on microstructural, mechanical and electrochemical evaluation of tungsten inert gas welding of low-nickel ASS. Anti-Corrosion Methods and Materials. 2018;**65**:605–615. DOI: 10.1108/ACMM-05-2018-1941

[26] Ahmed E, Ahmed R, EL-Nikhaily A, Essa ARS. Effect of heat input and filler metals on weld strength of gas tungsten arc welding of AISI 316 weldments. China Welding. 2020;**29**:8-16. DOI: 10.12073/j.cw.20200107001

[27] Merchant Samir Y. Investigation on effect of heat input on cooling rate and mechanical property (hardness) of mild steel weld joint by MMAW process. International Journal of Modern Engineering Research (IJMER). 2015;**5**: 34-41

[28] Kumar R, Arya HK, Saxena RK. Experimental determination of cooling rate and its effect on microhardness in submerged arc welding of mild steel plate (Grade c-25 as per IS 1570). Journal of Material Sciences & Engineering. 2014;**3**:1-4. DOI: 10.4172/2169-0022.1000138

[29] Ahmed R, Essa ARS, EL-Nikhaily A, Ahmed E. Effect of heat input and shielding gas on the performance of 316 stainless steel gas tungsten arc welding. Journal of Petroleum and Mining Engineering. 2020;**22**:9-15. DOI: 10.21608/jpme.2020.23038.1024

[30] Das D, Pratihar DK, Roy GG. Cooling rate predictions and its correlation with grain characteristics during electron beam welding of stainless steel. The International Journal of Advanced Manufacturing Technology. 2018;**97**:2241-2254. DOI: 10.1007/s00170-018-2095-6

[31] Ahmed R, Essa ARS, EL-Nikhaily A, Ahmed E. Effect of heat input and shielding gas on the performance of 316 stainless steel gas tungsten arc welding. Journal of Petroleum and Mining Engineering. 2020;**22**:9-15. DOI: 10.21608/jpme.2020.23038.1024

Chapter 3

Properties of Additively Manufactured Deposits of Alloy 718 Using CMT Process Depending on Wire Batch and Shielding Gas

Manuela Zinke, Stefan Burger and Sven Juettner

Abstract

Wire + arc additive manufacturing (WAAM®) is a versatile, low-cost, energy-efficient technology used in metal additive manufacturing (AM). This process uses arc welding to melt a wire and form a three-dimensional (3D) object using a layer-by-layer deposit. In the present study, the effect of heat input and shielding gas during CMT-WAAM welding on cooling time, mechanical properties at room temperature, and macro- and microstructure was investigated based on different part geometries (wall, block) using two S Ni 7718 wire batches. The heat input and consequently the cooling rate were varied by changing the wire feed and the travel speed. As expected, increasing the heat input leads to higher cooling times. Due to the 2D-heat conduction, the thin walls cool significantly slower than the multi-pass block welds. Nevertheless, the influence on mechanical properties is only marginal. Both the AM batch of S Ni 7718 with the lower Nb/C and the multi-pass block welds with the higher thermomechanical reactions exhibit a high susceptibility to unacceptable seam defects, such as hot cracks or lacks of fusion. But even the standard batch causes hot cracks. An influence of the shielding gas on microstructure, mechanical properties, and occurrence of the seam defects cannot be detected.

Keywords: additive manufacturing, alloy 718, wire batch, shielding gas, cooling time, seam defects, mechanical properties, hardness, microstructure

1. Introduction

Superalloy Alloy 718 (NiCr19NbMo, EN 2.4668) is a Ni-Cr-Fe-Mo alloy for service conditions up to 650°C, where hardening is achieved by additions of Nb, Ti, and Al resulting in controlled precipitation of mainly γ" (Ni_3Nb) and to a small extent γ' ($Ni_3(Al, Ti)$) [1]. Due to this, this alloy is widely used for static and rotating components in stationary gas turbines, rocket drives and spacecraft, motor vehicle turbo chargers, high-strength screws, springs and mounting elements, and for heat-resistant tools in forgeries, extruders, and separating shearers [2].

Ni-alloys have a high cost and are difficult to machine, with main reasons being their high temperature resistance, high initial work hardening rate, and presence of carbides in the microstructure [3]. Therefore, additive manufacturing (AM) techniques could result in cost, weight, and machining savings for a wide range of applications using Alloy 718 [4]. The AM with wire-based arc processes (WAAM®) offers the possibility of producing small and large-volume components of low and medium complexity at high deposition rates. A disadvantage is the poorer surface quality and accuracy and thus the need for mechanical finishing of functional surfaces [5–7]. A typical process for WAAM® is the energy-reduced Cold Metal Transfer (CMT) process, a variant of GMAW, which is widely used industrially for cladding with Ni-base solid wires [8–11]. This is due to the very low dilution with substrate, high deposition rate, very low heat input, less spatter, and low susceptibility to hot cracking [12], factors that are also advantageous for AM.

Recently, the number of publications on AM of Alloy 718 with the GMAW is increasing [13] published a study as early as 2007 on the use of the MIG process for AM of an internal flange of combustor outer casing of Alloy 718. The present study has highlighted the presence of deleterious Laves- and δ phases in the weld deposition structures encouraged by extended time at high temperature during either deposition or subsequent post-deposition heat treatment. In addition, associated discontinuities in the form of hot cracks and shrinkage porosity have been identified [14, 15] evaluated the effect of oxide, wire source, and heat treatment on the mechanical properties of additive plasma arc welding with cold wire feed of Alloy 718 with oscillating torch in inert shielding gas atmosphere. Results showed that oxides formed during deposition had no effect on the mechanical properties. Wires from different suppliers resulted in around 50 MPa difference in UTS. Standard heat treatment improved the strength from 824 MPa to 1110 MPa, but the average strength was lower than the wrought alloy and laser powder-based AM IN718. The microstructure of the WAAM material showed large columnar grains and numerous Laves phase [16] identified an aging effect in deposits of Alloy 718, which produced with CMT-WAAM® process. Deposits consisting of 10 layers were manufactured, where interpass time (0, 30 s, and 60 s) was the only process parameter varied. An aging effect was observed, which caused precipitation of the γ'' and γ' strengthening phases and the δ phase. The highest hardness value was measured in the middle section of each deposit due to precipitation of the strengthening phases [4] investigates the effects of processing parameters and heat treatments employed on CMT-WAAM® of Alloy 718. The process stability was analyzed by electrical transients and melt pool imaging, showing an opposite trend to the measured heat inputs. Laves length and carbide diameter decreased with travel speed, while the as-deposited hardness increased. These observations permitted a linear wall to be fabricated with a minimal heat input per layer of 181 J–185 J/mm. Compared with powder-based AM, the CMT welds exhibit a larger melt pool size and lower as-deposited hardness, but has been found to show satisfactory aging response and similar Laves phase area fraction.

In the present work, the influence of different heat inputs and shielding gases in CMT-WAAM with S Ni 7718 on the macrostructure, seam defects, chemical composition, mechanical properties, and hardness was investigated. The investigations were carried out with two different wire batches on two component geometries (wall and block). These different geometries generate different welding situations (single- or multi-bead welding) and stress conditions.

2. Experimental setup

For WAAM®, two welding equipment were used: a three-axis gantry machine with programmable logic controller (PLC) and a two-axis welding portal FCB 900–500/ML100 (**Figure 1**). To achieve a low heat input (E), the welds were made using the Cold Metal Transfer (CMT) process of the company Fronius. For logistical reasons, two power sources (PS) were used with the RCU 5000 control unit: CMT TPS 4000, first generation (PS 1) and CMT Advanced 4000, second generation (PS 2). However, the welds were made with the same synergy line (NIBAS 625/1.2mmM12 Ar + 2,5% CO_2). Accompanying the welds, the electrical welding parameters, the shielding gas flow, and the wire feed speed were recorded and monitored with external measuring system WeldAnalyst-S2 (HKS Prozesstechnik GmbH). During the welding process, infrared pyrometers were used to measure the temperature of the welded parts. Two batches of solid wire S Ni 7718 from different manufacturers with diameter 1.2 mm were used. One of them is a standard batch, while the other is made specifically for AM.

3. Deposition strategy

Since in additive welding, the parts geometry influences the cooling rate and thus the microstructure and mechanical properties, in a first step, two different part geometries (thin wall and multi-pass block) were initially investigated (**Figure 2**).

Depending on the geometry, different substrate sheets were used. These sheets were welded to a 40 mm thick carbon steel plate to counteract distortion and produce high thermomechanical reactions. For welding of both geometries, a stick-out of 15 mm and 30% He, 2% H_2, 0.05% CO_2, bal. Ar as shielding gas with a flow rate of 18 l/min were used. This gas is recommended for welding Ni alloys, as it ensures good flow and wetting behavior of the highly viscous melt. During the welding, the direction was changed after each layer. Between the beads or layers, there was no brushing. For all welds, the maximum interpass temperature (IT) was 150°C. To influence the deposition rate and heat input at welding the walls, the wire feed speed (WFS) was varied between 6.0 and 9.0 m/min and the travel speed (TS) between 0.6 and 1.0 m/min. To achieve a high deposition rate when manufacturing multi-pass block with stringer bead technique, a high wire feed speed (WFS: 9.0 resp. 9.5 m/min) and a low welding speed (TS: 0.6 m/min) were selected.

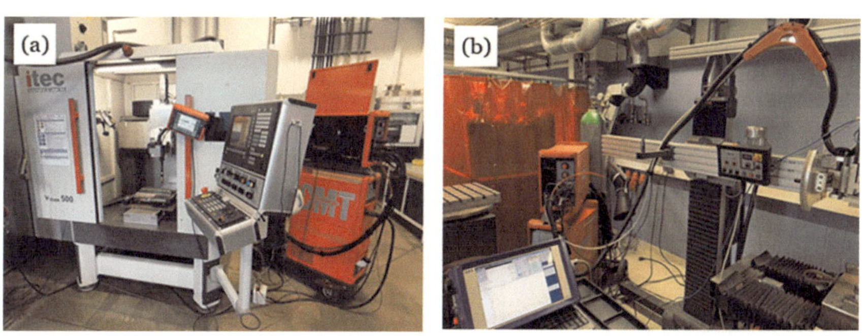

Figure 1.
Experimental setup for additively GMAW (a) three-axis gantry machine with CMT advanced 4000 and (b) two-axis welding portal with CMT TPS 4000.

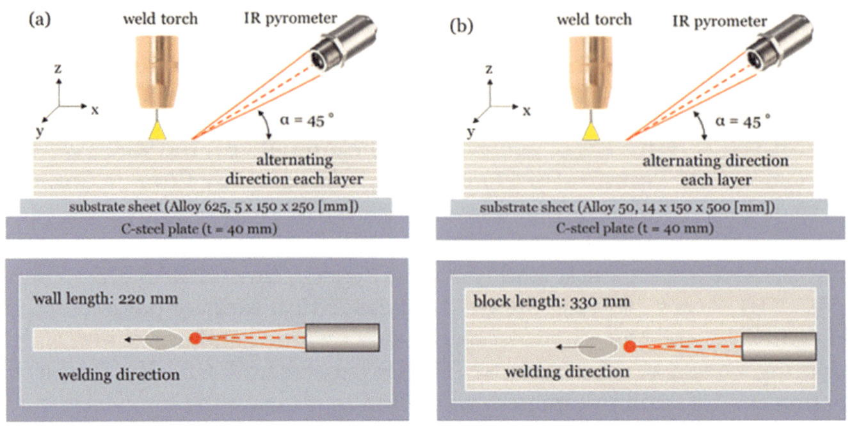

Figure 2.
Deposition strategy of (a) wall (b) block (above: side view, below: top view).

In second step, the influence of different shielding gases on the seam quality was investigated. Ni-alloys are normally welded with inert shielding gas. Additions of active gases (CO_2, O_2) in the ppm range and of He or H_2 improve the flow and wetting behavior of the melt. However, the presence of CO_2 or O_2 can also lead to a slight seam oxidation. This effect can increase the possibility of lack of fusion on multi-pass deposition. On the other hand, hydrogen has a reducing effect and can ensure an oxide-free seam appearance. The following shielding gases were used:

- bal. Ar/30% He/2% H_2/0,055%CO_2 (ArHeHC)
- bal. Ar/30% He/1% H_2 (ArHeH)
- bal. Ar/2% H_2 (ArH)
- bal. Ar/30% He/0,03% O_2 (ArHeO)

These investigations were also performed on walls and blocks with constant setting values (WFS: 9.0 resp. 9.5 m/min and TS: 0.6 m/min). The nominal dimensions were for walls 200 mm long and 60 mm high and for blocks 200 mm long and circa 25 mm–30 mm high and 30 mm–35 mm wide. In contrast to the previous blocks, the welding direction was changed after each stringer bead. The remaining welding conditions (stick-out, IT, etc.) corresponded to the previously tests.

In multi-pass deposition, maintaining the correct center distance d between stringer beads is of elementary importance. If this distance is too big, there is no connection to the neighboring bead and lack of fusion can occur. If it is too small, the beads overlap too much and no even seam surface is created. Therefore, it is important to ensure that the overlap areas and valleys are of equal size. In [17] it is proposed to calculate the ideal center distance depending on the ratio of WFS/TS. If WFS/TS > 12.5, the equations shown in **Figure 3** should be used.

4. Test methods

First the exact characterization of the wire electrodes took place. To determine the chemical composition, the wires were melted into buttons with the Arc Melter

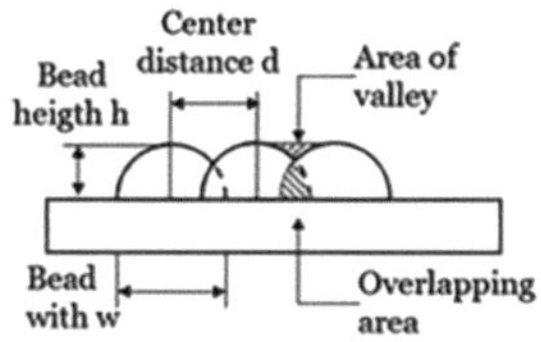

$$d = \frac{1}{2h}\left(2R^2 \arcsin\left(\frac{w}{2R}\right) + wh - wR\right) \quad \text{Eq. 1}$$

$$\text{with } R = \frac{\left(h^2 + \frac{w^2}{4}\right)}{2h} \quad \text{Eq. 2}$$

Figure 3.
Illustration of overlapping beads and equations for calculating the ideal center distance [16].

MAM-1 (Edmund Bühler GmbH) in pure argon atmosphere (99.996%) using a TIG arc. The methodology is described in [18]. The chemical composition of buttons was analyzed with the atom emission spectrometry (AES). In addition, the contents of oxygen, nitrogen, carbon, and sulfur were measured by means of carrier gas melt extraction (CGME) with combustion analysis (CA) directly on the solid wire in the delivery condition. Afterward, measurements of the wire diameter, the surface roughness (axial) with the tactile cut method [19], and tensile tests were carried out.

The welded components have been subjected to a visual and dye penetration test. In the first step, the metallographic cross sections (Y-Z plane) are taken. For the automatic preparation, it was necessary to cut the cross sections in the middle. All samples were processed with standard metallographic techniques with a final polish of 0.5 µm O.P.S (SiO_2). Subsequently, the central part of the walls and blocks were mechanically machined to allow high-quality X-ray inspection. While the image quality indicator according to [20] on the walls was 16 or 17, only an indicator of 10 was achieved on the blocks due to their higher thickness. For this reason, no radiographic testing was performed on the blocks. On walls and blocks, chemical analyses, tensile tests, and hardness measurements were performed. Standard tensile tests were performed to BS EN ISO 6892-1 [21] at room temperature. The tensile specimens with form E 5 x10 × 40 [22] were taken in welding direction. Vickers hardness testing was performed in compliance with BSEN ISO 6507-1 [23], using a load of 0.2 or 10 kg. Measurements were made at varying distances from the substrate. Only on the blocks the determination of the impact energy on Charpy V samples at room and low temperatures (−196°C) was carried out. The microstructure was investigated by light optical microscope (inverse incident light microscope Leica MeF4A, Leica), scanning electron microscope (XL30 FEG/ESEM, company FEI/Philips), and energy-dispersive spectrometer (EDAX Si (Li) detector).

5. Results and discussion

5.1 Characterization of wire batches

The chemical compositions of S Ni 7718 wire batched are listed in **Table 1**. The AM batch has lower contents of Cu and Co. Furthermore, it has slightly higher contents of C, Nb, and Fe than the standard variant. These elements influence the precipitation of the Laves phase, MC-type carbides, and TN in the weld metal [24–26]. Furthermore, the solidification cracking of austenitic materials is strongly dependent on the solidification temperature range (STR) and especially the solidification at the grain boundaries and interdendritic regions. The Nb/C ratio affects the amount and distribution of the γ + NbC eutectic and γ + Laves eutectic, which influence the STR. Addition of Nb at higher C levels promoted γ/NbC eutectic type constituent and at higher Si and Fe

Element	C	Si	S	Cr	Mo	Fe	Nb	Al	Ti	Cu	Co	Ni	Nb/C
Standard	0.04	0.06	0.001	19.1	2.6	17.9	5.67	0.53	1.12	0.05	0.07	52.6	149
AM	0.07	0.04	0.001	18.0	2.7	20.2	5.99	0.52	1.16	<0.01	<0.01	5.12	89

Table 1.
Chemical composition of different wire batches of S Ni 7718 (wt%).

Parameter	Ø [mm]	Surface	Rz [μm]	UTS [MPa]	A_{50mm} [%]
Standard	1.19	matt	1.38 ± 0.05	1407 ± 22	4.6 ± 0.4
AM	1.18	glossy	0.47 ± 0.04	1326 ± 30	9.8 ± 0.3

Table 2.
Mechanical properties at RT and roughness of solid wire electrodes.

levels promoted γ/laves [27]. When Nb/C ratio was increasing, the solidification temperature range (STR) was reduced [28].

A basic requirement of a stable gas-metal arc welding process is a permanently constant feed of the wire electrode, even with cable assembly of different lengths. The main factors influencing this are the surface condition of the wire electrode, mechanical properties, and also technological characteristics. Basically, the wire electrode for GMAW should have a high tensile strength (UTS) with sufficient elongation at rupture (A) to ensure high stiffness even in long cable assemblies [29]. Both wire variants guarantee these requirements (**Table 2**). For the sliding behavior in long hose assemblies, the basic rule is that with lower roughness (Rz) and greater pre-bending, the sliding ability of wire electrodes in the hose assembly increases [30]. The AM variant has a lower average roughness values, which also correlate with the appearance of the wire surface. Despite the different properties and characteristics, no significant influence on the wire feeding and thus on the process behavior was found under the selected test conditions.

5.2 Heat input and deposition rate

An arithmetic mean value for all weld beads of the wall or block was formed from the mean value of the respective stringer bead read off the power sources. The tolerance given is the average of the positive and negative error indicators. The heat input (E) is the quotient of arc power (P_{arc} = I x U) and travel speed (TS). The deposition rate (DR) is the amount of wire melted per unit time. Despite the same synergic line and setting parameters, differences in the measured values for current (I) and voltage (U) occurred for both current sources and consequently in the heat input (E) and deposition rate (DR) when producing the walls (**Table 3**). The large blocks were welded only with a high wire feed speed and a low welding speed to generate a high deposition rate (**Table 4**).

Moreover, this combination represents the critical case, since the heat input is comparatively high. Fourteen layers of nine stringer beads per layer each were made. Due to the long production time of 10.5 hours resulting from the interpass temperature, only one block was welded at a time. The welding time was only 70 minutes.

Tables 5 and **6** show the weld parameters of walls and blocks welds with the varied shielding gases. It can be seen that the shielding gas has a slight influence on these

Wire batch	Setting values		Mean values				
	WFS [m/min]	TS [m/min]	I [A]	U [V]	WFS [m/min]	E [J/mm]	DR [kg/h]
Standard/PS1	6.0	0.6	133 ± 2.0	15.3 ± 0.2	6.3 ± 1.2	203 ± 5.7	3.53
		1.0	144 ± 3.0	17.0 ± 1.4	6.8 ± 0.8	147 ± 9.1	3.83
	9.0	0.6	208 ± 3.0	19.9 ± 0.6	9.7 ± 1.2	414 ± 9.4	5.44
		1.0	204 ± 3.0	19.7 ± 1.0	9.8 ± 1.0	241 ± 11.1	5.50
AM/PS2	6.0	0.6	123 ± 2.3	15.6 ± 1.1	4.9 ± 0.8	192 ± 13.7	2.74
		1.0	120± 4.5	16.1 ± 0.6	5.0 ± 1.0	115 ± 6.1	2.78
	9.0	0.6	178 ± 2.5	21.8 ± 0.6	7.4 ± 1.2	388 ± 8.9	4.16
		1.0	167 ± 3.5	19.9 ± 0.5	8.2 ± 1.6	200 ± 6.0	4.58

Table 3.
Welding parameters for CMT-WAAM® of walls, welded with ArHeHC.

Wire batch	Setting values		Mean values				
	WFS [m/min]	TS [m/min]	I [A]	U [V]	WFS [m/min]	E [J/mm]	DR [kg/h]
Standard/PS2	9.0	0.6	159 ± 18.5	22.2 ± 7.6	8.6 ± 2.0	352 ± 49	4.82
Am/PS2	9.5	0.6	174 ± 22.0	22.0 ± 7.5	8.7 ± 1.1	382 ± 56	4.88

Table 4.
Welding parameters for CMT-WAAM® of blocks, welded with ArHeHC.

Shielding gas	Wall				Block			
	I [A]	U [V]	WFS [m/min]	E [J/mm]	I [A]	U [V]	WFS [m/min]	E [J/mm
ArHeHC	218 ± 4.4	21.1 ± 0.6	10.3 ± 0.4	461 ± 20	216 ± 9.7	20.9 ± 1.6	10.1 ± 0.7	451 ± 34
AeHeH	217 ± 3.6	20,5 ± 0.6	10.2 ± 0.3	445 ± 18	216 ± 5.0	19.8 ± 1.6	10.0 ± 0.5	429 ± 34
ArH	216 ± 1.7	19.6 ± 0.3	10.2 ± 0.2	423 ± 7	212 ± 6.2	19.3 ± 1.9	9.8 ± 0.8	409 ± 40
AeHeO	214 ± 3.4	19.2 ± 0.2	9.9 ± 0.2	416 ± 11	212 ± 7.2	19.2 ± 1.2	9.6 ± 0.7	407 ± 26

Table 5.
Mean values for CMT-WAAM® welds (standard batch, PS1, WFS: 9.0 m/min, TS: 0.6 m/min).

Shielding gas	wall				block			
	I [A	U [V	WFS [m/min]	E [J/mm]	I [A]	U [V]	WFS [m/min	E [J/mm]
ArHeHC	217 ± 3.8	20.3 ± 0.8	1.4 ± 10.5	439 ± 15	215 ± 6.6	19.6 ± 1.7	10.1 ± 1.0	422 ± 34
AeHeH	217 ± 4.0	19.8 ± 0.5	10.5 ± 0.8	430 ± 16	214 ± 7.4	18.9 ± 1.4	9.8 ± 0.6	405 ± 35
ArH	215 ± 4.1	18.8 ± 0.9	9.8 ± 0.7	404 ± 24	209 ± 6.8	18.0 ± 1.9	9.4 ± 0.8	375 ± 48
AeHeO	214 ± 3.8	18.3 ± 1.2	9.7 ± 0.3	392 ± 29	215 ± 7.1	17.8 ± 0.8	9.6 ± 0.4	383 ± 40

Table 6.
Mean values for CMT-WAAM® welds (AM batch, PS1, WFS: 9.5 m/min, TS: 0.6 m/min).

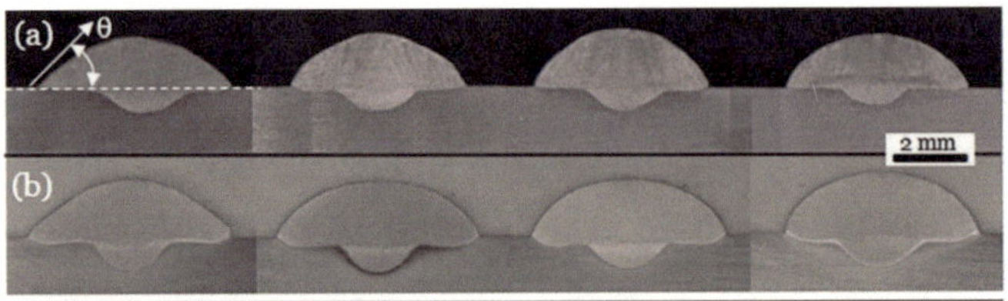

Shielding gas	ArHeHC	ArHeH	ArH	ArHeO
θ, ° (a) / (b)	60 / 61	70 / 69	68 / 73	71 / 79

Figure 4.
Cross sections of stringer beads depending on shielding gas: (a) standard batch (b) AM batch.

parameters. The lowest heat input is generated when ArHeO is used, while ArHeHC results in the highest values.

Cross sections of stringer beads as a function of shielding gas show the lowest wetting angles θ when ArHeHC is used (**Figure 4**). This gas caused the highest heat input. The wetting angle increases with all other shielding gases. A low wetting angle has a positive effect on reducing lack of fusion in multi-pass welds.

5.3 Cooling time

Infrared (IR) pyrometers were used to measure the temperatures directly on the seam surface for every second layer at the wall structures. Since there is no specific temperature interval for the cooling time for Ni-based materials, this was calculated for cooling from 1000–600°C. **Figure 5** shows the cooling times $t_{10/6}$ depending on heat input and layer for both batches when welding the walls. In the first layers, shorter cooling times occur due to the still possible heat conduction into the substrate. From about the 10th layer, the 3D changes to a 2D-heat conduction, which is why the cooling time remains almost constant. As expected, the cooling times increase with increasing heat input when welding the walls. At the highest heat energy of 414 J/mm, the cooling time is about 21 s. And at the lowest heat input of 105 J/mm, the cooling time is about 7 s.

The blocks cool significantly faster at comparable setting values and similar heat input due to the 2D-heat conduction. This is shown as an example in **Figure 6a** for the S Ni 7718 AM. The average $t_{10/6}$ times are approximately between 4 s and 5 s. The further layers then cool down somewhat more slowly. In addition, it can be observed that the outer beads of each layer in the upper layers also usually cool down somewhat slower than the inner beads (**Figure 6b**).

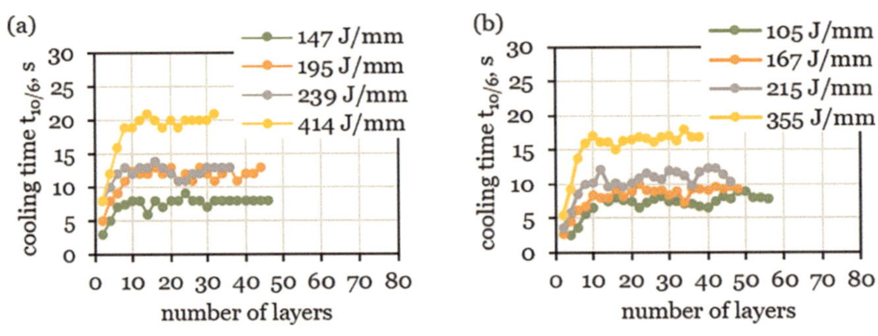

Figure 5.
$t_{10/6}$ cooling times of S Ni 7718 walls: (a) standard batch (b) AM batch depending on number of layers and heat input (shielding gas: ArHeHC).

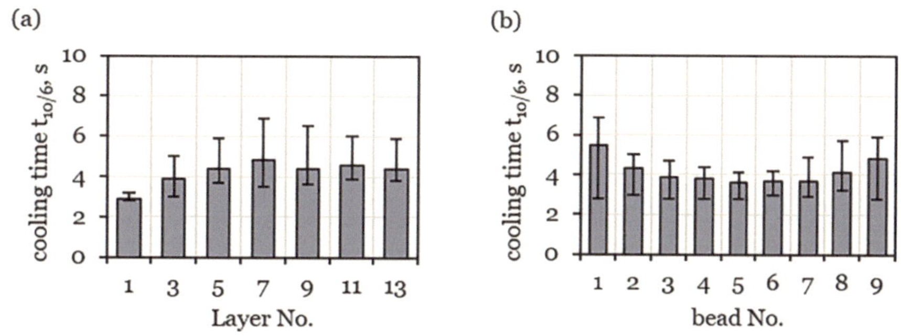

Figure 6.
Mean $t_{10/6}$ cooling times of S Ni 7718 AM blocks depending on (a) layer and (b) bead (shielding gas: ArHeHC).

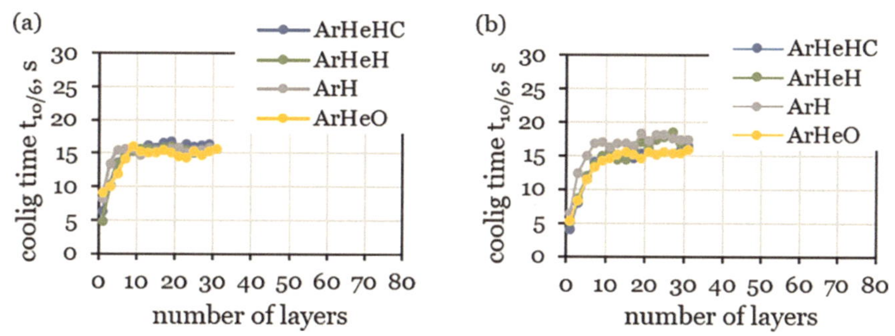

Figure 7.
$t_{10/6}$ cooling times in walls of S Ni 7718: (a) standard batch and (b) AM batch depending on number of layers and shielding gas.

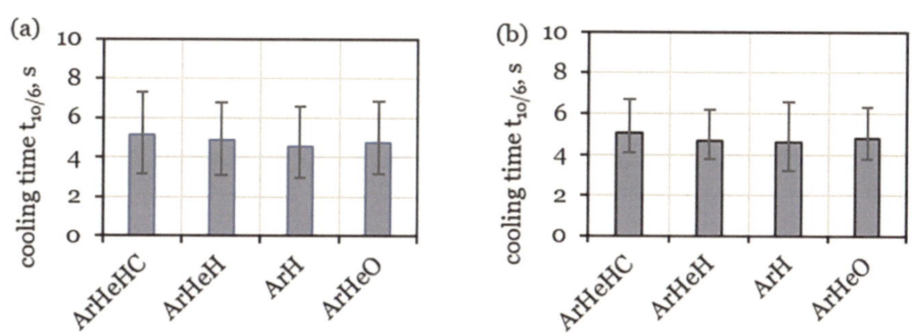

Figure 8.
Mean $t_{10/6}$ cooling times in blocks of S Ni 7718 depending on shielding gas: (a) standard batch (b) AM batch.

The shielding gas exerts only a minor influence on the cooling time, since the amounts for the heat input do not differ significantly (**Figures** 7 and **8**).

5.4 Nondestructive testing

As a result of the visual inspection, no external defects such as cracks, lack of fusion or pores were observed. Only some welds of the standard charge of S Ni 7718 with shielding gases containing active gas components (CO_2 and O_2) showed dark particles on the weld surfaces (**Figure 9**).

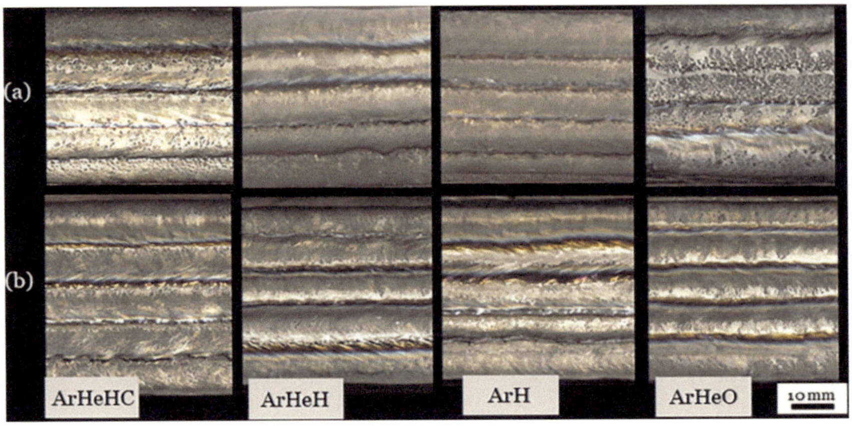

Figure 9.
Surfaces of S Ni 7718 blocks depending on shielding gas: (a) standard batch (b) AM batch.

It is assumed that metallurgical reactions with the oxygen-affine elements Al or Cr occur due to the size of the molten pool and slow cooling. On the stringer beads' surfaces of welded walls fabricated with higher heat input, these particles also occurred, but in both wire batches. However, no deterioration of the CMT process stability was observed.

The penetration tests on walls and blocks resulted in some red indications independent of wire and welding speed (**Figure 10**).

On the cross sections of walls, it was determined that these defects always occur in the last welded layer of walls and in some cases extend to the surface (**Figure 11**). Due to their structure and dimensions, these phenomena are not classified as solidification cracks or pores but as micro blowholes. Since the components are subsequently machined, these defects are not significant.

The X-ray examinations showed no or very small porosity of less than 0.001%. The porosity is the quotient of the sum of the areas of all pores and the X-rayed weld metal area. **Figure 12** shows an X-ray image of a wall with one pore. No radiographic tests were performed on the blocks due to the low image quality number, as they did not provide sufficient information.

Figure 10.
Example of penetration test of last layer surface with red indications.

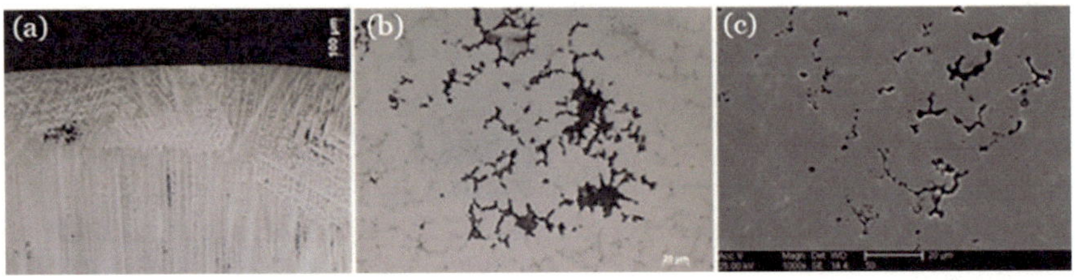

Figure 11.
Cross sections (Y-Z plane) with examples for micro blowholes: (a) 100x magnification (b) and (c) 1000fach.

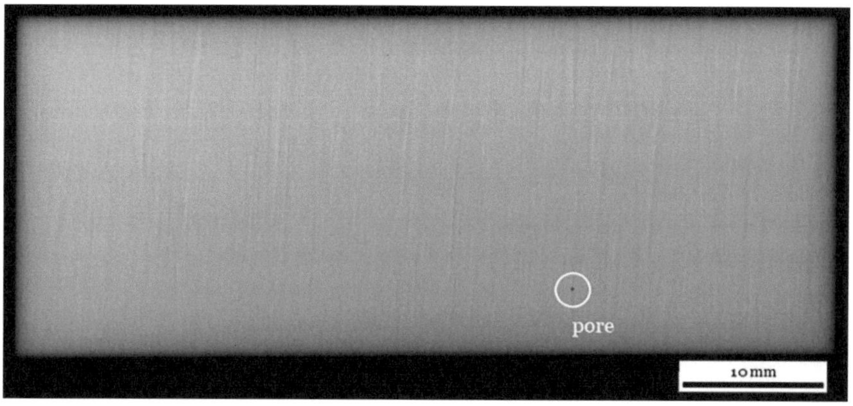

Figure 12.
X-ray test on the example of S N 7718 AM wall welded with ArH.

5.5 Chemical analysis

Since a shielding gas with low active and reducing shielding gas components was used and no brushing between the individual layers took place, the chemical composition of welds was determined. **Table 7** shows the values of the additive welded wall as a function of the different shielding gases in comparison with the element contents of wire batch.

An influence of shielding gas or component geometry and heat input on the chemical compositions cannot be proven. If deviations occur, they are more likely to be due to measurement inaccuracies resulting from calibration tolerances. Thus, the weld metals also show the same tendency with respect to the Nb/C ratio.

The O- and N-contents of part geometries of both batches welded with ArHeHC show a slight increase compared with the wire batches (**Figure 13**). A similar behavior is shown for the blocks produced with the different shielding gases (**Figure 14**).

However, the weld metal of the standard batch dissolves slightly more oxygen and nitrogen compared with the AM batch. In addition, the reducing effect is observed for the shielding gas with the highest H content (Ar with 2% H_2). The low O_2 and CO_2 contents do not cause oxygen pickup by the weld metal.

Element	C	Si	S	Cr	Mo	Fe	Nb	Al	Ti	Cu	Co	Ni	Nb/C
standard	*0.04*	*0.06*	*0.001*	*19.1*	*2.6*	*17.9*	*5.67*	*0.53*	*1.12*	*0.05*	*0.07*	*52.6*	*149*
ArHeHC	0.04	0.06	0.001	19.0	2.6	17.5	5.65	0.56	1.14	0.06	0.06	53.2	145
ArHeH	0.04	0.07	0.001	18.9	2.5	17.7	5.65	0.56	1.14	0.06	0.06	53.2	149
ArH	0.04	0.07	0.001	18.9	2.5	17.7	5.63	0.56	1.14	0.06	0.07	53.2	148
ArHeO	0.04	0.07	0.001	18.9	2.5	17.7	5.60	0.55	1.14	0.06	0.07	53.4	147
AM	*0.07*	*0.04*	*0.001*	*18.0*	*2.7*	*20.2*	*5.99*	*0.52*	*1.16*	*<0,01*	*<0.001*	*51.2*	*89*
ArHeHC	0.07	0.03	0.001	17.8	2.9	20.2	5.80	0.54	1.08	<0.01	<0.001	51.5	85
ArHeH	0.07	0.03	0.001	17.9	2.9	20.2	5.86	0.53	1.08	<0.01	<0.001	51.3	83
ArH	0.07	0.03	0.001	18.0	2.9	20.2	5.80	0.53	1.08	<0.01	<0.001	51.2	83
ArHeO	0.07	0.03	0.001	18.0	2.9	20.2	5.83	0.53	1.08	<0.01	<0.001	51.3	84

Table 7.
Chemical composition of wire batches of S Ni 7718 blocks depending on shielding gas (wt%).

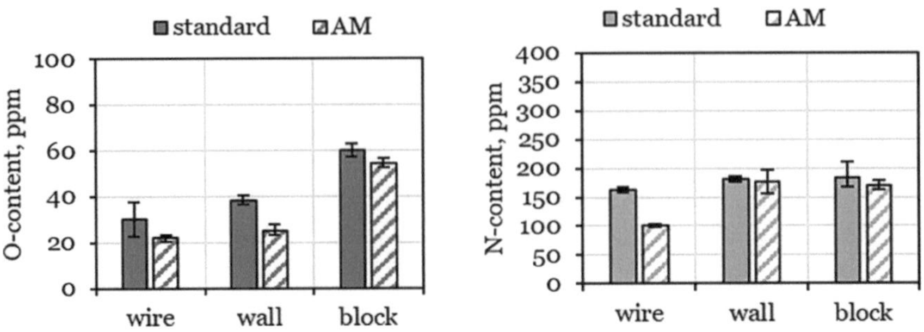

Figure 13.
ON-contents of S Ni 7718 walls and blocks depending on wire batch (shielding gas: ArHeHC).

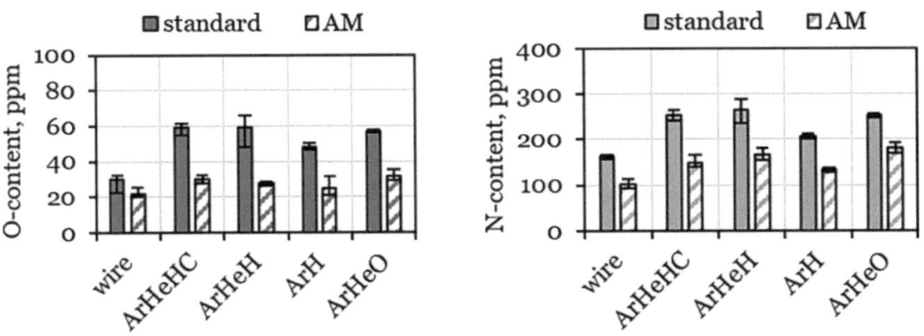

Figure 14.
ON contents of S Ni 7718 blocks depending on shielding gas.

5.6 Macrostructure and defects

Figure 15 shows the macro cross sections (Y-Z plane) of the walls of different wire batches. At constant wall height, the weld width increases with rising deposition rate and the number of layers reduces. As a result of the lower deposition rate when welding with the AM batch, more plies had to be welded to achieve equal weld heights. In addition, comparatively large thickness reductions are recorded in the lower wall structure area when welding with 6.0 m/min wire feed and 1.0 m/min travel speed. To avoid this effect, it would be necessary to adjust the wire feed speed or, if necessary, to preheat the substrate sheet, but this was not the focus of these investigations. In order to obtain an evaluation of the lateral surface waviness, the end contour proximity (ECP) was determined according to Eq. 3.

$$ECP = \frac{W_{inside}}{W_{outside}} \cdot 100\ [\%] \tag{3}$$

Since the sections have been split, the lower value of the nominal wall width and the higher value of the actual wall width are relevant. During the evaluation, the upper rounded and the lower, partially constricted areas of the samples were not considered. All wall structures except one wall achieved an ECP of ≥75%, which is according to [31] a good value. Since the image quality indicator of the X-ray inspection is not sufficient to detect micro hot cracks, the cross sections of the walls and blocks were inspected for internal seam defect. With one exception, there were no hot cracks in the walls. Only in one cross section of the AM batch of S Ni 7718 hot cracks were separated and visible in the second and third layers (**Figure 16**).

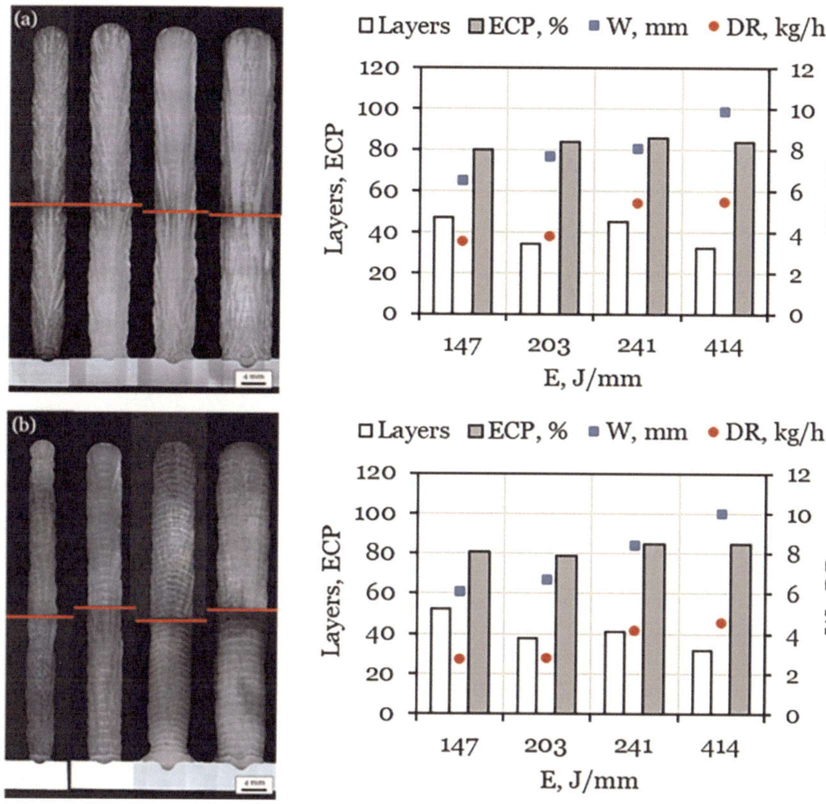

Figure 15.
Cross sections (Y-Z plane) of S Ni 7718 walls: (a) standard (b) AM welded with ArHeNC.

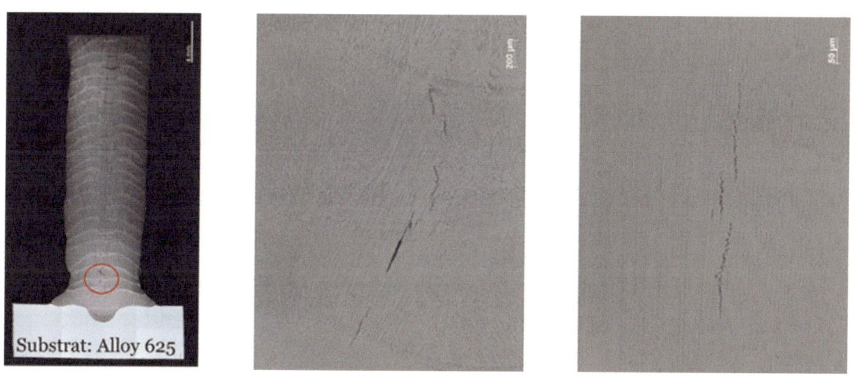

Figure 16.
Hot cracks in the lower layers of S Ni 7718 AM wall (WFS:9.0 m/min, TS: 1.0 m/min, ArHeHC).

For the blocks, the higher welding feed speed in a wider block with the same number of layers and beads (**Figure 17**) can also be seen. The machining allowances are comparable to those of the walls. However, the ECP is very high due to the large width of the block. When the blocks are welded, lateral tracking of the beads can be observed on each side in each position. Due to this, the process is more unstable and a lot of weld spatter occurs. In principle, the wire feed speed would have to be adjusted for these weld beads, but this has not been done.

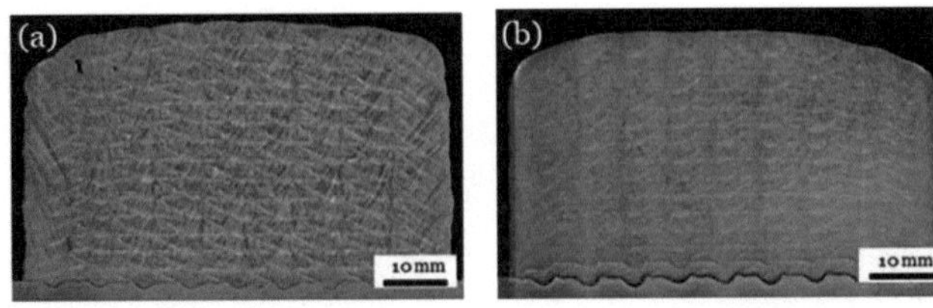

		Standard batch	AM batch
Layers / beads	[-]	9 / 126	9 / 126
H_{block}	[mm]	42.0	39.0
W_{block}	[mm]	57.5	60.9
ECP	[%]	97	98

Figure 17.
Cross sections (Y-Z plane) of S Ni 7718 blocks: (a) standard, (b) AM welded with ArHeHC.

Defect	Hot crack		Lack of fusion
	Number	CL, μm/mm²	Number
Standard	17	2.3	2
AM	83	33.3	1

Table 8.
Defects in S Ni 7718 blocks, welded with ArHeHC.

Due to the significantly larger number of weld beads in large blocks, the thermal-mechanical reactions were higher than for the walls, so that hot cracks occurred in both blocks despite faster cooling. In addition to the hot cracks, also few lacks of fusion were also found. A quantitative evaluation of the number of cracks and the weld metal area-related crack length (Eq. 4) shows a significantly higher hot crack sensitivity of the AM batch of S Ni 7718 (**Table 8**). This confirms the investigation results of [27, 28] that a higher Nb/C ratio results in greater susceptibility to hot cracking.

$$CL = \frac{\text{Total length of cracks}}{\text{weld metal area}} \ [\mu m/mm2] \qquad (4)$$

Figure 18 shows the blocks produced with different shielding gases. Eq. 1 and Eq. 2 (**Figure 3**) were used here to determine the center distance d between the beads. At the side block edges, there is always a lateral flow of the melt, which leads to an uneven deposit. While the calculated center distance was set for the blocks of the AM batch, a fixed spacing of 5.5 mm was set for welding with the standard batch. This, together with the higher wire feed speed of 9.5 m/min, resulted in wider blocks with lower heights for the same number of beads per layer in the AM batch.

The evaluation of the macro sections showed for the blocks from the AM batch not only many hot cracks but also a large number of lacks of fusion (**Figure 19**, **Table 9**). But these defects also occurred in the blocks of the standard batch. No internal defects were visible on the walls of S Ni 7718.

5.7 Mechanical properties at room temperature and hardness

Tables 10 and **11** show the mechanical properties at room temperature of S Ni 7718 walls and blocks welded with ArHeHC.

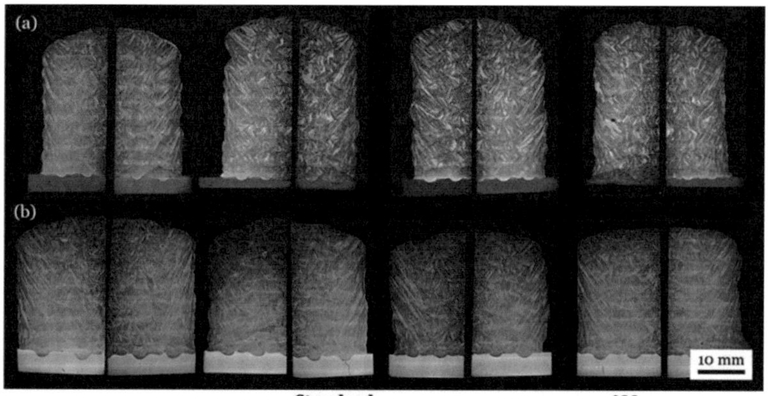

		Standard				AM			
		ArHeHC	ArHeH	ArH	ArHeO	ArHeHC	ArHeH	ArH	ArHeO
d	mm	6,4	6,1	6.1	6,1	7.3	6.5	6.5	6.8
Layers / beads	-	11 / 55	12 / 60	12 / 60	12 / 60	12 / 65	11 / 55	10 / 50	11 / 55
H_{block}	mm	33.0	33.4	33.2	32.3	27.0	25.3	25.2	26.9
W_{block}	mm	32.4	30.2	31.8	32.2	39.2	35.1	34.1	36.2
ECP	%	88	79	89	79	86	83	82	86

Figure 18.
Cross sections (Y-Z plane) of the block structures of S Ni 7718 batches: (a) standard (b) AM.

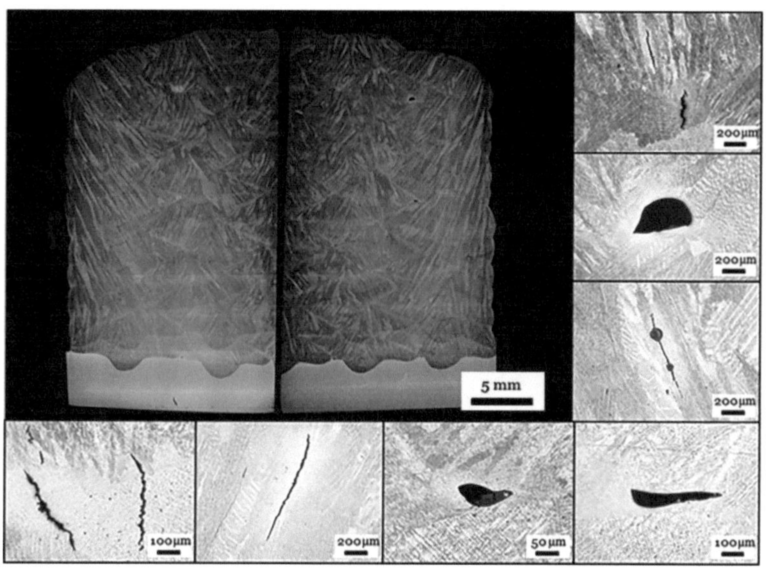

Figure 19.
Example of unacceptable defects in S Ni 7718 blocks of AM batch, welded with ArHeHC.

	Standard				AM			
	ArHeHC	AeHeH	ArH	ArHeO	ArHeHC	ArHeH	ArH	ArHeO
No. hot crack	6	5	5	6	20	11	7	13
CL, $\mu m/mm^2$	2,5	1.4	1.6	3.0	5,8	3,8	2,3	5,1
No. lack of fusion	0	0	1	3	15	13	12	8

Table 9.
Defects in S Ni 7718 blocks, welded with different shielding gases.

		Standard				AM			
E	[J/mm]	**147 ± 9.1**	**203 ± 5.7**	**414 ± 9.4**	**241 ± 11.1**	**115 ± 6.1**	**192 ± 13.7**	**200 ± 6.0**	**388 ± 8.9**
UTS	[MPa]	767 ± 10	765 ± 10	762 ± 6	754 ± 1	759 ± 11	775 ± 8	771 ± 4	785 ± 1
$YS_{0.2}$	[MPa]	470 ± 10	488 ± 10	454 ± 4	464 ± 9	409 ± 16	499 ± 13	440 ± 15	439 ± 10
A	[%]	42,0 ± 1.9	39,7 ± 6.0	38,9 ± 1.8	42,1 ± 5.7	39,7 ± 0.3	40,1 ± 0.7	35,5 ± 4.6	34,3 ± 0.2
HV10	[−]	237 ± 24	244 ± 17	245 ± 22	237 ± 21	242 ± 24	249 ± 17	247 ± 16	242 ± 19

Table 10.
Summary of tensile properties and hardness measurements of S Ni 7718 walls (shielding gas: ArHeHC).

	UTS [MPa]	$YS_{0.2}$ [MPa]	A [%]	HV10	KV, 20°C [J]	KV, −196°C [J]
Standard	792 ± 2	487 ± 7	36.1 ± 3.9	271 ± 19	88.5 ± 3.5	77.5 ± 2.5
AM	792 ± 33	528 ± 14	23.6 ± 3.9	275 ± 34	65.0 ± 5.0	49.0 ± 1.5

Table 11.
Results of tensile and charpy tests as well as hardness measurements of S Ni 7718 blocks (shielding gas: ArHeHC).

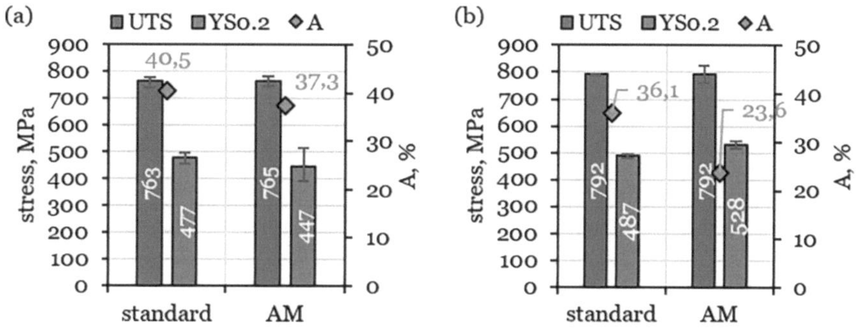

Figure 20.
Mean values of tensile properties of S Ni 7718 walls (a) and blocks (b), welded with ArHeHC.

If the influence of wire batch is considered independently of the geometry, the AM batch leads to a lower elongation at rupture (A) compared with the standard batch (**Figure 20**). The values for the 0.2% yield strength, on the other hand, behave divergently, while no significant difference occurs for the tensile strength. Considering the influence of the geometry, the blocks show a slightly higher tensile strength, 0.2% yield strength and hardness, while the elongation at rupture is significantly lower. The reasons for the higher strengths of the blocks compared with the walls are probably due to the faster cooling.

An influence of the shielding gases on the strength is not detectable (**Table 12**). The impact energy values determined scatter slightly, but a correlation with the shielding gases cannot be demonstrated (**Table 13**).

For the identification of an aging effect on the deposit hardness, local hardness maps (HV0.2) were performed over 2–3 layers in the center of the walls and blocks welded with the highest heat input (**Figure 21**). The results of the walls show no significant changes in hardness, indicating an age hardening effect.

5.8 Microstructure of deposit weld metals

Already on the polished cross sections of walls and blocks, a large number of irregularly distributed cubic particles are visible, which are either TiN or NbC (**Figure 22**). The nitrides containing a certain amount of Ti are already formed in the melt at a later stage of solidification [24].

The Laves phase is already clearly visible in the etched sections at the light microscope at sufficiently high magnification (**Figure 23**).

The scanning electron microscope (SEM) images demonstrate the precipitated phases in the microstructure of samples, such as Laves Phase and complexe NbC, Ti N-particles (**Figure 24**).

The precipitation of these phases appeared for all walls and blocks. The brittle Laves phase is generally considered to be the major microstructural segregation

		Standard				AM			
		ArHeHC	ArHeH	ArH	ArHeO	ArHeHC	ArHeH	ArH	ArHeO
UTS	[MPa]	729 ± 5	731 ± 5	738 ± 4	729 ± 4	765 ± 6	753 ± 7	756 ± 5	758 ± 16
0.2 YS	[MPa]	440 ± 2	434 ± 14	438 ± 9	410 ± 22	452 ± 20	453 ± 3	452 ± 20	445 ± 17
A	[%]	39.9 ± 2.7	39,2 ± 2.3	38.2 ± 1.6	41.0 ± 1.4	39.3 ± 0.9	36.2 ± 1.6	39.4 ± 2.3	39.3 ± 2.3
HV10	[−]	245 ± 23	243 ± 17	248 ± 25	258 ± 14	244 ± 22	242 ± 17	247 ± 16	247 ± 16

Table 12.
Summary of tensile properties and hardness measurements of S Ni 7718 walls depending on shielding gas.

		Standard				AM			
		ArHeHC	ArHeH	ArH	ArHeO	ArHeHC	ArHeH	ArH	ArHeO
HV10	[–]	275 ± 24	278 ± 25	282 ± 13	279 ± 15	264 ± 22	263 ± 26	266 ± 26	274 ± 36
KV, 20°C	[J]	77.5 ± 0.5	80.0 ± 2.0	84.0 ± 0	81.0 ± 1.0	75.0 ± 0	68.5 ± 1.5	80.0 ± 2.0	77.0 ± 1.0
KV, −196°C	[J]	63.5 ± 1.5	67.5 ± 1.0	66.0 ± 1.5	61.0 ± 1.0	62.0 ± 2.0	63.5 ± 0.5	61.5 ± 1.5	65.5 ± 2.5

Table 13.
Results of Charpy V tests and hardness measurements of S Ni 7718 blocks dependent on shielding gas.

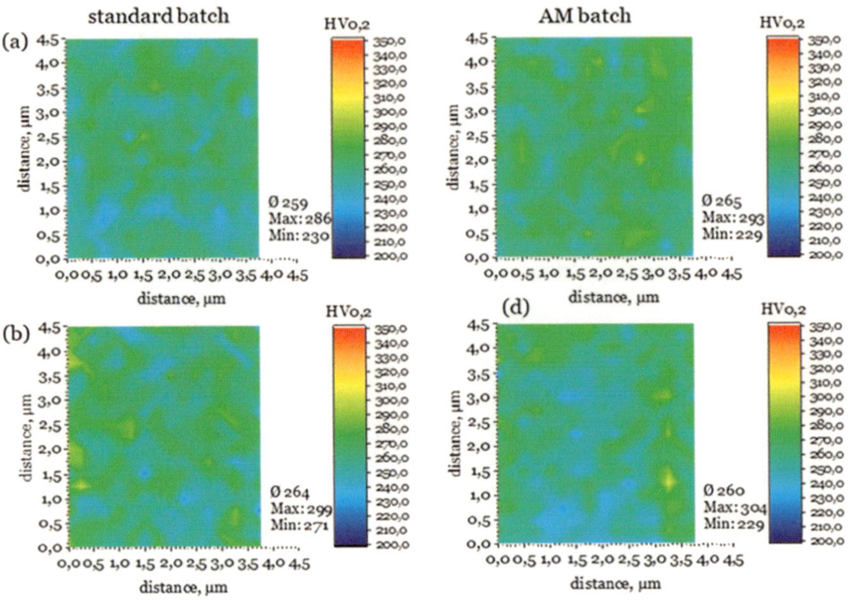

Figure 21.
Hardness maps an (a) walls and (b) blocks of different wire batches of S Ni 7718.

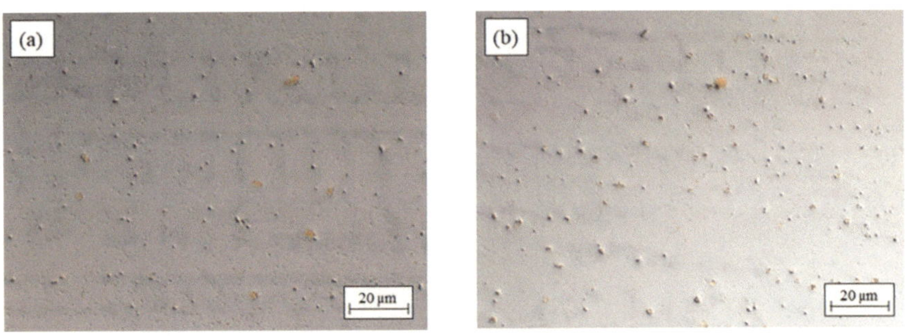

Figure 22.
Light microscopic images of polished walls of S Ni 7718: (a) standard (b) AM.

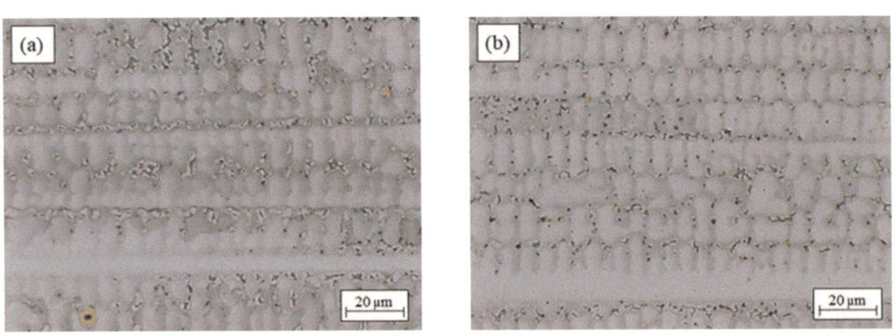

Figure 23.
Light microscopic images of etched walls of S Ni 7718: (a) standard (b) AM (etchant: NaOH).

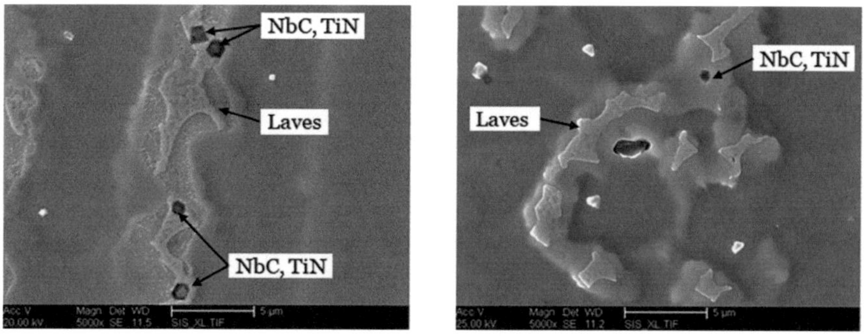

Figure 24.
SEM of etched walls of S Ni 7718: (a) standard (b) AM batch.

Elements	Nb	Mo	Ti	Cr	Fe	Ni	Phase
Standard	21.84 ± 1.01	5.23 ± 0.42	1.63 ± 0.09	13.06 ± 0.42	12.10 ± 0.28	43.14 ± 1.38	Laves
	4,42 ± 0.98	3,42 ± 0.34	1,04 ± 0.17	18,32 ± 0.36	18,0.4 ± 0.62	53,97 ± 0.50	Matrix
AM	22,27 ± 1.79	5.63 ± 0.34	1.62 ± 0.15	12.64 ± 0.17	13.29 ± 0.29	44.55 ± 0.82	Laves
	3.93 ± 0.84	3.40 ± 0.24	0.97 ± 0.14	17.75 ± 0.21	20.60 ± 0.53	52.57 ± 0.66	Matrix

Table 14.
SEM with secondary electron detector on walls of S Ni 7718: (a) standard (b) AM (wt%).

constituent of welded alloy 718. Nb and Mo are enriched in the Laves phase, whereas Al, Fe, Cr, and Ni are depleted (**Table 14**).

6. Conclusions

1. The chapter shows that the part geometry and the wire composition of Alloy 718 significantly affect the hot cracking tendency in CMT-WAAM weld metal. The influence of geometry is greater in this case.

2. In multilayer WAAM welds using the energy-reduced CMT process, the risk of occurrence of unacceptable lack of fusion is very high despite useful calculation tools due to the narrow beads, large wetting angles, and uneven deposition in the edge areas of components.

3. Mechanical properties of the WAAM weld metal of both batches differ only slightly. The AM batch generates slightly higher tensile strength and yield strength, but lower elongation at rupture and notched bar impact work.

4. The composition of the shielding gas mixture in CMT-WAAM welding is rather redundant in terms of improved wetting behavior of molten metal. An influence on the mechanical properties is not visible.

5. The hardness of blocks is about 30 HV 10 higher than that of walls, probably caused by increased precipitation of the brittle and hard phases (Laves, NbC, TiN). In the solution-annealed condition, the hardness is only 200 HV10.

6. The precipitation of the γ'' and γ' strengthening phases in as-welded condition was not observed. But precipitates, such as Laves, NbC, TiN, which are often described in the literature, appear in the weld metal.

Acknowledgements

The present contribution is a part of the AiF project IGF-No. 20.616 B/DVS-No. 01.3145 of the German Research Association on Welding and Allied Processes (DVS). It was kindly supported by the AiF (German Federation of Industrial Research Associations) within the program for promoting the Industrial Collective Research (IGF) of the Federal Ministry for Economic Affairs and Energy (BMWi), based on a decision by the German Bundestag.

Conflict of interest

Manuela Zinke, Stefan Burger, and Sven Juettner declare that they have no conflict of interest.

Author details

Manuela Zinke*, Stefan Burger and Sven Juettner
Institute of Materials and Joining Technology, Otto von Guericke University, Magdeburg, Germany

*Address all correspondence to: manuela.zinke@ovgu.de

References

[1] Debarbadillo JJ, Mannan SK. Alloy 718 for oilfield applications. JOM. 2012;**64**: 265-270. DOI: 10.1007/s11837-012-0238-z

[2] Paulonis DF, Schirra JJ. Alloy 718 at Pratt & Whitney–Historical perspective and future challenges. Superalloys 718, 625, 706 and various derivatives. 2001: 13-23. DOI: 10.7449/2001/Superalloys_2001_13_23

[3] Ezugwu E, Wang Z, Machado A. Themachinability of nickel-based alloys: A review. Journal of Materials Processing Technology. 1999;**86**:1-16. DOI: 10.1016/S0924-0136(98)00314-8

[4] Kindermann RM, Roy MJ, Morana R, Prangnell PB. Process response of Inconel 718 to wire + arc additive manufacturing with cold metal transfer. Materials & Design. 2020;**195**:109031. DOI: 10.1016/j.matdes.2020.109031

[5] Yilmaz O, Ugla AA. Shaped metal deposition technique in additive manufacturing: A review. Journal of Engineering Manufacture. 2016;**230**(10): 1781-1798. DOI: 10.1177/0954405416640181

[6] Rodrigues TA, Duarte V, Miranda RM, Santos TG, Oliveira JP. Current Status and Perspectives on Wire and Arc Additive Manufacturing (WAAM). Materials. 2019;**12**:1121. DOI: 10.3390/ma1207112

[7] Ding D, Pan Z, Cuiuri D, Li H. Wire-feed additive manufacturing of metal components: Technologies, developments and future interests. International Journal of Advanced Manufacturing Technology. 2015;**81**:465-481. DOI: 10.1007/s00170-015-7077-3

[8] Näkki J. Properties of alloy 625 claddings made with laser and CMT methods. (Tampere University of Technology. Publication; Vol. 1593). Tampere, Finland: Tampere University of Technology; 2018

[9] Solecka M, Kopia A, Radziszewska A, Rutkowski B. Microstructure, microsegregation and nanohardness of CMT clad layers of Ni-base alloy on 16Mo3 steel. Journal of Alloys and Compounds. 2018;**751**:86-95. DOI: 10.1016/j.jallcom.2018.04.102

[10] Ola T, Doern FE. A study of cold metal transfer clads in nickel-base INCONEL 718 superalloy. Materials & Design. 2014;**57**:51-59. DOI: 10.1016/j.matdes.2013.12.060

[11] Benoit A, Jobez S, Paillard P, Klosek V, Baudin T. Study of Inconel 718 weldability using MIG CMT process. Sci Technol Weld Joi. 2011; **16**(6):477-482. DOI: 10.1179/1362171811Y.0000000031

[12] Selvi S, Vishvaksenan A, Rajasekar E. Cold metal transfer (CMT) technology - an overview. Defence Technology. 2018; **14**(1):28-44. DOI: 10.1016/j.dt.2017.08.002

[13] Clark D, Bache M, Whittaker M. Shaped metal deposition of a nickel alloy for aeroengine applications. Journal of Materials Processing Technology. 2008; **203**:439-448. DOI: 10.1016/J.JMATPROTEC.2007.10.051

[14] Xu X, Ding J, Ganguly S, Williams S. Investigation of process factors affecting mechanical properties of INCONEL 718 superalloy in wire + arc additive manufacture process. Journal of Materials Processing Technology. 2019; **265**:201-209. DOI: 10.1016/J.JMATPROTEC.2018.10.023

[15] Seow CE, Coules HE, Wu G, Khan RHU, Xu X, Williams S. Wire + Arc Additively Manufactured Inconel 718: Effect of post-deposition heat treatments on microstructure and tensile properties. Materials & Design. 2019;**183**:108157. DOI: 10.1016/j.matdes.2019.108157

[16] Van D, Dinda GP, Park J, Mazumder J, Lee SH. Enhancing hardness of Inconel 718 deposits using the aging effects of coldmetal transfer-based additive manufacturing. Materials Science and Engineering A. 2020;**776**: 139005. DOI: 10.1016/j.msea.2020. 139005

[17] Xiong J, Zhang G, Gao H und Wu L. Modeling of bead section profile and overlapping beads with experimental validation for robotic GMAW-based rapid manufacturing, Robotics and Computer-Integrated Manufacturing 2013;**29**(2):417–423. DOI: 10.1016/j.rcim.2012.09.011

[18] Wittig B, Zinke M, Jüttner S, Keil D. Experimental simulation of dissimilar weld metal of high manganese steels by arc melting technique. Weld World. 2017;**61**:249-256. DOI: 10.1007/s40194-017-0427-z

[19] ISO 21920-3:2021-12. Geometrical product specifications (GPS) - Surface texture: Profile - Part 3: Specification operators. Berlin: Beuth-Verlag GmbH; 2021

[20] DIN EN ISO 17636-1:2013. Non-destructive testing of welds - Radiographic testing - Part 1: X- and gamma-ray techniques with film. Berlin: Beuth-Verlag GmbH; 2013

[21] DIN EN ISO 6892-1:2020. Metallic materials - Tensile testing - Part 1: Method of test at room temperature. Berlin: Beuth-Verlag GmbH; 2020

[22] DIN 50125:2016-12. Testing of metallic materials - Tensile test pieces. Berlin: Beuth-Verlag GmbH; 2012

[23] DIN EN ISO 6507-1:2018. Metallic Materials. Vickers Hardness Test. Test Method. Berlin: Beuth-Verlag GmbH; 2018

[24] Knorovsky GA, Cieslak MJ, Headley TJ, Romig AD, Hammetter WF. INCONEL 718: a solidification diagram. Metallurgical Transactions A. Oct. 1989; **20**(10):2149-2158

[25] Antonsson T, Fredriksson H. The effect of cooling rate on the solidification of INCONEL 718. Metallurgical and Materials Transactions B: Process Metallurgy and Materials Processing Science. Feb. 2005;**36**(1):85-96. DOI: 10.1007/s11663-005-0009-0

[26] Radavich JF. The physical metallurgy of cast and wrought alloy 718, SuperAlloy 718s 718. Metall Appl. 1989: 229-240

[27] DuPont JN, Notis MR, Marder AR, et al. Solidification of Nb-bearing superalloys: Part I Reaction sequences. Metallurgical and Materials Transactions A: Physical Metallurgy and Materials Science. 1998;**29**:2785-2796. DOI: 10.1007/s11661-998-0319-3

[28] Nakao Y, Ohshig H, Koga S, Nishihara H, Sugitani J. Effect of Nb/C on the sensitivity of liquation cracking in 24Cr-24Ni-1. SNbFe-base heat resisting alloy. Journal of the Japan Welding Society. 1982;**51**:989-995

[29] Hauck G. Qualitätsanforderungen an Drahtelektroden beim Schutzgasschweißen mit Robotern. DVS Berichte. 1989;**118**:71-76

[30] Tammi V, Dios M. Reibkraftmessung in Schutzgas-

Schweißbrennern – die optimale Drahtförderung. Binzel, Gießen: Sonderdruck A; 1989

[31] Witzel JMF. Qualifizierung des Laserstrahl-Auftragschweißens zur generativen Fertigung von Luftfahrtkomponenten [Dissertation]. Aachen, Germany: RWTH Aachen University; 2015. Available from: http://publications.rwth-aachen.de/record/463044/files/463044.pdf

Chapter 4

Resistance Spot Welding: Principles and Its Applications

Das Tanmoy

Abstract

Resistance spot welding is one of the primary welding techniques extensively utilized in the automotive and aviation industries. Some 2000–3000 spots are made in a single body of automobiles, which are numerically controlled nowadays. Resistance spot welding works on the principle of Joule's law of heating, where the heat generated is directly proportional to the square of the welding current. This welding technique is generally used to join thin sheets of steel, titanium, aluminum, magnesium, etc. The welding of non-ferrous metals like aluminum and magnesium is quite tedious owing to their high thermal conductivities and the oxide formation on their surfaces. Thus extensive surface preparation is required before welding. Numerous limitations are also there in this welding technique which includes low strength of the joints and thickness limitation.

Keywords: resistance spot welding, welding current, welding time, automobile, Joule's law

1. Introduction

Resistance spot welding (RSW) or simply spot welding is widely employed in automotive and aeronautical industries. This type of welding employs a tremendous amount of current and a very low voltage. Numerous types of metals like different grades of steels, aluminum (Al), magnesium (Mg), titanium (Ti), copper (Cu), and their alloys. Generally, thin sheets of similar or dissimilar metals are joined by RSW in lap joint configuration. RSW also suffers from a few limitations like liquation cracking, voids, misalignment, electrode wear, etc. Thicker sheets of metals are generally difficult to weld by RSW because the heat flows into the surrounding metal very easily.

2. Principles of resistance spot welding

2.1 Theory

Resistance spot welding is a fusion welding process that works on the principle of Joule's law of heating, which states that: $Q = I^2Rt$, where 'Q' is the amount of heat generated during RSW, 'I' denotes the welding current used, 'R' is the resistance setup at the interface of the metal sheets, and 't' is the welding time employed. RSW technique uses two truncated cone/dome-shaped copper alloy electrodes to concentrate the welding current into a fixed small spot and to simultaneously clamp the sheets together without any misalignment. Thin sheets of metals used as workpieces are held together under pressure exerted by the electrodes. The thickness of the metal sheets generally varies between 0.5 and 3 mm. The enforcement of a large amount of current through the spot will melt the metal and form a weld. RSW allows a large amount of energy to be distributed to a specific location in a short period (approximately 10–100 milliseconds). This allows the welding to take place without overheating the rest of the metal sheet.

The resistance between the electrodes and between the electrodes and metal sheets, as well as the amplitude and duration of the welding current, control the amount of heat energy transferred to the produced spot. The amount of energy is chosen to match the sheets' material properties like thermal conductivities, coefficient of thermal expansion, electrical conductivity, etc. Applying too little energy will not melt the localized region and sufficient strength will not be developed. Whereas, applying too much energy will melt too much metal, eject molten material and make a void rather than a spot [1]. **Figure 1** depicts a schematic of a resistance spot welding process.

2.2 Technique and equipment used

There are generally three stages in the resistance welding process which are stated as follows: (a) the electrodes are being brought to the surface of the metal sheets being welded and a slight amount of pressure is applied and (b) the welding current from the electrodes is then applied for a very short time after which the current is removed but the electrodes maintain the pressure to allow the weld metal to cool and solidify. The applied weld times normally range from 0.01 to 0.8 s depending on the thickness of the metal, the electrode force, and the electrode tip diameter [2, 3].

The resistance spot welding setup mainly consists of tool holders and copper alloy electrodes. The tool holders act as a mechanism to hold the electrodes firmly in place

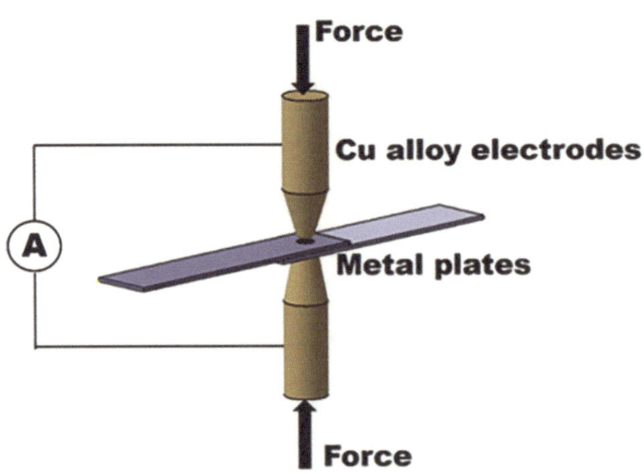

Figure 1.
Schematic of resistance spot welding.

and also to support the cooling water hoses that are used for cooling the electrodes. Tool holding techniques generally include paddle-type light-duty, universal and regular offset. The electrodes are made up of low resistant highly conductive metals like copper and are manufactured in numerous designs such as truncated cone, dome, flat shapes depending on the application needed.

The metal sheets to be welded together are known as workpieces and should be a good conductor of electricity. The width of the metal sheets is limited by the throat length of the welding equipment and ranges typically from 5 to 50 inches (13–130 cm). The thickness can vary between 0.20 and 3 mm.

In the case of RSW, there are two critical components of the tooling system whose characteristics have a significant impact on the entire process: the gun and its kind, as well as the size and form of the electrode. The C-type gun is commonly employed in applications where the gun layout must be as stiff as possible due to large applied forces. This design provides great stiffness and tooling flexibility, as well as collinear electrode motion. The X-type arrangement, like the C type, provides minimal stiffness, even though the reachable workspace is significantly greater than the C-type. As a result, this architecture is highly frequent where tin and flat objects are processed. However, low flexibility is provided in terms of tooling, because the paths of the moving electrodes are not collinear, hence a dome-shaped electrode tip should be used.

Electrodes that are used in spot welding also vary in terms of their uses. For high heat applications, radial type electrodes are used, truncated tip electrodes are used for high pressure, eccentric electrodes for welding corners, offset eccentric tips for reaching into corners and narrow places, and lastly offset truncated into the workpiece itself.

2.3 Features of resistance spot welding

Resistance spot welding tends to work harden the material during the application of electrode force causing it to warp. This phenomenon leads to the reduction of the materials fatigue strength and may stretch the material as well as anneal it. The various defects of spot welding include internal cracking, liquation cracking at the interior of the weld nugget, and a bad appearance. The chemical properties affected include the metal's internal resistance and its corrosive properties.

The welding times used are very short, which can cause electrode wear- they cannot move fast enough to keep the material clamped. During the first pulse, the electrode contact may not be able to make a good weld. The first pulse will soften the metal. During the interval between the two pulses, the electrodes will come closer and make better contact. Also, a higher welding current creates a huge magnetic field, and when the electric current and magnetic field intersect, a large magnetic force field is produced, which causes the melted metal to move very quickly, up to 0.5 m/s. As a result, the fast motion of the melted metal could substantially alter the heat energy distribution in spot welding. A high-speed camera can be used to observe the rapid motion of spot welding [4–6].

2.4 Power supply

The basic spot welding setup consists of a power supply, an energy storage unit (e.g., a capacitor bank), a switch, a welding transformer, and the welding electrodes. The capacitor bank acts as a supplier of high instantaneous power levels. The accumulated energy is dumped into the welding transformer when the switch is pressed. This transformer then reduces the voltage while increasing the current. The transformer's main feature is that it reduces the amount of electricity that the switch can tolerate. The transformer's secondary circuit includes the welding electrode. A control box is also present, which controls the switch and may also monitor the welding electrode voltage or current.

A large number of resistances are being set up in different regions, thus making the resistance offered quite intricate. Secondary winding, cables, and welding electrodes all have their resistances. The contact resistance between the welding electrodes and the workpiece is also a factor. There's also the resistance of the workpieces and the resistance of the workpieces' contact. Because contact resistances are typically high at first, the majority of the energy is wasted there. The heat created by the clamping force softens and smoothens the material at the electrode-material interface, resulting in better contact and lower contact resistance. As a result, more electrical energy will be transferred into the workpiece, and the junction resistance between the two workpieces will increase. The electrodes and the workpiece conduct the heat away as electrical energy is provided to the weld and the temperature rises. The most important need is to provide enough energy to melt a piece of the material within the spot without melting the entire spot. The perimeter of the spot will channel considerable heat away and keep the perimeter at a lower temperature due to thermal conductivity. Because less heat is transferred away from the inside of the spot, it is the first to melt. When a significant welding current is used, the entire spot melts, the material pours out, and a hole instead of a weld is formed.

The working voltage needed for welding is dependent on the resistance of the material to be welded, the sheet thickness, and the desired size of the nugget. When welding a 2 mm lapped joint, the voltage between the electrodes is only about 1.5 V at the start of the weld but can fall as low as 1 V at the end of the weld. This voltage reduction is due to the reduction in resistance owing to the workpiece melting. The open-circuit voltage from the transformer is higher than this which ranges from 5 to 22 V typically.

3. Literature review

3.1 Resistance spot welding of similar and dissimilar advanced high strength steels (AHSS)

Advanced High Strength Steels (AHSS) are generally have been used in the automotive body structures in automotive industries owing to the reduced vehicle weight, high strength safety requirements, good corrosion resistance, and improved crash resistance [7–10]. The numerous AHSSs under consideration are dual-phase (DP) steels, transformation-induced plasticity steel, complex phase steels, and martensitic steels. Pouranvari [11] investigated the failure mode transition from interfacial to pullout mode of DP600 steel and low carbon steel in both tensile-shear and cross-tension loading conditions. They reported that increasing the carbon equivalent decreases the ductility ratio. Wang et al. [12] investigated the effect of martensite volume fraction and morphology on the dynamic mechanical properties of three DP steel sheets (DP600, DP800, and DP1000) and one MS steel (M1200). They found out that the increment of tensile strength decrease with the increase of martensite volume fraction. The ferritic plastic deformation dominated the fracture mode. **Figure 2** shows the obtained deformed microstructures and holes at various distances from the substrate surface at the strain rates of 10^{-3} s^{-1} and 10^{3} s^{-1}.

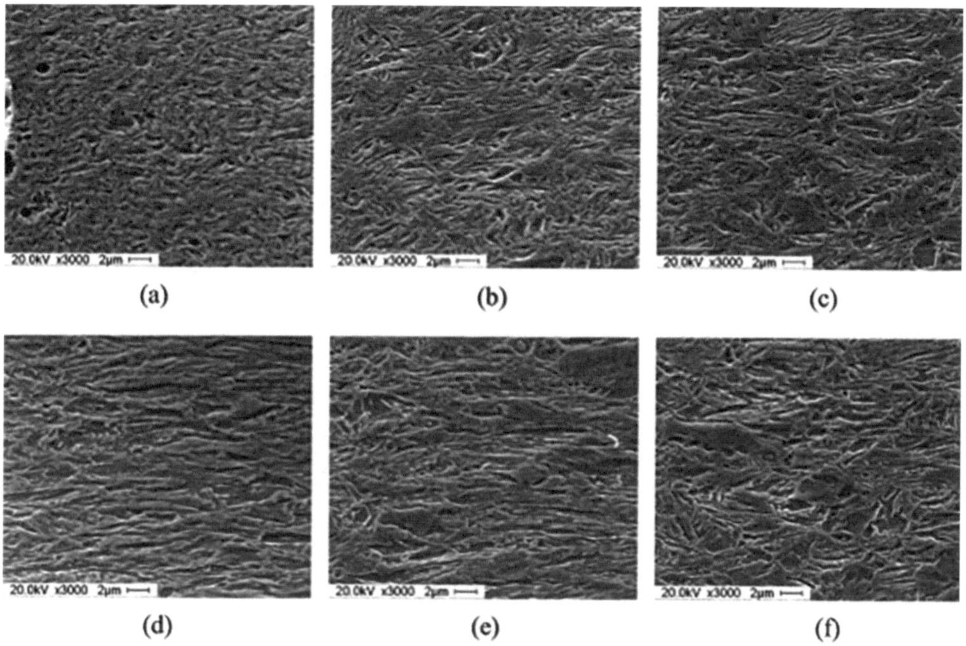

Figure 2.
The deformed microstructure and the holes at various distances from the surface of the M1200 at the strain rates of 10^{-3} s^{-1} and 10^3 s^{-1}. (a) 0.1 mm (10^{-3} s^{-1}). (b) 0.3 mm (10^{-3} s^{-1}). (c) 0.5 mm (10^{-3} s^{-1}). (d) 0.1 mm (10^3 s^{-1}). (e) 0.3 mm (10^3 s^{-1}). (f) 0.5 mm (10^3 s^{-1}) [12].

Hayat and Sevim [13] carried out the spot-welded joints of galvanized DP600 steel and found out that the fracture toughness of the welded joint varies with the welding current and the welding time. Also, the fracture toughness of spot weld is not only dependent on the nugget diameter but also on the sheet thickness, tensile rupture force, welding time, and current. Pal and Bhowmick [14] investigated the RSW characteristics of DP780 steel and found out that the maximum load-carrying capacity is affected by the mode of fracture, that is, interfacial fracture attributes lower load-carrying capability compared to plug and hole type fracture. Zhao et al. [15] carried out the RSW of similar DP600 joints and concluded that the electrode force has an obvious effect on the weld nugget size of the weld joint. And there is a critical electrode force with which the weld nugget size attains its maximum value. The mechanical properties enhanced with the increase in the welding current and time. The variation of the effect of electrode force on the nugget diameter and penetration rate and the relationship between the tensile shear load, and absorbed energy with the nugget diameter has been depicted in **Figure 3a** and **b**.

Banerjee et al. [16] evaluated the fatigue characteristics of resistance spot welded DP590 steel sheets and reported that the fatigue life of the joints depends on the nugget size, notch sensitivity, load regime, and associated shear and nominal stress conditions. In the high and intermediate load regimes, the fatigue performance can be correlated to the nugget diameter with the larger nuggets exhibiting better performance. Matlock et al. [17] studied the recent developments in AHSSs for automotive applications. Early dual-phase and TRIP steel research in the late 1970s and early 1980s evolved into the core ideas for these breakthroughs. Controlling austenite stability and volume fraction to make highly ductile TRIP steels was highlighted as a significant factor in the development of new third-generation AHSS. Khan et al. [18] carried out

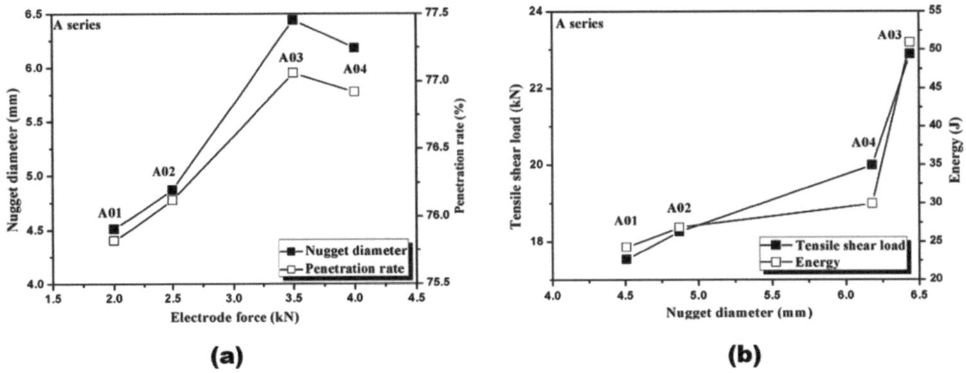

Figure 3.
(a) The electrode force effect on nugget diameter and penetration rate and (b) the relationship between tensile shear load, energy with nugget diameter [15].

the resistance spot weldability study of AHSSs like DP600, DP780, TRIP780, and 590R. They found out from the study that the typical inter-critical heat affected zone (ICHAZ) microstructure comprised of undissolved ferrite and dispersed martensitic islands. The TRIP steel exhibited some retained austenite within the ICHAZ. Also, AHSS produced superior tensile failure loads relative to HSLA. The interfacial fracture was observed during tensile testing of DP600, while button pullout failure modes occurred for HSLA 590R, DP780, and TRIP780. Shojaee et al. [19] investigated the mechanical properties and failure behavior of RSW joints in third-generation 980 and 1180 sheets of steel. They concluded that the tensile shear results (TSS) exhibited that welds can show IF mode and possess high load-bearing capabilities. This suggests that using failure mode as the primary criterion for gauging TSS tests is inaccurate and that load-bearing capacity is a better indicator of weldment performance under shear loads. **Figure 4** shows the Vickers microhardness maps across the weld cross section of (a) 3G-980 steel welded at 9.3 kA and 3G-1180 steel welded at 9.1 kA. And **Figure 4c** shows the hardness profiles extracted from the maps.

3.2 Techniques to improve the RSW joints

Numerous techniques have been developed to enhance the joint strength of spot-welded joints of both similar and dissimilar metals. The techniques are listed as below:

3.2.1 Use of double pulsed current

Soomro and Pedapati [20] studied the effect of second pulse current on the microstructure and mechanical behavior of RSW HSLA350 steel and concluded that the introduction of a second pulse current enhanced the energy absorption capability and tensile shear strength of the weld. Also, shear dimples with a low fraction of micro-cracks were observed compared with tearing ridges with a high fraction of micro-cracks ass seen in a single pulse weld. Soomro et al. [21] carried out both single-pulse and double pulse welding of DP590 steel and observed the maximum improvement of 62% in tensile peak load and 62.3% failure energy in double pulse welds compared with single pulse welds. Also, an increment of 3.7% at heat input (Q) = 0.25 and 13.8% at Q = 1 was observed in weld nugget size. Jahandideh et al. [22] investigated the effect of post-heating time and post-heating current on the weld quality

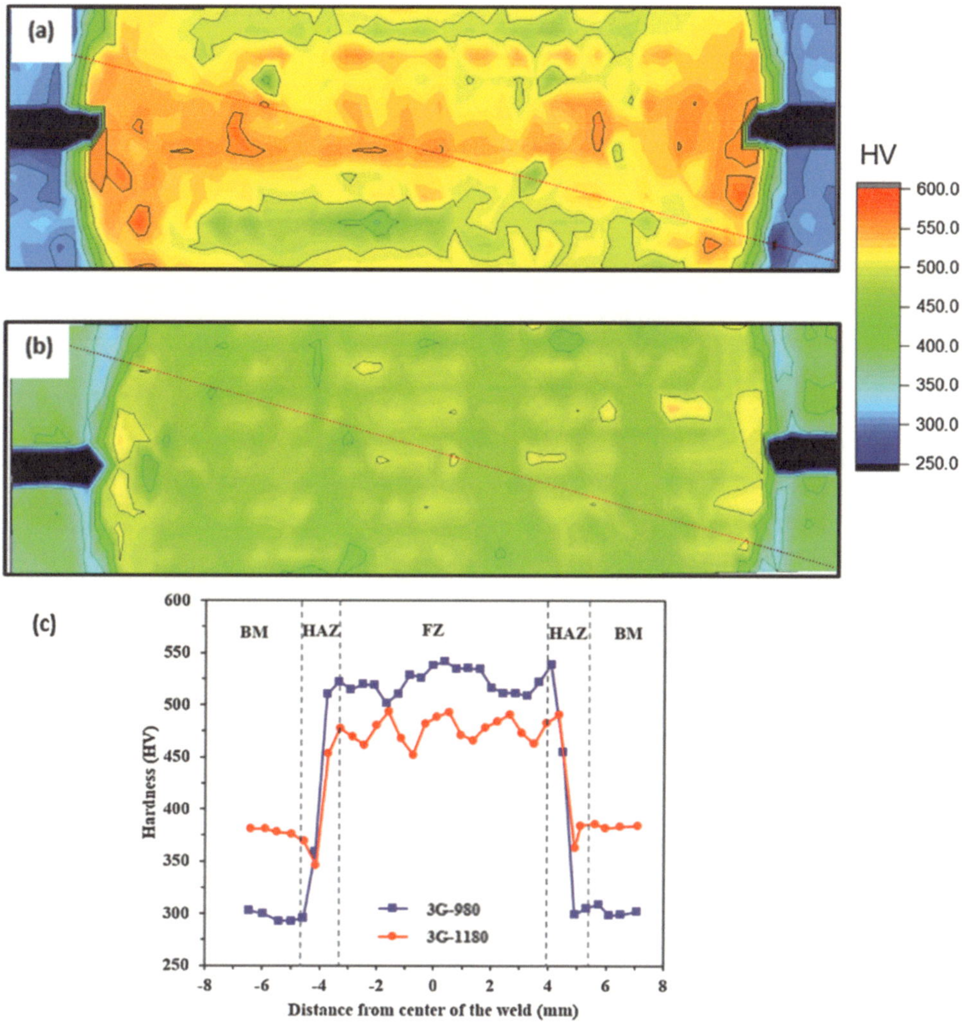

Figure 4.
(a) Vickers microhardness maps across the cross-section of (a) 3G-980 welded at 9.3 kA and (b) 3G-1180 welded at 9.1 kA. (c) Hardness profiles extracted from maps [19].

of SAPH40 steel and concluded that post-heating time reduces interfacial fractures but with a lower rate. The post-heating stage does not have a significant effect on the shear-tensile strength of the welded joints and the failure mode for the tensile shear lap tests. **Figure 5a** shows the nugget hardness and failure modes at various weld parameters and **Figure 5b** and **c** depicts the typical fracture modes.

Eftekharimilani et al. [23] investigated the effects of single and double pulse RSW on the microstructures of an AHSS. The elemental distribution of phosphorous at the primary weld nugget edge of the double pulse welds is more uniform, according to the researchers. When the area is heated to a higher temperature, the distribution improves (i.e., a second current pulse of equal magnitude to the first). The mechanical properties also enhanced due to double pulsing and welds subjected to two equal current pulses show the highest maximum cross tensile strength and tensile shear strength and a favorable plug failure. Liu et al. [24] studied the effect of double-pulse

RSW on the mechanical properties and fracture process of Q&P980 steel. They observed that martensite was the predominant microstructure in the weldment and this steel was susceptible to liquation crack formation. Also, the application of higher secondary current improved the tensile-shear strength and failure mode, while a medium value-enhanced cross-tensile strength and ductility ratio. **Figure 6** portrays the typical fracture modes in TSS tests and their corresponding fracture surfaces and cross-sections.

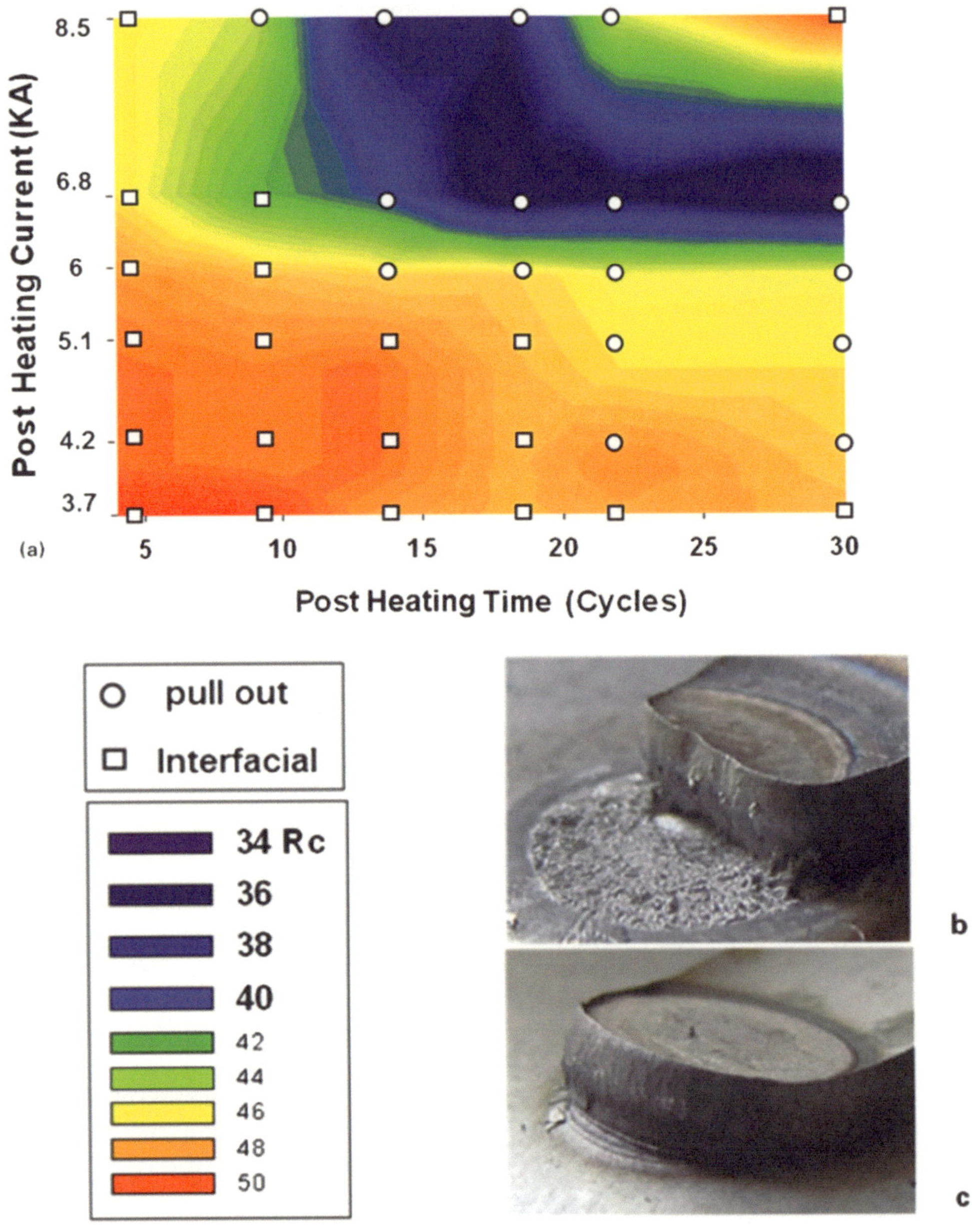

Figure 5.
(a) Hardness of nugget and failure mode at various post-welding currents and times, (b) typical interfacial fracture, and (c) typical pullout fracture [22].

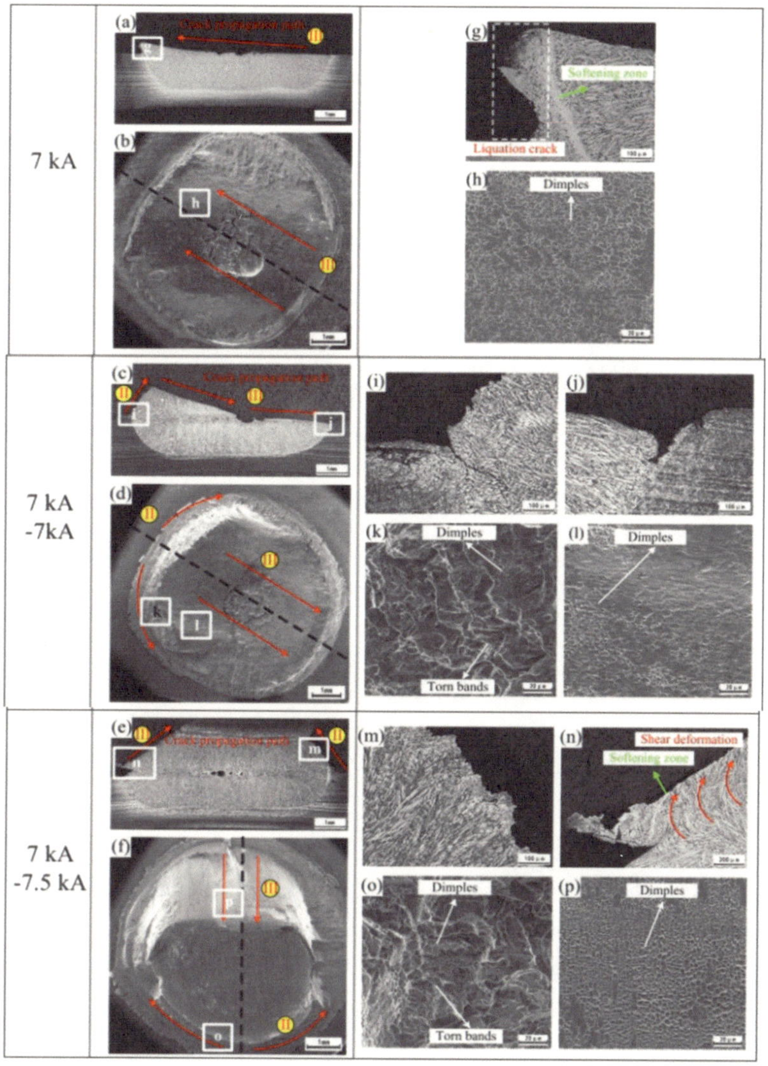

Figure 6.
Typical fracture modes in TSS tests: (a)–(f) are fracture surfaces and cross-sections; (g)–(p) are magnified images located at locations (g)–(p), respectively [24].

3.2.2 Use of interlayers

Ibrahim et al. [25] investigated the weldability study of A6061-T6 sheet to SS304 using Al-Mg alloy as an interlayer. They concluded that Al/steel dissimilar welds with an interlayer exhibited higher tensile shear force than those without interlayer. The tensile shear and fatigue strengths of RSW Al/steel dissimilar welds were higher than those of FSSW ones fabricated using a scroll grooved tool without a probe. Plug, shear, and upper Al sheet fracture were dominant at high, medium, and low load levels, respectively. Zhang et al. [26] carried out the thermo-compensated RSW of AA5052-H12 Al alloy and AZ31B Mg alloy using Zn as an interlayer. The addition of a Zn interlayer between the sheets does not affect the tensile properties of the Mg/Al dissimilar joints, and the tensile shear force of the weld joint was improved to 219 N using a thermos-compensated method, whereas the peak load of the Mg/Al RSW joints and the Mg/Al with Zn interlayer RSW joints was only 33 and 727 N, respectively. Das et al. [27] carried out the RSW of AISI-1008 steel to Al-1100 alloy using graphene nanoplatelets (GNPs) coating as an interlayer. They reported an enhancement of ~124% in the weld strength in one of the welding parameters. There was also an increment in the hardness owing to the interplay of different strengthening mechanisms. Intermetallic compounds (IMCs) of Al-Fe like $FeAl_3$, Fe_2Al_5, and Fe_4Al_{13} were formed at the interfacial region of Al/Fe, which were brittle. **Figure 7a–d** presents the load vs. displacement plots of the bare and GNP coated specimens and also the percentage enhancement owing to the GNP addition.

Penner et al. [28] investigated the effect of gold-coated nickel interlayer on the mechanical and microstructural behavior of dissimilar Al-Mg resistance spot welds. They reported that no joints were produced using a bare Ni interlayer. The welds made with 24 kA current had an average peak load of 4.69 kN, which was as high as 88% of the optimized similar AZ31B welds. And the formation of Al-Mg IMCs was completely suppressed using a gold-coated nickel interlayer. Thus gold-coated nickel represented a promising approach in dissimilar RSW. Sun et al. [29] carried out the dissimilar RSW of AA5052 to AZ31 alloys with Sn-coated steel interlayer.

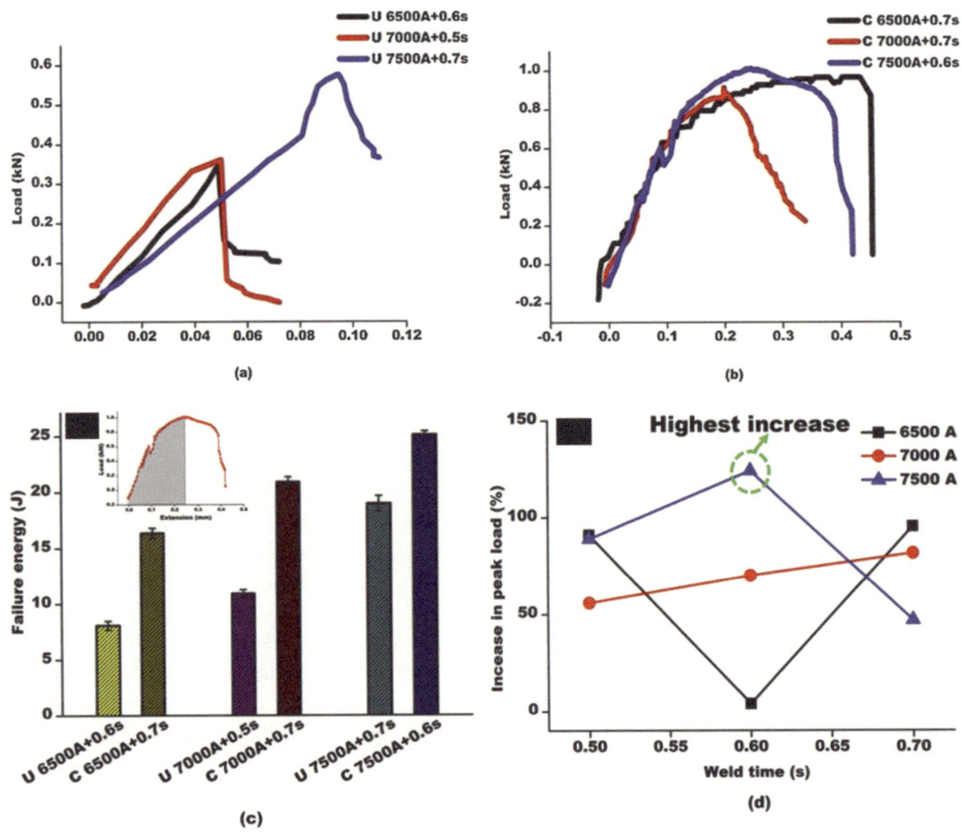

Figure 7.
Load vs. extension plots of the (a) uncoated specimens, (b) graphene-coated samples processed at the best welding parameters, (c) failure energy of the samples, and (d) percentage increase in the peak load by graphene addition as compared to bare samples [27].

They reported that strong joints were achieved using the interlayer and it reached 88% of the maximum value of AZ31 similar RSW joints. The thickness of the Al-Mg IMCs reduced and also the voids reduced due to the long downslope time and maybe also the high boiling temperature of Sn. Das et al. [30, 31] studied the effect of multi-walled carbon nanotubes (MWCNTs) on the RSW of AISI-1008 steel joints. They concluded that an enhancement of ~45% in the joint strength was observed owing to

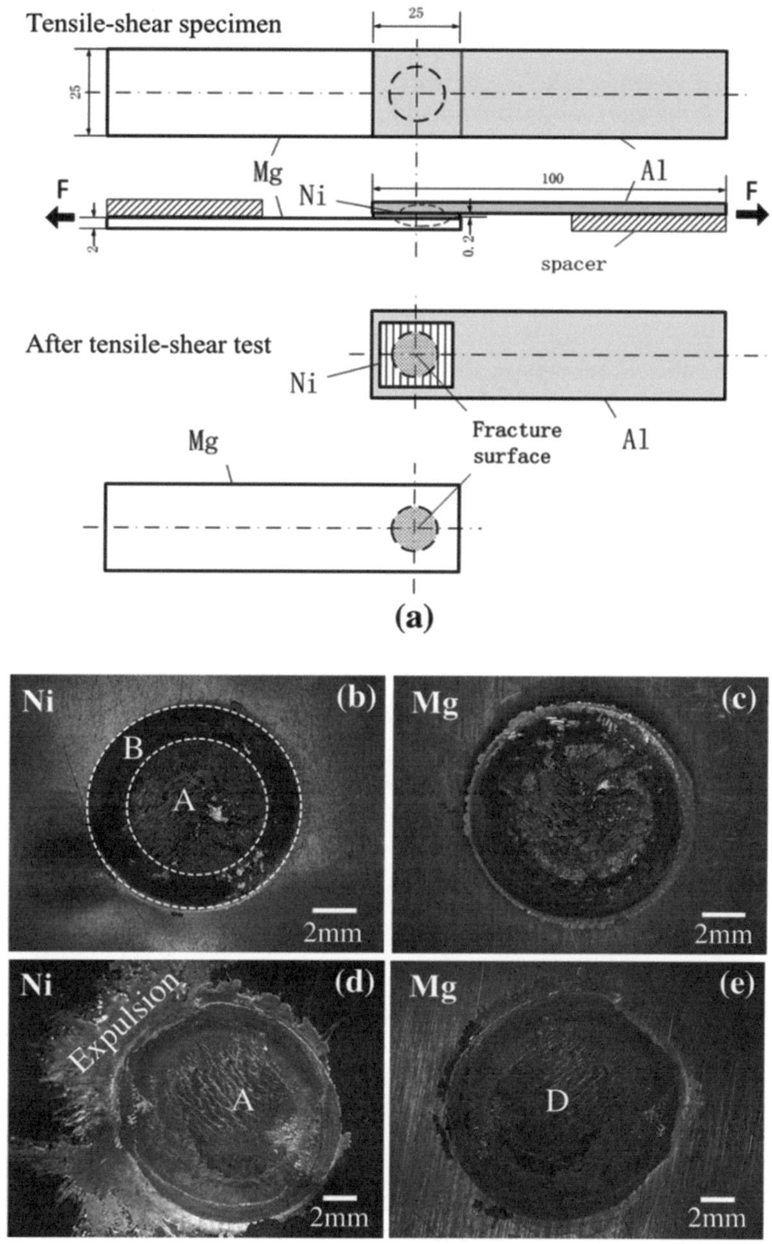

Figure 8.
(a) schematic of the tensile shear specimen and fracture surfaces; fracture surface of the nugget on Ni (Al) and Mg side at welding current of 32 kA (b) and (c) and 42 kA (d) and (e) [32].

the incorporation of MWCNTs interlayer. The failure energy is also enhanced with the increase of welding current and with the use of an interlayer. Sun et al. [32] carried out the RSW of dissimilar AZ31 Mg alloy to aluminum AA5754 with a commercially pure Ni as an interlayer. They summarized that increasing the welding current increased the nugget diameter and hence the joint strength increased to 36 kA. But defects such as cracks and porosity formed in the Mg/Ni interfacial region were excessively high at 42 kA. This led to the early fracture at the Mg/Ni interfacial region and a reduction in the joint strength. **Figure 8a–e** presents the schematic of the tensile shear specimen and fracture surfaces depicting clearly the interfacial mode of failure and metal expulsion.

Das et al. [33] investigated the effect of graphene nanoplatelets on the RSW of similar AISI-1008 steel joints and concluded that an enhancement of ~63% at a welding parameter was observed. Microhardness studies also reported an increase in the hardness with the incorporation of GNPs interlayer and also with the increase of welding current.

4. Applications of resistance spot welding

Resistance spot welding is typically used when joining particular types of sheet metal, welded wire mesh, or bare wire mesh. Thicker sheets are more difficult to spot weld owing to the dissipation of heat into the surrounding metal more easily. Spot welding is possible with aluminum alloys, but their higher thermal and electrical conductivities necessitate larger welding currents. This necessitates the use of larger, more powerful, and more costly welding transformers.

Spot welding is most commonly used in the automobile manufacturing business, where it is nearly routinely used to join the sheet metals of car frames. Spot welders can also be fully automated and many of the industrial robots found on assembly lines are spot welders.

Spot welding is also utilized in orthodontist clinics to resize metal “molar bands” used in orthodontics with small-scale spot welding equipment.

To create batteries, spot welding is also used to attach straps to nickel-cadmium, nickel-metal hydride, or lithium-ion battery cells. Spot welding thin nickel bands to the battery terminals connect the cells. This method prevents the battery from overheating, which could occur if traditional soldering was used.

Some design practices must be followed for spot welding like connecting surfaces should be free of contaminants such as scale, oil, and dirt to ensure quality welds.

5. Modifications of resistance spot welding

A modified version of resistance spot welding has been developed which is known as projection welding. The weld is localized in projection welding by using raised areas, or projections, on one or both of the metal sheets to be connected. Heat is concentrated at these projections, allowing for the welding of heavier parts or closer weld spacing. The projections can also be used to position the workpieces so that they are balanced. Studs, nuts, and other threaded machine parts are frequently welded to a metal plate using projection welding. Crossed wires and bars are typically joined with it. Multiple projection welds can be arranged by appropriate planning and jigging in this high-production technique [34].

6. Conclusions

This chapter dealt with the resistance spot welding technique and their various working principles and applications. Resistance spot welding is mainly used for the joining of thin metal sheets and their alloys. The difficulties associated with the joining of highly thermally conductive metals like aluminum and magnesium were also discussed. The power supply used was also discussed. Various techniques have also been discovered by researchers to enhance the joint strength of the weldments. The techniques included the use of double-pulse welding current and various metallic or non-metallic interlayers. Also, the broad field of applications of resistance spot welding was mentioned, and also a modified RSW technique which is the projection welding was also elaborated in this chapter.

Appendices and nomenclature

RSW	resistance spot welding
GNPs	graphene nanoplatelets
MWCNTs	multi-walled carbon nanotubes
IMCs	intermetallic compounds
AHSS	advanced high strength steels
DP	dual-phase
HSLA	high strength low alloy
TRIP	transformation induced plasticity
Q&P	quench and partitioning
ICHAZ	inter-critical heat affected zone

Author details

Das Tanmoy
Indian Institute of Technology Kharagpur, Kharagpur, India

*Address all correspondence to: tanmoy.nerist@gmail.com

References

[1] Schumacher BW, Dilay W. Adaptive schedule selective weld control. 1984

[2] Pouranvari M, Khorramifar M, Marashi SPH. Ferritic-austenitic stainless steels dissimilar resistance spot welds: Metallurgical and failure characteristics. Science and Technology of Welding and Joining. 2016;**21**:438-445. DOI: 10.1080/13621718.2015.1124491

[3] Larsson H, Chamberlain A, Walin S, Schouri S, Nilsson L, Myrsell E, et al. Welding methods for electrical connections in battery systems. 2019

[4] Li YB, Lin ZQ, Shen Q, Lai XM. Numerical analysis of transport phenomena in resistance spot welding process. Journal of Manufacturing Science and Engineering Transaction of the ASME. 2011;**133**:1-8. DOI: 10.1115/1.4004319

[5] Li YB, Lin ZQ, Hu SJ, Chen GL. Numerical analysis of magnetic fluid dynamics behaviors during resistance spot welding. Journal of Applied Physics. 2007;**101**:053506-1 to 053506-10. DOI: 10.1063/1.2472279

[6] Li Y, Wei Z, Li Y, Shen Q, Lin Z. Effects of cone angle of truncated electrode on heat and mass transfer in resistance spot welding. International Journal of Heat and Mass Transfer. 2013;**65**:400-408. DOI: 10.1016/j.ijheatmasstransfer.2013.06.012

[7] Kleiner M, Chatti S, Klaus A. Metal forming techniques for lightweight construction. Journal of Materials Processing Technology. 2006;**177**:2-7. DOI: 10.1016/j.jmatprotec.2006.04.085

[8] Zhang XQ, Chen GL, Zhang YS. Characteristics of electrode wear in resistance spot welding dual-phase steels. Materials and Design. 2008;**29**:279-283. DOI: 10.1016/j.matdes.2006.10.025

[9] Ozturk F, Toros S, Kilic S. Tensile and spring-back behavior of DP600 advanced high strength steel at warm temperatures. Journal of Iron and Steel Research International. 2009;**16**:41-46. DOI: 10.1016/S1006-706X(10)60025-8

[10] Ma C, Chen DL, Bhole SD, Boudreau G, Lee A, Biro E. Microstructure and fracture characteristics of spot-welded DP600 steel. Materials Science and Engineering A. 2008;**485**:334-346. DOI: 10.1016/j.msea.2007.08.010

[11] Pouranvari M. Susceptibility to interfacial failure mode in similar and dissimilar resistance spot welds of DP600 dual phase steel and low carbon steel during cross-tension and tensile-shear loading conditions. Materials Science and Engineering A. 2012;**546**:129-138. DOI: 10.1016/j.msea.2012.03.040

[12] Wang W, Li M, He C, Wei X, Wang D, Du H. Experimental study on high strain rate behavior of high strength 600-1000MPa dual phase steels and 1200MPa fully martensitic steels. Materials and Design. 2013;**47**:510-521. DOI: 10.1016/j.matdes.2012.12.068

[13] Hayat F, Sevim I. The effect of welding parameters on fracture toughness of resistance spot-welded galvanized DP600 automotive steel sheets. International Journal of Advanced Manufacturing Technology. 2012;**58**:1043-1050. DOI: 10.1007/s00170-011-3428-x

[14] Pal TK, Bhowmick K. Resistance spot welding characteristics and high cycle

fatigue behavior of DP 780 steel sheet. Journal of Materials Engineering and Performance. 2012;**21**:280-285. DOI: 10.1007/s11665-011-9850-2

[15] Zhao DW, Wang YX, Zhang L, Zhang P. Effects of electrode force on microstructure and mechanical behavior of the resistance spot welded DP600 joint. Materials and Design. 2013;**50**:72-77. DOI: 10.1016/j.matdes.2013.02.016

[16] Banerjee P, Sarkar R, Pal TK, Shome M. Effect of nugget size and notch geometry on the high cycle fatigue performance of resistance spot welded DP590 steel sheets. Journal of Materials Processing Technology. 2016;**238**:226-243. DOI: 10.1016/j.jmatprotec.2016.07.023

[17] Matlock D, Speer J, De Moor E, Gibbs P. Recent developments in advanced high strength sheet steels for automotive applications: An overview. Journal of Engineering Science and Technology. 2012;**15**:1-12

[18] Khan MI, Kuntz ML, Biro E, Zhou Y. Microstructure and mechanical properties of resistance spot welded advanced high strength steels. Materials Transactions. 2008;**49**:1629-1637. DOI: 10.2320/matertrans.MRA2008031

[19] Shojaee M, Midawi ARH, Barber B, Ghassemi-Armaki H, Worswick M, Biro E. Mechanical properties and failure behavior of resistance spot welded third-generation advanced high strength steels. Journal of Manufacturing Processes. 2021;**65**:364-372. DOI: 10.1016/j.jmapro.2021.03.047

[20] Soomro IA, Pedapati SR. Application of in situ post weld heat treatment using double pulse technology and its effect on microstructure and mechanical performance of resistance spot welded HSLA350 steel. International Journal of Advanced Manufacturing Technology. 2019;**105**:3249-3260. DOI: 10.1007/s00170-019-04569-2

[21] Soomro IA, Pedapati SR, Awang M. Optimization of postweld tempering pulse parameters for maximum load bearing and failure energy absorption in dual phase (DP590) steel resistance spot welds. Materials Science and Engineering A. 2021;**803**:140713. DOI: 10.1016/j.msea.2020.140713

[22] Jahandideh A, Hamedi M, Mansourzadeh SA, Rahi A. An experimental study on effects of post-heating parameters on resistance spot welding of SAPH440 steel. Science and Technology of Welding and Joining. 2011;**16**:669-675. DOI: 10.1179/1362171811Y.0000000053

[23] Eftekharimilani P, van der Aa EM, Hermans MJM, Richardson IM. Microstructural characterisation of double pulse resistance spot welded advanced high strength steel. Science and Technology of Welding and Joining. 2017;**22**:545-554. DOI: 10.1080/13621718.2016.1274848

[24] Liu XD, Xu YB, Misra RDK, Peng F, Wang Y, Du YB. Mechanical properties in double pulse resistance spot welding of Q&P 980 steel. Journal of Materials Processing Technology. 2019;**263**:186-197. DOI: 10.1016/j.jmatprotec.2018.08.018

[25] Ibrahim I, Ito R, Kakiuchi T, Uematsu Y, Yun K, Matsuda C. Fatigue behaviour of Al/steel dissimilar resistance spot welds fabricated using Al–Mg interlayer. Science and Technology of Welding and Joining. 2016;**21**:223-233. DOI: 10.1179/1362171815Y.0000000086

[26] Zhang Y, Luo Z, Li Y, Liu Z, Huang Z. Microstructure characterization and tensile properties of Mg/Al dissimilar joints manufactured by

thermo-compensated resistance spot welding with Zn interlayer. Materials and Design. 2015;**75**:166-173. DOI: 10.1016/j.matdes.2015.03.030

[27] Das T, Das R, Paul J. Resistance spot welding of dissimilar AISI-1008 steel/Al-1100 alloy lap joints with a graphene interlayer. Journal of Manufacturing Processes. 2020;**53**:260-274. DOI: 10.1016/j.jmapro.2020.02.032

[28] Penner P, Liu L, Gerlich A, Zhou Y. Feasibility study of resistance spot welding of dissimilar Al/Mg combinations with Ni based interlayers. Science and Technology of Welding and Joining. 2013;**18**:541-550. DOI: 10.1179/1362171813Y.0000000129

[29] Sun M, Niknejad ST, Gao H, Wu L, Zhou Y. Mechanical properties of dissimilar resistance spot welds of aluminum to magnesium with Sn-coated steel interlayer. Materials and Design. 2016;**91**:331-339. DOI: 10.1016/j.matdes.2015.11.121

[30] Das T, Rawal S, Panda SK, Paul J. Resistance spot-welding of AISI-1008 steel joints with MWCNT coating interlayer. Materials and Manufacturing Processes. 2021;**36**:448-456. DOI: 10.1080/10426914.2020.1843667

[31] Das T, Panda SK, Paul J. Microstructure and mechanical properties of resistance-spot-welded AISI-1008 steel lap joints using multiwalled carbon nanotubes as an interlayer. Journal of Materials Engineering and Performance. 2021;**30**:3333-3341. DOI: 10.1007/s11665-021-05687-3

[32] Sun M, Niknejad ST, Zhang G, Lee MK, Wu L, Zhou Y. Microstructure and mechanical properties of resistance spot welded AZ31/AA5754 using a nickel interlayer. Materials and Design. 2015;**87**:905-913. DOI: 10.1016/j.matdes.2015.08.097

[33] Das T, Sahoo B, Kumar P, Paul J. Effect of graphene interlayer on resistance spot welded AISI-1008 steel joints. Material Research Express. 2019;**6**:0865c3

[34] Zhang X, Yao F, Ren Z, Yu H. Effect of welding current on weld formation, microstructure, and mechanical properties in resistance spot welding of CR590T/340Y galvanized dual phase steel. Materials (Basel). 2018;**11**:2310. DOI: 10.3390/ma11112310

Chapter 5

Engineering Challenges Associated with Welding Field Repairs

Tyler J. McPheron and Robert M. Stwalley III

Abstract

Welding as technology exists in two worlds. Manufacturers execute designs typically based on professional society-backed standards. Repair service centers that administer field repairs where welding applications are required can sometimes have staff members with little formal education. The challenges of a technical manager seeking welded field repairs to equipment are significant and numerous. This chapter will seek to outline the process of executing a successful welding field repair by breaking down the analysis into three parts—(1) the identification of the engineering challenges associated with a specific job, including significant stresses, difficult materials or locations, and adequate piece preparation to ensure of weld integrity; (2) the ability to properly specify the type of repair, including knowledge of the types of weld junctions and preparations, the various types of welding processes and their features, weld types and associated drawing symbols, and the repair design and repair support process; and (3) the challenges for field engineers and technical managers in identifying weld defects, executing measures, and providing adequate examination and evaluation of weld quality in the field. This chapter tries to bridge the gap between the formal, engineered welds used in manufacturing and the sometimes-needed expediency of fieldwork.

Keywords: welding, repairs, MIG, TIG, arc, weld defects

1. Introduction

Welded connections play a substantial role in the manufacturing processes of many types of parts, structures, equipment, and materials. Equipment manufacturers for vehicles and implements use a considerable amount of advanced welding techniques and applications throughout the design of their manufactured products. In the industrialized world, frequent breakdowns, failures, and the necessity of repairs are a part of ongoing operations. Mechanical and structural component failures are inevitable, particularly with mobile agricultural and construction equipment. Therefore, it is essential for engineers, field technicians, and those in similar roles to be educated on the characteristics of a successful weld repair, welding fundamentals, and the associated challenges of executing field repairs.

Identifying the engineering challenges of field repairs will be examined first. Properly executed welding techniques incorporate an extensive amount of engineering fundamentals to maintain the original assembly's design and ensure adequate structural integrity once complete. There are many types of welding repairs that can be categorized by location, component type, design criteria, and the degree of critical quality requirements. Understanding basic material properties and identifying the material to be repaired is important for the welding process selection and the structural analysis of the proposed repair design. Certain repairs may not be accessible enough to clean the mill-scale, corrosion, or debris away, thereby narrowing the choice of possible welding processes, so repair preparation may also influence the welding process selection.

Understanding the various types of welding processes is vital to a manager seeking repairs. Each welding process features unique characteristics, often making one process more suited for a specific repair than other potential processes, based on the technical characteristics of the welding process equipment. Identifying the base material properties and measuring the thickness of pieces are both easily quantifiable metrics that are used to determine the proper weld size. For economic reasons, this analysis is always performed in manufacturing applications, but it is also necessary for quantifying the strength of a repair weldment to ensure that enough weld material has been applied to a given joint. Weldment strength mathematical field calculations will be examined in the chapter and illustrated with appropriate figures.

Finally, challenges for field engineers and technical managers will be summarized with emphasis on the identification of weld defects, preventative methods, and suggestions that can be used to minimize the occurrence of weld defects when conducting field repairs. Technical procedures for weld quality examination in field repairs should be useful for managers when instructing repairmen on alleviating defects or challenges within a job. Finally, the chapter concludes with a discussion of the importance of repairs to business enterprises. There is value in having basic welding technical knowledge for an array of industry specialists. This chapter seeks to promote further education and interest in the engineering properties of welding repairs, their application to practical problems, and provide some value to operational managers utilizing heavy field equipment.

A variety of primary sources have been used for this chapter. Various sections of Blodgett's *Design of Welded Structures* (1972) are used heavily throughout the chapter [1]. This classic textbook incorporates a significant amount of published welding calculations, and its primary focus is on welding in industrial and manufacturing environments. However, many of the calculations and formulas are valid across all fields of welding technology and application. Information regarding material properties and analysis comes from the *American Society of Mechanical Engineers* (ASME), the *American Institute of Steel Construction* (AISC), and the *American Society for Testing Materials* (ASTM) publications. These organizations are responsible for writing many of the technical standards and recommended protocols for design, material testing, and property specification in the industry. Codes and practical definitions regarding welding standards are from the *American Welding Society's* (AWS) *Structural Welding Code* [2]. The AWS provides detailed information in its standards for a diverse array of welding applications. Various standards include code requirements, measurements, strength, specific practice recommendations, given constants for equations, education, safety practices, and other facets of welding. Where appropriate, the reader will be directed to primary sources for specific information.

2. Identification of engineering challenges in repairs

There is a multitude of jobs and types of repairs requiring welding applications. Challenges can often be categorized by the location, component type, design criteria, and the degree of critical quality needed in the repair. Repair location presents its own challenges from a logistical sense. Outdoor welding applications require mobilized welding equipment to access the repair. Mobilized welding equipment can require a fuel-based power source, due to a remote location or conditions. Due to the space constraint of certain repairs, it is common for mobile welding equipment to feature lengthy leads, which can also create unique circumstances. Outdoor welding repairs can present additional challenges derived from environmental factors, such as weather, terrain, and less than ideal base material circumstances. Not all welding processes are well-suited for this type of environment, limiting available repair options. These challenges call for strategic planning and an understanding of certain technical properties for repairs made in outdoor environments. There are fewer constraints for indoor welding repairs, where a variety of processes are available. Indoor welding environments found in machine shops, manufacturing facilities, prototype shops, and fabrication workshops are well equipped for operator comfort and process quality, without exposure to adverse environmental conditions.

2.1 Repair preparations

Pipeline and structural welding are common examples of strictly outdoor welding applications. Professional welders of this category are highly skilled, due to the criticalness of their weldments and their ability to produce quality welds in difficult or uncomfortable body positions. This may consist of laying under a pipe in the mud to bevel, grind, mate, and weld pipe together, or it may mean being suspended hundreds of feet in the air, welding steel beams for structural applications. Critical preparation techniques are always used during the preliminary repair preparation process to ensure the quality of the weldment because, in many circumstances, time in the repair zone or position is limited.

Metal cleanliness and preparation are the most critical features for any field repair. Insufficient attention to base metal preparation will lead to unwanted imperfections in the repair. While certain welding processes are capable of penetrating through some surface rust, failing to clean the intended weld area of contaminants will always yield certain weld defects, such as slag inclusions, porosity, and craters. Weld defects lead to unwanted future cracks in the weldment, causing the component to lose structural integrity and have progressively weaker joint strength. Metal preparation is usually conducted using an electric angle grinder with a variety of disc attachments. Standard grinding discs are used to remove material quickly, and they perform well for removing large amounts of surface rust, paint, and other contaminates. Sanding disks, also referred to as "flap disks," are effective at removing mill-scale on the base material and polishing the metal surface.

Material fitment to maintain proper dimensions and alignment with other components during the welding process is nearly as vital as material preparation. Repairs due to a defect or flaw causing a component or related component to be displaced from their original state must be approached with caution. These conditions require the fitment of material to be returned to a near-original state prior to the repair process. However, the removal of old material often includes the removal of the original weldment or the removal of material near the damaged area. Taking measurements beforehand or using a secondary, mirrored part for dimensioning can be helpful for reestablishing the original location of a displaced component in need of repair. When proper material fit-up is completed, weld material can be added to the removed areas by making a series of weld passes to build up the filler material. Adding an amount of weld material more than what is necessary is generally recommended, so later grinding can remove the excess.

After achieving proper fitment of the base pieces, constructing a jig or welding additional structures to the member may be necessary to maintain the position of the pieces while welding. Welding induces rapid temperature changes with the heat concentrated in a small area. This thermal transfer of energy causes the metal to expand slightly, and under extreme conditions, it can cause warpage and deformation. High temperatures cause a crystalline structural rearrangement and reduction in tensile and yield strength in most metal materials [3]. Metal warpage is more likely to occur when applying a large amount of weld material to thinner materials with thicknesses of less than 30 mm. Part of the preparation process for the repairman is evaluating and mapping out the intended weldments in an effort to evenly distribute heat applied to the material during the repair. If the material movement cannot be avoided by distributing weld material and heat evenly, a temporarily fixed member can be welded in place to prevent movement from occurring. The temporary support can later be removed and grinded away.

2.2 Material analysis

Welding repairs may be necessary on a variety of metal components that can be made from ferrous materials, such as carbon steel, stainless steel, and cast iron, as well as non-ferrous materials, such as aluminum and titanium alloys. Identification of the material is essential to executing a welding repair. Repairmen need to accurately be able to access which materials they are dealing with. All raw materials have specific properties associated with the type of material. Material properties are defined as measurable, quantifiable properties associated with the material. The material properties help categorize different materials and ease the process of material selection. Evaluating categories of mechanical properties of the material is an effective way to identify material for a field repair. Documentation is always best, but in the field, it seldom exists. When confronted with an unknown material, investigation should include at a minimum, a chemical test, and a hardness test. These two property indicators will help qualify the weldability of the material. Other properties can help narrow uncertainty in the base material. There are generally considered to be five categories of mechanical properties for common building materials [4], which are as follows:

- physical properties;
- mechanical properties;
- thermal properties;
- electrical properties; and
- chemical properties.

Physical properties are perhaps the most easily identifiable material characteristics when conducting field repairs. These properties include the shape, size, color, texture, finish, porosity, and luster of the subject material [4]. Technological properties are also referred to as basic mechanical properties for the metal, and these include hardness, malleability, machinability, weldability, and formability. It is recommended that the material's physical properties be evaluated first. This will allow an easier understanding and identification of mechanical properties, once known. In a repair situation, material identification is important for understanding the behavior of the component's base metal and how it is likely to react to different welding processes.

Mechanical properties are critical to understanding structural repair applications, particularly when there are critical zones of stress and strength maintenance requirements. Material properties commonly found in engineering material references are as follows [3]:

- ultimate tensile strength;
- elongation;
- modulus of elasticity;
- compressive strength;
- shear strength; and
- fatigue strength.

Tensile strength for different types of material is experimentally determined by a standard testing method, conducted using a tensile test machine. The selected material is marked at two locations, 50.8 mm apart. Once the selected material is placed on the machine, an axial load is applied by pulling the material in opposite directions at a constant rate. As the test progresses, the load divided by the original cross-sectional area of the material within the marked area represents the resistance that the material has to the tensile load being applied [1].

The stress (σ) unit is in force per area, while the strain (ϵ) unit is formally dimensionless, it is expressed as displacement in length per original length. The maximum load applied before failure of the material, divided by the static cross-sectional area of the material being tested is equal to the ultimate tensile strength (σ) of the material. From stress and strain values, the modulus of elasticity of a material can be calculated as [1]:

$$Modulus\ of\ Elasticity\ (E) = \frac{Stress\ (\sigma)}{Strain\ (\epsilon)} \tag{1}$$

Modulus of elasticity (E) is a way to quantify the springiness of a material or the stress value of a given material as it is deformed by a force in one direction. It is also commonly referred to as Young's Modulus, after English physicist, Thomas Young. The AISC states that the standard for all low-carbon steel is a modulus of elasticity of 200,000 mPa [5]. Section area is an important metric, used when calculating the stress and strength of materials with loads applied in compression, tension, and shear configurations. If the member is not symmetric throughout the length of the applied load, then the section at which the material or structure will induce the most stress is used in the calculation. Once the desired cross section is found, the neutral axis must be located. The neutral axis of a section represents the plane of zero strain and zero stress, and it can be a good place to locate a spot weld during the fitment process [1].

Material hardness is a well-correlated property with many other physical properties, and it is determined using a Brinell hardness test. The test is conducted by applying a known load to the surface of the material using a hardened steel ball. The diameter of the impression that the ball leaves on the tested material is the measured result of the test. The diameter of the impression can be converted to the Brinell number as follows [6]:

$$BHN = \frac{2P}{\left(\pi D\left(D - \left(D^2 - d^2\right)^{0.50}\right)\right)} \tag{2}$$

where *BHN* = Brinell Hardness Number;
P = Load on indenting tool (kg);
D = Diameter of hardened steel ball (mm); and.
d = measured diameter at the rim of the impression (mm).

Fortunately, the Brinell Hardness Number does not typically need to be calculated. For most materials, the number can be found using various Brinell charts. One might be exposed to materials with a high Brinell hardness utilized in high wear environments, as in abrasive situations, due to contact with other moving components. The Brinell hardness can be increased using a thermodynamic hardening process. There are various methods of hardening materials, but in the simplest form, hardening is achieved by increasing the temperature of the material to a modest degree and then rapidly cooling it by quenching the material. The quick change from a high temperature to a cold temperature hardens the material by locking-in elevated temperature crystal structures.

Tool steel used for drill bits, mill cutters, and hand tools is typically hardened, along with other critical mechanical components, such as shafts, bearings, and gears. Hardened materials can be a challenge for welding repairs, due to their impenetrable nature of the material. The heat applied to the material during the welding process, along with the rapid cooling typically present in welding, can make the hardened steel base material brittle and cause cracking along the joint. Heating the material slowly and evenly with an oxygen-acetylene torch, while monitoring the temperature of the joint before welding, will soften the material and allow the weld process to penetrate deeper. After the weld is complete, cooling the material around the joint slowly will maintain the material's hardness, but make the material less brittle. This softening process is also referred to as annealing [7].

The ASTM is an organization established to produce standards for material properties of all sorts. For nearly 120 years, ASTM has written technical standards for materials, products, and other systems [8]. Physical properties are defined by characteristics, such as corrosive resistance, hardness, density, and thermal conductivity, to name a few. When choosing metal material or evaluating an existing component for repair, corrosive resistance is an important factor to consider. Understanding the environment that the material is exposed to aids the welder in selecting preparation requirements and the welding process. In general, the welder should determine the following material properties before starting a repair:

- the material hardness;
- the moisture exposure;
- will paint need to be removed; and

- whether is it outside or inside the application.

It is commonly known that mild steel is corrosive, but when painted, the structural life of the steel is lengthened significantly. The corrosion rate is simply measured by the millimeters of corrosive penetration into the material per year. A common alternative to painting is the plating of the metal. Materials can be plated with a variety of different plating materials. Zinc-plated bolts are a common example of a component that receives plating for increased durability. It might be best to use stainless steel for structures exposed to salt water or high moisture atmospheric environments. Stainless steel is highly resistant to oxidation and a popular choice to use for marine structures, along with many industrial food processing applications having health code precautions. Materials with any of these anti-corrosion features must be treated specially, and repair welders must take these elements into account when planning repairs. Additionally, a repair technician should almost always consult a specialist, if they spot a highly stressed area, find corrosion or hydrogen embrittlement, or have a high-temperature operational environment [9]. Finally, if chemical information about the base material is available, levels of carbon and other alloying materials provide critical information to the repair technician. Certain stainless steels cannot be effectively welded, and in general, a 0.35% carbon level is typically considered the upper limit for welding. Material chemistry mainly identifies what cannot be welded or where extreme caution should be applied when making repairs. There are now test kits available that can determine a metal's chemical content within a matter of hours [10]. There is simply no excuse for not knowing what material you are dealing with anymore.

3. Welding fundamentals and equipment

There are three primary welding processes—gas metal arc welding (GMAW/MIG), gas tungsten arc welding (GTAW/TIG), and shielded metal arc welding (SMAW/Stick/Arc). This section will describe the characteristics and advantages of each process. Various welding positions, the different weld joint types, weldment preparations to the base materials, the calculation of weld fillet size, and weld strength will additionally be reviewed.

3.1 Arc welding

Shielded metal arc welding (SMAW/Stick/Arc) is chosen for a variety of applications due to its uniquely robust properties, allowing it to be used in many environments. Stick welding is heavily used in structural and industrial heavy metal applications when less than ideal conditions are present. Common SMAW/Stick/Arc welding joints include the assembly of structural frameworks for buildings and the joining of pipe segments together in pipeline applications. The connections of high-pressure pipeline segments are generally made by SMAW/Stick/Arc welding. The pipe ends are typically beveled, mated, and welded with an E6010 electrode. This is called a root pass. There can be several root passes depending on the application. Root passes are then overlaid and capped by several additional weld passes, building additional filler material with each pass. This kind of weld is durable and self-healing around small weld defects. It is the oldest form of electrode welding and is still preferred when deep penetration is required.

There is an assortment of consumable electrodes available to the repairman. These are also referred to as "rods." Electrodes are 30 cm long sticks of filler metal, wrapped with a flux coating. As shown in **Figure 1**, they are referenced with a four-digit numerical code with each number referencing a specification for the electrode. The first two numbers reference the tensile strength of the weld material that the rod will produce. The third number refers to the position(s) in the electrode that can be used for welding. One means that it can be used in all positions, and two means it can only be used in the flat and horizontal positions. The fourth number refers to the type of flux coating, the hydrogen content of the electrode, and if it can be used with alternating current, direct current, or both. An advantage of the stick welding process is its ability to make high penetrating welds on imperfectly prepared material with oxidation and mill scale, the oxidized scaly substance on the surface of most hot-rolled material.

This makes SMAW/Stick/Arc welding an excellent choice for field repairs. The stick welding process is commonly used by repair shops for agricultural, off-road, and construction equipment repairs. With the correct rod, rusty and dirty metal can be welded effectively with minimal preparation. This process produces a very strong weldment with high material penetration and can produce an esthetically pleasing weld bead using certain electrodes. Unfortunately, as shown in **Figure 2**, it generates far more splatter and slag peel than other welding processes. However, those other processes require clean metal with minimal mill scale to make solid weldments. Stick welding can operate with either direct current or alternating current. Stick welding machines use a ground clamp and an electrode holder, commonly referred to as a "stinger" to make the necessary electric circuit.

SMAW/Stick/Arc welding is the generally preferred choice of the process by many repairmen for field repairs. As previously mentioned, field repairs often require the use of mobile welding equipment. Notable welding equipment manufacturers, such as Lincoln Electric and Miller Electric, manufacture gasoline and diesel-powered welders for this purpose. Test engineers, pipefitters, construction crews, and other field repairmen will generally have their service vehicles equipped with machines of this category. Mobile welding rigs typically have electrode holders and ground leads in lengths of 15–30 m for repairs and jobs, where only limited vehicle access is available. If the stick welding process is utilized for repair work to an appearance-critical piece of equipment, extreme care must be taken to effectively tarp and shield the balance of the unit from the inevitable splatter of the process.

3.2 MIG welding

Gas metal arc welding (GMAW/MIG) is the predominate joining method in manufacturing and industrial settings. This process is preferred, because of its weld

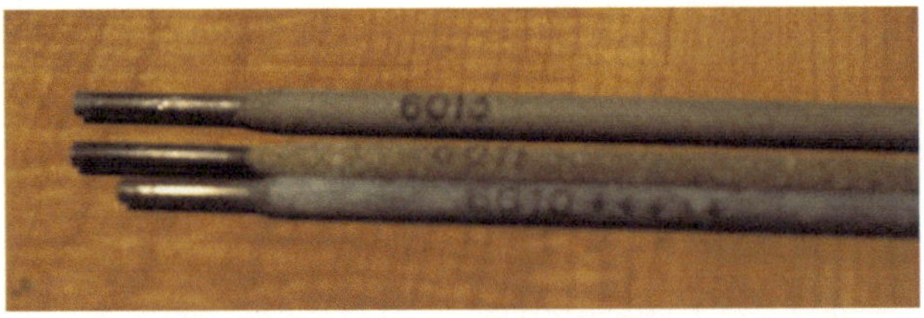

Figure 1.
Electrode specification stamped on the flux coating of a typical welding rod.

Figure 2.
Stick welded components showing the splatter form the process.

quality and speed, making it less expensive to implement in production situations. GMAW/MIG welding produces very little slag and creates a very clean welded connection when welding clean material. MIG welding is often referred to as "wire feed welding." As shown in **Figure 3**, its nickname is due to the spool of wire fed through the MIG gun by the machine. When the trigger on the gun is actuated by the operator, electrified wire is fed through the end and shrouded by a gas cloud. A typical combination of shielding gas for welding mild steel is 75% argon and 25% CO_2. As the welding wire is pushed toward the base material, it strikes an electric arc and melts the target material and filler materials together. Filler metals are rated by their tensile strength. A tensile strength rating of 482 mPa is common for most spools of filler wire. MIG welding is not as effective as SMAW/Stick/Arc for use on oxidized material, without significant clean-up and material preparation. However, it is more time-efficient than other processes for clean assemblies, making it an excellent process in production environments. As shown in **Figure 4**, it makes a beautiful, clean weld on clean material. Materials such as steel, aluminum, magnesium, carbon steel, nickel, and other alloy metals can be welded with the GMAW/MIG process, making this a

Figure 3.
Spool of 0.58 mm filler wire mounted on MIG welding machine.

versatile industrial tool, when multiple materials are in use. The ability to weld a variety of materials makes it a top choice in high production manufacturing [11].

3.3 TIG welding

Gas tungsten arc welding (GTAW/TIG) is a process commonly used for specialty welding repairs and applications, where additional craftsmanship is necessary. The GTAW/TIG welding process is commonly used on components that are made of stainless steel and other non-ferrous metals. TIG welding is expensive and time-consuming, due to its precision, but as shown in **Figure 5**, the quality of the GTAW/TIG weld is unsurpassed. Consequently, it is not often used for facilitating field repairs, but may be used for smaller repairs, especially for stainless-steel repairs. Stainless steel is an alloy with unique properties that do not allow it to be welded easily with other welding processes or without specific equipment configurations. Stainless steel material is subject to cracking at high welding temperatures from other welding processes, causing additional weld defects, which result in weld quality issues. However, stainless can be welded nicely with the use of a TIG welder, due to its low heat application from the tungsten rod in tandem with an argon shielding gas. The user has the ability to control the welding heat with great precision. TIG welding uses a foot pedal to control the machine amperage. It can operate on direct current or alternating current, but the tungsten electrode rod is not actually consumed very rapidly. The process requires filler metal to be fed manually by the welder into the arc that is struck by the tungsten rod attached to what is referred to as a "TIG torch." TIG welding is commonly used in applications where bead esthetics are preferred because it produces no weld slag. These applications include stainless-steel exhaust headers, pipe/tubing, and roll cages. TIG welding produces a high tensile strength weld and is a very good choice for a variety of applications requiring clean-finished beads or the repair of exotic materials. It is a well-suited process for applications requiring clean, good-looking welds, and it is a highly regarded choice for critical stainless-steel joints and connections.

Figure 4.
Horizontal fillet weld on clean material done by GMAW/MIG welding.

Figure 5.
GTAW/TIG fillet weld on aluminum material.

3.4 Welding positions

The American Welding Society identifies four basic welding positions. These positions characterize the position of the electrode in relation to the object being welded. The four welding positions are described as the horizontal position, flat position, vertical position, and overhead position. These welding position definitions have been established by the AWS, due to the subtly, but critically differing, techniques associated with each position. Professional welders must pass welding application tests in each position to prove their competence for all of the established welding positions [3]. Weld metal behaves differently in all positions, creating challenges to consider when conducting a field repair. Identifying where the welding material will travel due to gravity, is a major consideration when making repairs in each welding position. **Table 1** summarizes several of the welding symbols associated with different welding positions and types of weld. Well-specified repair instructions identify the welding position, weld type, and joint preparation to be used.

3.4.1 Flat position

The flat position is the simplest position and the easiest to execute by novice welders. Most welders find welding in this position to be natural and comfortable, typically traveling from left to right. It is generally recommended that the component being welded be turned to the flat position for easier constructability, if possible. **Figure 6** shows the location of the material and fillet in a flat position weld. Molten weld material travels downward naturally in the flat position, penetrating the material adequately. The flat welding position is executed similarly for both fillet and groove joint configurations [12].

3.4.2 Horizontal position

The horizontal welding position is similar to the flat position with the joint rotated at 90°. A fillet weldment in the horizontal position can be slightly more challenging to execute than weldments made in the flat position. Horizontal groove welds typically require more attention and control of the weld pool to establish an equal penetration

Welding positions and symbols		
Position	**Weld type**	**Symbol**
Flat	Fillet	1F
Flat	Groove	1G
Horizontal	Fillet	2F
Horizontal	Groove	2G
Vertical	Fillet	3F
Vertical	Groove	3G
Overhead	Fillet	4F
Overhead	Groove	4G

Table 1.
Welding positions and weld types identified by welding symbols for technical drawings.

Figure 6.
Base material orientation and working position for flat position fillet weld (1F).

Figure 7.
Base material orientation and working location for horizontal position fillet weld (2F).

into both base metals. Again, left to right travel is generally preferred. **Figure 7** shows the location of the material and the fillet in a horizontal position weld.

3.4.3 Vertical position

The vertical position requires traveling uphill or downhill when welding. The vertical welding position is illustrated in **Figure 8**. It can be difficult to perform, due to the downward gravitational pull on the weld material before it cools. This can cause the weld material to build up when traveling uphill, and it can extinguish the arc when traveling downhill. To decrease the occurrence of material build-up, welders typically lower the amperage on the machine slightly. This allows the weld material to cool faster, decreasing its ability to move gravitationally. Traveling uphill is the preferred direction of travel for vertical position welding. This gives the welder more weld pool visibility, without the weld material blocking the welder's view of the arc. The weld pool will cool away from the arc, so the proper speed of travel is critical to executing a proper vertical weld.

DOI: http://dx.doi.org/10.5772/intechopen.104263

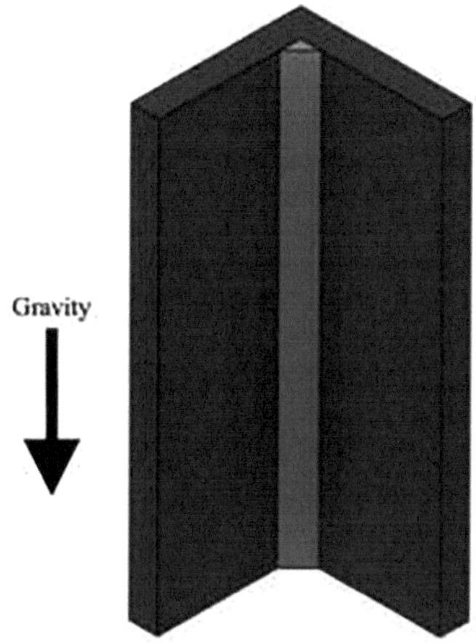

Figure 8.
Base material orientation and working location for vertical position fillet weld (3F).

3.4.4 Overhead position

The overhead welding position is the most difficult, even for the most skilled tradesman. **Figure 9** illustrates some of the challenges posed by the gravitational vector in executing an overhead weld. In the overhead position, the weld pool wants to naturally flow in the opposite direction of the joint. This will cause weld material to sag away from the joint as it cools, creating a crown [12]. A crown is an excess concavity of the face of the weld material that sags away from the joint [13]. Poor

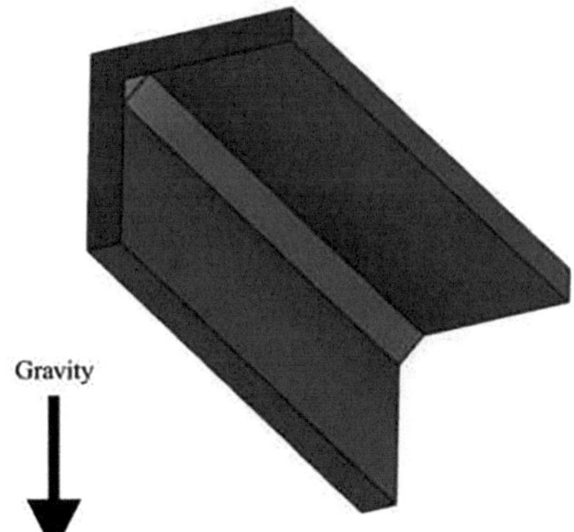

Figure 9.
Base material orientation and working position for overhead position fillet weld (4F).

overhead welds can have void spaces, so keeping the arc tight and the diameter of the weld pool small helps avoid this defect. Welding in the overhead position is generally avoided if possible, but if required, additional care must be taken to identify problems and correct deficiencies. A project manager must understand that this weld is extremely challenging and to be approached with caution in a repair.

3.5 Weld joint types

In a welding repair scenario, there will have been a joint already constructed. Identifying the joint type is critical to know what type of weld should be applied to repair the member. There are five primary joint configurations established by the AWS shown in **Figure 10**. The five weld joint configurations are described by the physical intersection between two joining materials. The shape of the base pieces is irrelevant to the joint classification. All metal joints can be classified under one of these five categories, whether the joint is constructed on a plate, pipe, or pre-made structural shapes. Weld joint classification is different from the weld type. The former depends on the relative position of the base pieces to one another, while the latter depends primarily on the joint preparation. In general, there are three weld types: fillet, groove, and slot types. The various weld types are illustrated in **Figure 11**.

3.5.1 Fillet welds

The fillet weld is a type of weld used in most applications and is notable for being the most common type of welded joint connections. Fillet welds are used to make lap joints, corner joints, and T-joints [14]. The basic shape of a fillet weld is a right

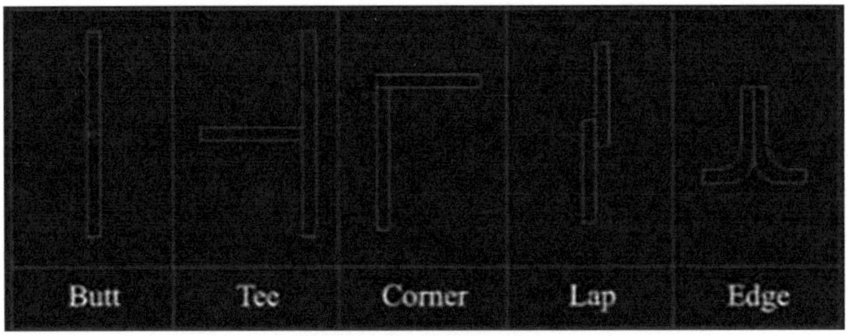

Figure 10.
The five established weld joint configurations.

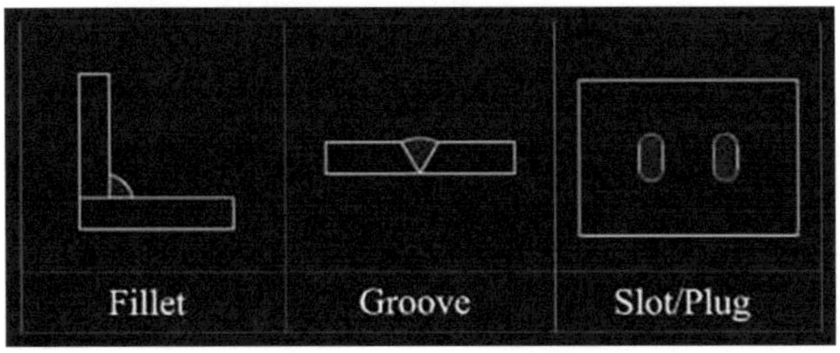

Figure 11.
The three established weld types.

triangle with two equal leg sizes. However, it is difficult to make a fillet weld that formulates a perfect right triangle. There is a reasonably generous tolerance for the shape of a fillet weld. The weld throat is assumed to be the length of the hypotenuse, across the weld shape. Fillet welds must maintain a certain size relative to the thickness of the base metal. In other words, the throat of the weld cannot be too convex or too concave. A fillet weld with a throat that is too convex means there is not enough penetration into the joining material. A fillet weld with a throat that is too concave results from not enough weld metal applied. Either condition creates a weak weldment, and both should be flagged for repair, at present.

3.5.2 Groove welds

Groove welds are classified by the various shapes and sizes of the preparation on the base materials. The types of groove weld classifications are shown in **Figure 12** and listed below:

- square groove;
- V-groove;
- bevel groove;
- U-groove;
- J-groove;
- flare V-groove; and
- flare bevel groove.

Groove welds can be used in corner joints, T-joints, and butt joints. Groove welds are often used for tight fitment applications. Each of the groove weld types listed in the text above has its own symbol and associated dimensioning method per the AWS symbol standard. There are significant material preparation techniques associated with groove weldments, and they typically concern the construction of the specific groove. These details are specified within the weld symbol structure. Beveling the ends of the material in a specific manner is necessary to form the groove. The beveling

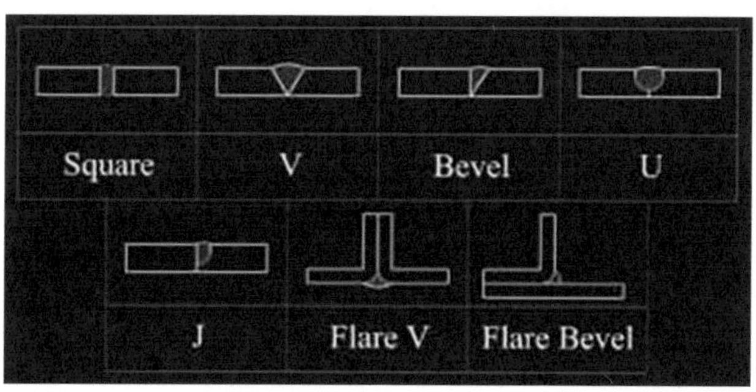

Figure 12.
Groove weld preparations in base material to be joined.

extent is specified by different bevel angles. Bevels allow for weld metal to fill the beveled-out area more easily and increase the strength of the weld.

Figure 13 above represents the material preparation of two steel plates for a 60° V-groove weldment. The angular dimension for the groove is specified within the weld symbol structure. After the weld has been applied between the pieces of base metal, the excess can be ground-away to present a smooth surface of the steel plates, as if the plate is one piece. If executed properly, groove welds can provide superior penetration and strength.

3.5.3 Plug and slot welds

Plug welds and slot welds are similar and are often used to join two overlapping plates. One of the two plates will have drilled holes or milled-out elongated slots. This internal opening to the second piece is where the weldment will be produced. When dimensioning a plug or slot weld, the hole diameter will be shown to the left of the plug symbol, and the plug spacing (distance between each hole) is shown to the right of the symbol. For slot welds, the width of the slot is shown to the left of the symbol, while the length and pitch are shown to the right of the symbol. A drawing detail is often referenced in the tail of the weld symbol structure. If the slot or plug is not to be filled completely with weld metal, the fill depth will be specified [14].

3.6 Determining weld size

Determining the proper size of a weld is essential in conducting a welding repair. Over-welding by repairmen does not generally enhance the chances of success for a repair. A common misconception is that applying an excess of weld material provides additional strength to the joint. This is false, and over-welding can actually weaken the joint by the additional application of heat transferred to the material around the joint. Over-welding is quantifiable and can be calculated in manufacturing applications to minimize production costs. For repairs, it adds time to the job without any benefit to the customer, and it may be counterproductive. The depth of weld penetration is a more important factor in creating a good weld, and it can be properly gauged by correctly sizing the thickness of the weld.

The thickness of the weld is controlled by the depth of the material being welded. It is critical to understand that the material thickness input must be the thinnest material associated with the joint that is being welded. This is especially important in weldment strength calculations. *Design of Welded Structures* define the recommended size of a fillet weld using Eqs. (3) and (4) [1]:

$$L_s = \frac{3}{4} * (M_t) \tag{3}$$

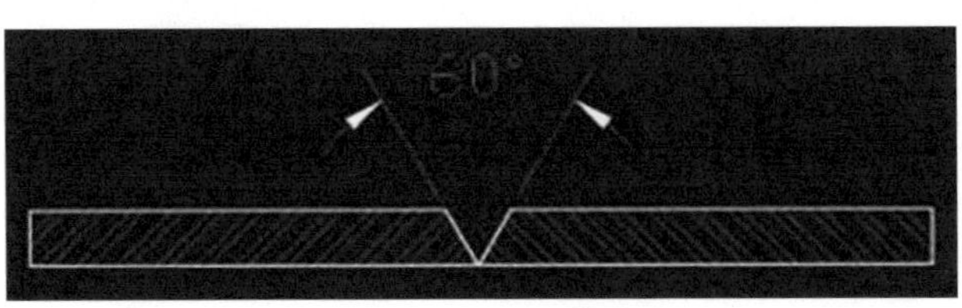

Figure 13.
Material preparation for V-groove weldment.

$$W_t = \cos(45°)(L_s) \tag{4}$$

where M_t is the base material thickness;
L_s is the weld leg size; and.
W_t is the weld throat size.

Figure 14 illustrates these variables on an illustration of a cross section of a tee joint. Following this recommendation for the fillet size will create the proper amount of weld penetration into the base material for a solid joint.

When calculating the throat size, a 0.71 scale factor sometimes appears in the equation for calculating throat size. This is the rough equivalent of $\cos(45°)$. Depending on the structural application, type of stress on the effective weldment, and how much stress the weld will endure, different welds will require different lengths of run. Determining the proper length for a weld minimizes the potential for over-welding. For instance, if two plates needed to be joined, but no load or minimal load was applied in shear, then the most effective way to join the two plates would be to apply a series of small stitch welds across the connection. Stitch welds are intermittent, rather than continuous welds. Fillet welds used to transmit forces should not be less than four times its leg size in length or 3.8 cm, whichever is greater [1]. Neither should the effective weld size of the fillet weld exceed ¼ of the length of the weldment [1]. **Figure 15** presents these fillet length constraints graphically.

Stitch welding is a technique used to connect base material pieces that will see little stress and fatigue in their duty cycles. Additional weld fillet length beyond the structural need is wasteful and costly. Stitch welding is generally a great technique for field repairs that require lengthy connections, eliminating the need to weld a solid bead that would introduce a significant amount of heat into the base material, potentially causing distortion and warpage. Engineering drawings feature unique callouts for stitch welding that include dimensions referencing the length of the stitch weld, the distance between each stitch weld, the weld type, and the size of the weld [15]. The drawings will also feature callouts for the intermittence of the weldment, such as intermittent welds on one or both sides of the joint and the staggered intermittent interval for the weldments [15]. In field repairs, it is usually up to the repairman to evaluate the stitch weld dimensions and the weld runs intermittently, depending on

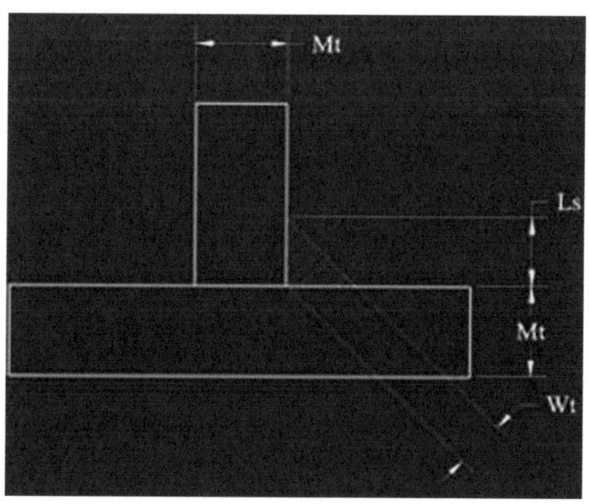

Figure 14.
An illustration of fillet weld size variables used in weld strength calculations.

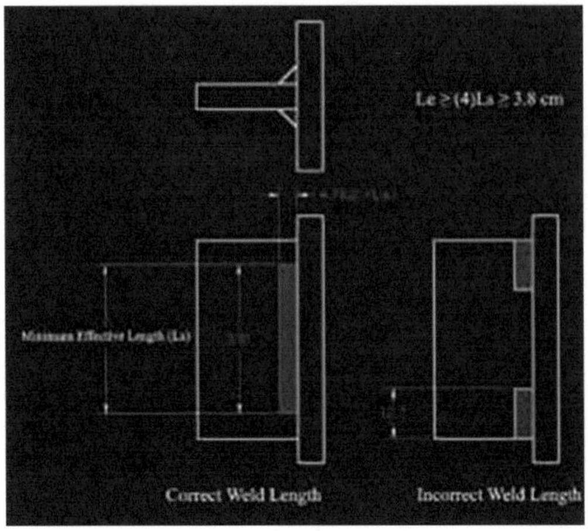

Figure 15.
Minimum effective weld length of run.

the stress of the joint. Significant guidance for the design of the weld can be had through the proper estimation of the needed weld strength.

3.7 Weld strength

Weld metal has tremendous strength, generally higher than the raw strength of the base metals that are being joined. This strength can be calculated with the type of load the material will face. Welded components may endure multiple different types of stress in unison. Fortunately, the superposition of loads principle allows the calculations of each type of stress remain separate. Transverse and shear loads represent the two most common load orientations.

The fillet welds represented by the right triangles shown in **Figure 16** are very resistant to the transverse type of loading. However, if the load follows the direction of the arrow and only one fillet weld existed on the right-hand side of the plate, the connection would be weakened and would likely bend over itself and break off at the joint, toward the right-hand direction. In general, the best practice would be to weld both sides. However, if the application required only one fillet weld, due to a low load application, then the weld should be placed on the side of the plate that places the weld in tension, not compression. In this case, it would be the arrow side of the plate. If the load was in the opposite direction, then the preferred weld location would move to the other side.

Transverse loading is only when the load applied is perpendicular to the weldment. If the load is present in a direction parallel to the weld, as shown in **Figure 17**, then the

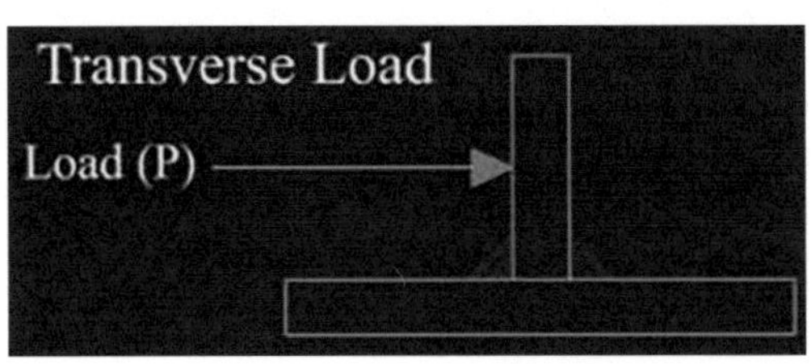

Figure 16.
Illustration of transverse loading on weldment.

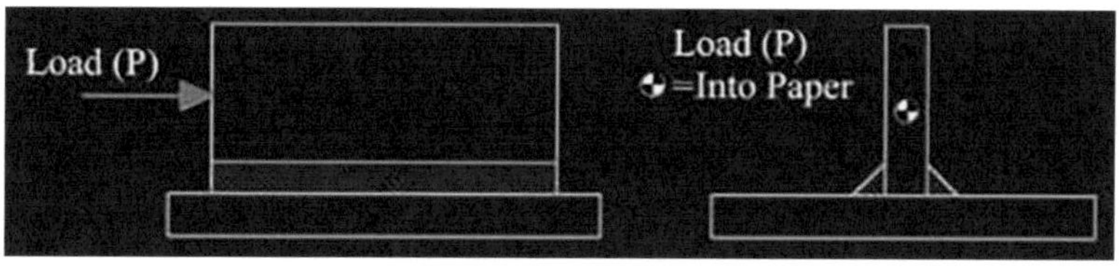

Figure 17.
Illustration of shear loading on weldment.

strength of the weld must be calculated in shear. Shear loading is more detrimental to the structural integrity of a weldment than a transverse load. The decreased structural strength of the weld means that the weld strength cannot be calculated using the full tensile strength of the filler metal. Instead, the AWS recommends that the tensile strength of the filler material be reduced by multiplying by a value of 0.30, reducing the tensile strength by 70% [2]. Otherwise, the strength calculation process is the same.

The calculation of weld strength for transverse load is straightforward, but it is important to emphasize that this is the calculated static strength of the weldment, without measurement against an applied load. Transverse loading requires three basic inputs for determination: filler tensile strength (T_F), the thickness of base material (M_t), and length of weld applied (L_W). Eqs. (3) and (4), along with Eqs. (5) and (6), are used to successfully calculate the strength of a weld under transverse load [1]:

$$A = W_t * L_W \tag{5}$$

$$F = T_F * A \tag{6}$$

where A is the sectional area of the weld;
W_t is the weld throat size;
L_W is the length of the applied weld;
T_F is the tensile strength of the weld filler material; and.
F is the resistive force the weld is capable of supporting.

Table 2 shows the application of this model in calculating the weld strength for material and weldment length using specific values. The tensile strength for the most common MIG filler wire is 483 mPa, and that has been used here. The calculations show that the maximum strength of the weldments under transversal load is 200 and 198 kN in shear load. This number would be an upper bound for loading because the calculation lacks any accounting for imperfections. Any weld imperfections, defects, undercut, material imperfections, material fit-up problems, or dynamic or thermal loading during welding will reduce the strength of a welded joint. This calculation merely provides the theoretic strength of the specific design under specific conditions. In manufacturing, weldment specifications are properly simulated using finite element analysis, tested, and inspected to ensure that the applied loads do not exceed the factor of safety in the design. However, this level of analysis and design is typically not available to repair specialists. They simply tend to increase the factor of safety within their designs, accordingly.

All of the necessary parameters for the basic weldment strength calculations can be easily obtained when conducting field repairs. A tape measure and a phone calculator are the only tools truly needed to execute this calculation in the field. The tensile strength of the filler wire can be found on the wire spool attached to the welding

Name	Value	Unit	Description
Strength of weldments			
Inputs			
TF	482	mPa	Tensile strength of filler
TM	3.18	mm	Thickness of material
LW	1092.2	mm	Length of weld applied
Transverse loading			
LS	2.38	mm	Leg size
TS	1.25	mm	Throat size of weld
A	1366.26	cm^2	Effective area of weld
F	658,536	N	Strength of weld
STF	482	mPa	Tensile strength of filler
STM	3.18	cm	Thickness of material
SLW	1092.2	cm	Length of weld applied
Shear loading			
SF	197,561	N	Strength of weld

Table 2.
Spreadsheet calculation for weld strength under transverse and shear loading.

machine. For arc welding applications, the tensile strength is called-out on the electrode. Proper weldment design in repairs is just as critical, perhaps more so than in original manufacture.

Stitch welding is a technique used to connect base material pieces that will see little stress and fatigue in their duty cycles. Additional weld fillet length beyond the structural need is wasteful and costly. Stitch welding is generally a great technique for field repairs that require lengthy connections, eliminating the need to weld a solid bead that would introduce a significant amount of heat into the base material, potentially causing distortion and warpage. Engineering drawings feature unique callouts for stitch welding that include dimensions referencing the length of the stitch weld, the distance between each stitch weld, the weld type, and the size of the weld [15]. The drawings will also feature callouts for the intermittence of the weldment, such as intermittent welds on one or both sides of the joint and the staggered intermittent interval for the weldments [15]. In field repairs, it is usually up to the repairman to evaluate the stitch weld dimensions and the weld run intermittence, depending on the stress of the joint. Significant guidance for the design of the weld can be had through the proper estimation of the needed weld strength.

4. Challenges for engineers and technical managers

One of the more difficult tasks for an equipment manager is signing off on the acceptance of repair work when the work falls within the domain of a specialist. Welding repairs certainly meet these criteria for most individuals. This section will examine weld defects and how to recognize them. It will cover the repair of defective welds and then go into detail about more extensive examination methods necessary to verify the quality of a critical weld.

4.1 Weld defects

Defects in repair welds can have calamitous results. Therefore, it is necessary to identify defects and imperfections within a repair weld, before returning equipment to duty. Understanding the allowable variations in weld material's physical and mechanical structure will provide a reasonable basis for inspecting welds for imperfections. Pockets of impurities and variations in the weld that are within the acceptable tolerance range are called "discontinuities." Discontinuities that exceed the acceptable tolerance are called weld "defects" [16]. Defects must be ground away and repaired. Defects are generally caused by a poor welding technique, poor joint fit-up, or both. In the world of repair, poor joint fit-up is the most common cause of defective weldments. Repair jobs often consist of components that have separated and are no longer in their original position, making it difficult for the repairman to achieve the original geometry. This can be accomplished for many repairs, but sometimes it requires significant attention to detail and a bit of creativity to manipulate the original material into an acceptable repair position.

Defective welding techniques can cause impurities that affect the shape and size of the weldment, cause imperfections in the internal structure of the weld, and create other defects that adversely affect the weld's strength. The basic categories of weld defects are overlapping, undercutting, distortion and warpage, cracks, craters, porosity, and inclusions. Overlap is excess weld metal that reaches beyond the joint onto the base metal. The primary cause of overlap is an incorrect angle of the electrode with respect to the base material. If the electrode is not angled away from the weld pool on the leading edge, the arc will manipulate the weld pool, causing notches to form between the overlap and the base metal.

Undercutting defects occur when electrode travel speed is too fast, causing grooves or gaps on the toe of the weldment between the base metal and weld metal. The rapid travel speed does not allow enough time for the weld metal to fill in after penetration has occurred into the base metal. A thin or narrow weld bead also indicates excessive travel speeds. Good technique and weld speed will eliminate this problem.

The misconception that over-welding creates stronger joints was mentioned previously. Over-welding can also cause distortion and warpage of the base metal by the excessive application of heat. This occurs due to an improper joint design from the base metal thickness and dimensions. Applying a large amount of weldment in several passes can induce a large amount of heat, and it will also cause thin materials to flex and warp. This weld defect is extremely difficult to correct after the fact, due to the alteration of the base metal's natural shape. Heavy clamping in critical locations, before welding materials are likely to deform, is usually best. Once a piece has deformed into a complex shape, corrective choices become limited and generally involve minimizing further damage.

Cracks in the weld metal are often caused by other weld defects or excess stress induced at the joint. There are several classifications of cracks. Some cracks are internal and cannot be identified from the surface or at the throat of the weld. The stress-induced area at the toe of weldment is a common origination site for weld cracks [16]. Craters can form in the weld bead, causing a gap or hole in the weld metal due to the lack of filler metal deposited. Craters are easily corrected by revisiting the defect area after the weldment is complete. A new electrode should be used, beginning the tie-in 2.5 cm or so in front of the crater, allowing weld metal to melt and deposit into the crater, and extinguishing the arc after the proper amount of weld metal has filled the depression.

Porosity occurs when gas bubbles become trapped within the weld metal. This creates an uneven distribution of foamy weld metal. Porosity weakens the weld and can generally be identified at the surface of the weldment. Visible porosity is an indication that there is also porosity located below the throat of the weld. Inadequate weld preparation and contamination are the most common cause of weld porosity. The trapped gases are usually do not shield gases from the welding process, but gases from the oxidation and vaporization of the contaminates on the surface of the base metal [16].

Slag inclusions can result from trapped welding slag within the weld metal. This occurs only when using the SMAW/Stick/Arc welding process since GTAW/TIG and GMAW/MIG processes produce little slag peel. Slag inclusions can happen when an electrode has been consumed, and the welder ties into the old weldment with a new electrode, before removing the slag peel on the surface. Defects caused by improper welding techniques can be easily eliminated by following established welding procedures and maintaining the proper travel speed, arc length, and electrode angle recommended for the specific welding process.

4.2 Repairing weld defects

Welding defects must be repaired by the complete removal of the defect using either torch removal or grinding methods. Portable grinders with stone wheels are an excellent tool for the removal of weld defects. There is a selection of sizes in grinding wheels, and some wheels are designed to use the surface of the wheel, while others are designed to be used only on the edge. Grinding wheels designed for grinding with the outer edge of the wheel are a popular choice for removing joints, where a groove or bevel is present. Once the defect has been ground away completely, another welding attempt can be tried. For long stretches of removal, an oxygen-acetenyl torch may be used to gouge out the defect. This can be a more efficient method of removal in certain situations. Torch gouging may also be the only practical choice, if the joint is space-constrained, with little room to reach the defect area with a grinding wheel.

4.3 Weld quality and examination

It is important for field engineers and welders to evaluate the quality of their weldments. Each kind of defect displays unique characteristics, simplifying the identification and analysis process. Some defects are visually obvious, but a surprising number of repairmen do not thoroughly inspect their work after completion. Visually inspecting the weldment is vitally important to ensure the integrity of the repair, and it does not require significant effort. Cleaning the weldment and the area around the joint by removing slag with a wire brush is an easy process that results in improved visibility. The visual examination of a weld can usually be conducted with a flashlight, looking closely for the undesirable characteristics associated with the defects previously discussed.

The best practice for larger repairs and welding jobs is to have the inspection process functioning throughout the repair, conducted by a trained inspector at each step of the way. This preventative inspection protocol eliminates the continuance of defects during a repair [1]. However, an independent inspector for field repairs is not usually feasible. Therefore, it is vital for the welder to be capable of identifying defects as they occur. Quality control can be improved by having two or more qualified persons provide inspections. A fillet weld gauge is a handy tool for determining the size and quality of welded field repairs. This device is a measuring tool for checking the leg size and throat size of the weld [17].

In specialty applications where structural integrity is of utmost importance, inspections must be regular, and welds must meet the acceptable defect tolerance level. Specialty applications, such as underwater bridge welding, ship welding, and oil-rig applications must adhere to strict guidelines, codes, and are classified as a Class A weld. Welders in these fields must be licensed by passing a series of weld tests for joints configured in the 3F and 4F positions, administered and evaluated by AWS certified inspectors before beginning a new job, regardless of the prior experience of the welder. This requirement is due to the potential ramifications of a failed weldment [18].

4.4 Filling gaps and repairing cracked components

Cracked components are common in the mechanical industry and are a regular failure mode for many pieces of equipment. These repairs are some of the toughest to execute, and this section will illustrate the overall process, so the equipment manager has an idea of the complexity of the operation. To repair a crack in carbon steel, a bevel directly on the crack must be ground to allow the weld metal to penetrate the base material, creating a tight joint at the crack. An electrode diameter size close to the width of the groove made by the grinder must be used. A standard size filler of 0.89 mm wire is generally sufficient if using a MIG welding process. Depending on the size and characteristics of the crack, an E6010 electrode and capped with a low-hydrogen E7018 electrode would be recommended to eliminate any undercut or hydrogen deposits created by the E6010 electrode, if using a GAW/Stick/Arc process. Finally, the weldment should be ground flush with the base metal to maintain its original appearance.

A field technician might encounter a situation where a gap needs to be bridged between two base pieces. A gap is when two materials should be welded together, but the pieces should not contact each other, as in **Figure 18**. The gap must be filled with weld metal, as in **Figure 19**. This technique should be considered the second stage in repairing a crack. A break that needs to be fixed by bridging a gap often originates from a hairline fracture or crack. When the material separates, it cannot always be drawn back together for repair. Filling gaps can be done in a variety of ways with different welding processes. Large gaps can be filled by using a backing strip. A

Figure 18.
Tack welds applied to establish a 0.3 cm gap between the separated pieces.

Figure 19.
Multiple large tack welds used to fill the gap between the separated pieces, where a continuous weld would likely blow through the gap.

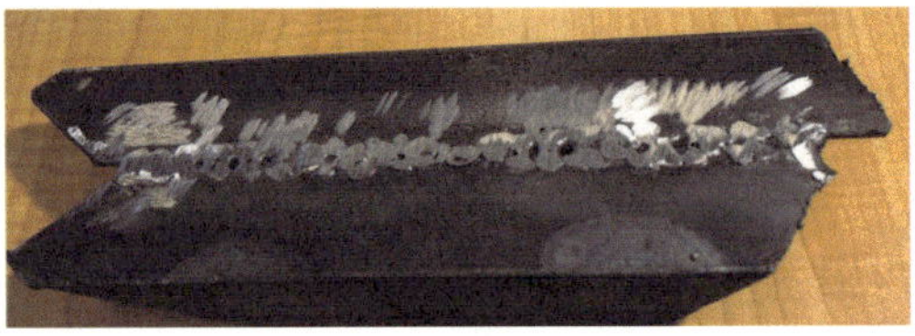

Figure 20.
Welds ground-down to remove defects that occurred during the tack welding process joining the two separated pieces.

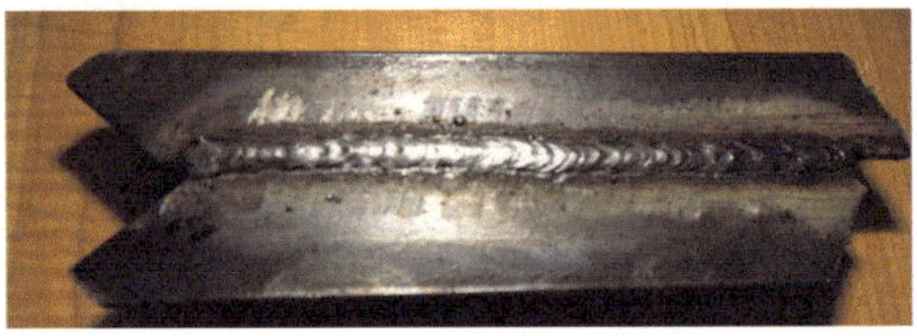

Figure 21.
A fine cap fillet pass used to finish joining two separated pieces, which may be ground flush for a solid appearance.

backing strip is a section of material tack welded or stitch welded onto the base metal. This provides a surface for multiple welding passes without burning through. If a backing strip is used, it can later be grinded off and removed, or it can be left in place, depending on the criticality of the original dimensions. One simple technique to execute a gap fill is by making a series of tack welds along the joint, evenly spaced out. This allows the joint to be welded solid, while the tack welds provide enough weld metal to not blow through the joint. The fillet is typically ground flush after sufficient penetration and connection between the pieces has been established, as in **Figure 20**. Finally, a finish fillet can be applied to cap the weld, as in **Figure 21**.

5. Closure and value of welding technical knowledge

Field repairs are frequently needed and often encountered by many technical professionals. Certified welders in manufacturing, pipeline, industrial, and fabrication shops gain a large amount of theoretical and application-based welding knowledge during their skilled-trade education. On the other hand, engineers and technologists are typically provided with very basic knowledge of welding processes and applications and have minimal field experience. These career areas may experience the need for welding repairs, especially when prototyping and testing. Gaining additional knowledge to further understand the theoretical process and mathematics behind welding techniques will increase the ability to execute, facilitate, and manage repairs of this nature. Management professionals will likely encounter the need to retain welding services, and they need to be able to competently specify the work to be done and be able to inspect it.

Challenges are present in all welding repairs. Critical considerations of these challenges are often not made, due to a lack of knowledge surrounding some of the more advanced topics. It can be difficult to evaluate all the factors in-field repair environments when time is constrained. Material properties, hardness of the material, and stress characteristics should really be thoroughly investigated when the repair is critical. Welding processes, equipment, and weldment preparation can provide the basis for conducting a proper field repair or general welding task, and it is incumbent upon a good equipment manager to understand the basics of welding. A breakdown of fundamental welding positions and types of welds has been shown. It has been demonstrated that the repairman's physical position in relation to the weld is important to constructing a proper joint and executing a sufficient weldment. It was shown that the calculation for the correct weld size in relation to the base material is a simple field calculation. In more advanced major repairs, where stress-induced members are present, further field calculations for structural analysis of the weldment can be performed without excessive computation. These calculations provide quantitative metrics to determine the integrity of the repair weldment design if it were to be applied properly. The repairman can then judge whether the proposed weldment is adequate and sufficient for the repair.

The identification of welding defects, imperfections, and allowable tolerances for each is imperative in evaluating welds and managing quality repairs. The difference between an acceptable weld and an unacceptable weld can be minimal depending on the defect, and it is critical to be able to differentiate between the two. Finally, knowledge regarding the repairs and the correction of welding defects is perhaps the most valuable skill an equipment manager can have, as the correction of these deficiencies is frequently among the most commonly needed field repairs.

Acknowledgements

We would like to acknowledge our classmates from the Spring 2021 Power Units and Power Trains class at Purdue University's School of Agricultural and Biological Engineering for their contributions to the structure and content of this technical chapter.

Conflicts of interest

The authors declare no conflict of interest.

Author details

Tyler J. McPheron and Robert M. Stwalley III*
Purdue University Agricultural and Biological Engineering, West Lafayette, Indiana, USA

*Address all correspondence to: rms3@purdue.edu

References

[1] Blodgett OW. Design of Welded Structures. Cleveland: The James F. Lincoln Arc Welding Foundation; 1972

[2] American Welding Society. Structural Welding Code—Steel (AWS D1.1). In: Structural Welding Code. Miami, Florida, United States: The American Welding Society; 2020

[3] Phillips DH. Welding Engineering: An Introduction. New York: John Wiley & Sons; 2016

[4] Kakani SL, Kakani A. Material Science. New Delhi: New Age International Publishers, Ltd.; 2012

[5] M June. Mechanics of Materials-Steel. Washington, D.C.: Learn Civil Engineering; 2014. Available from: http://www.learncivilengineering.com/wp-content/themes/thesis/images/structural-engineering/Structural-steel-structural-light-gage-reinforcing.pdf [Accessed: December 18, 2021]

[6] Engineering Toolbox. BHN—Brinell Hardness Number. Austin, TX: Engineering ToolBox; 2008. Available from: https://www.engineeringtoolbox.com/bhn-brinell-hardness-number-d_1365.html [Accessed: December 18, 2021]

[7] University of Illinois. Metals, Champaign, IL: University of Illinois Urbana-Champaign, 1995. Available from: http://matse1.matse.illinois.edu/metals/metals.html [Accessed December 18, 2021]

[8] American Society for Testing and Materials. About Us—STM. West Conshohocken, PA: ASTM; 1996. Available from: https://www.astm.org/ABOUT/overview.html [Accessed: December 18, 2021]

[9] National Board of Boiler and Pressure Vessel Inspectors. Identifying Existing Materials. Columbus, OH: NBBI; 2014. Available from: https://www.nationalboard.org/SiteDocuments/Members%20Only/Technical%20Presentations/2014-8_TechPresentation_Beach_Scribner.pdf [Accessed: December 18, 2021]

[10] MetalTekInternational. How to Evaluate Materials. Waukesha, WI: MetalTek International; 2020. Available from: https://www.metaltek.com/blog/how-to-elevate-materials-properties-to-consider/ [Accessed December 18, 2021]

[11] O Nguyen. 3 Most Common Industries for MIG Welding. Tulsa, OK: Tulsa Welding School, 2018; Available from: https://www.tws.edu/blog/welding/3-most-common-industries-for-mig-welding/ [Accessed: December 18, 2021]

[12] Tulsa Welding School. What Are the Different Welding Positions?. Tulsa, OK: Tulsa Welding School; 2020. Available from: https://www.tws.edu/blog/welding/what-are-the-different-welding-positions/ [Accessed: December 18, 2021]

[13] Lincoln Global, Inc. Parts of a Weld Poster (WC-482). Cleveland, OH: Lincoln Electric Company; 2015. Available from: https://www.lincolnelectric.com/assets/US/EN/literature/WC482.pdf [Accessed: December 18, 2021]

[14] Miller Electric Manufacturing, LLC. Deciphering Weld Symbols. Appleton, WI: Miller Electric Manufacturing Company; 2007. Available from: https://www.millerwelds.com/resources/article-library/deciphering-weld-symbols [Accessed: December 18, 2021]

[15] Moran CD. Interpreting Metal Fab Drawings. Salem, Oregon: Open Oregon Educational Resources; 2021

[16] Industrial Training Partners Ltd.. Weld Defects: Causes and Corrections. Dublin, OH: WorldCat, 1983

[17] Ohio Department of Transportation. Field Welding Inspection Guide. Columbus, OH: Ohio Department of Transportation, 2011. Available from: https://www.dot.state.oh.us/Divisions/ConstructionMgt/Materials/Miscellaneous/Field-Welding-Inspection-Guide.pdf [Accessed: December 18, 2021]

[18] American Welding Society. Specifications for Underwater Welding (AWS D3.6M), Miami, FL: American Welding Society, 2017

Section 2

Microstructural Evolution and Residual Stresses

Chapter 6

Study on Microstructure Evolution and Mechanical Properties of Al5083 Joint Obtained from Friction Stir Spot Welding: Effect of Vibration and Plunge Depth

Behrouz Bagheri, Mahmoud Abbasi and Farzaneh Sharifi

Abstract

In this investigation, the vibration of the workpiece is accompanied by the rotating movement of the tool during friction stir spot welding. The method is entitled to friction stir spot vibration welding (FSSVW). Al5083 alloy samples are joined by two welding methods, friction stir spot welding (FSSW), and FSSVW under different plunge depths (DP). The microstructures and fracture surface of the welded zones were analyzed by optical microscopy (OM) and scanning electron microscopy (SEM), respectively. Analyzing the microstructures and mechanical properties of welded samples in both methods revealed that weld region grain size reduced and its hardness increased as the mechanical vibration and high plunge depth have been applied. In addition, the strength and ductility values of FSSV welded specimens with high plunge depth were higher than those produced by FSSW under low plunge depth. Furthermore, it was concluded that the effect of vibration on microstructure and mechanical properties of welded specimens increase as vibration frequency is increased.

Keywords: friction stir spot welding, mechanical vibration, mechanical properties, microstructure, grain size

1. Introduction

The welding process is a significant issue in metal industrial applications such as automotive, aerospace, electronics, and medical functions. Friction stir spot welding (FSSW) is known as a solid-state metal joining process that is applied to join two dissimilar metal plates. It is most often utilized when the combination of metals through fusion welding is not applicable. In this process, a welding tool, including a shoulder and a pin, is rotated, and then contact with workpieces is going to be joined. This process includes three main steps: plunging, stirring, and drawing out [1–3]. In the plunging step, the rotating tool with the pin plunges and penetrates the joining

IntechOpen

workpieces until the shoulder contacts the surface of the up workpiece and reaches the desired depth. Due to friction between the tool and workpiece heat is generated. The heated and softened material around the pin deforms plastically in the stirring step and the two workpieces are mixed. In the drawn-out step, the solid-state bonding between the upper and lower workpieces is achieved. The most important parameters of FSSW include rotational speed (RS) of the tool, dwell time, plunging rate, and plunge depth. These parameters are the key parameters for the size of the stirred zone during welding and weld outcome.

Several welding procedures and several applications for FSSW have been reported [4–8]. Czechowski [9] studied the effect of corrosion cracking on various aluminum alloys during the friction stir welding process. The heat generation analysis in the FSSW process of aluminum alloys was carried out by Awang and Muncino [10]. The results showed that most of the heat (around 96.84%) was produced from friction at the interface of the tool and workpiece. Matori et al. [11] investigated the corrosion properties of AA6061-T6 joint obtained using FSSW. The effect of pin length on the FSSW process for dissimilar aluminum and steel joints was studied by Piccini and Svoboda [12]. It was found that the maximum tool load value increases as tool penetration depth increases and the pin length decreases. Mechanical properties of the weld produced between AA6082-T6 sheets via FSSW were investigated by Aydin et al. [13]. The results showed that the tensile shear load increased almost linearly with increasing plunge depth. Jedrasiak et al. [14] determined the thermal modeling of FSSW for Al-Al and Al-steel joints. It was reported that the results gave a noticeable quantitative prediction of the radial diversity of the thickness of the intermetallic layer. FSSW for automotive applications was studied by Capar et al. [15].

Chevan and Shete [16] studied the optimization of FSSW parameters by using of Artificial Neural Network (ANN). It was found that the FSSW provided the maximum lap shear strength of 3.749 N/mm^2 on the tool rotation speed of 900 rpm and dwell time of 30 sec for the taper cylindrical pin. Astarita et al. [17] studied the stress corrosion behavior of joints made by FSSW between aluminum alloys for aeronautic applications. It was reported that the weld area has lower resistance to intergranular and pitting corrosion compared to other areas. Shekhawat and Nadakuduru [18] analyzed the bonding zones during the FSSW in underwater and normal conditions for Al6061-T6 alloys.

Rostamiyan et al. [19] mixed two welding methods, including FSSW, and ultrasonic welding to improve the weld quality. In their research, FSSW was performed by the ultrasonic vibration of the tool. The impact of process parameters namely vibration, tool rotary speed, tool plunge depth, and dwell time on mechanical properties such as lap-shear force and hardness were examined. It was reported that introduce of vibration enhanced the lap shear force and hardness. Ji et al. [20] introduced the "ultrasonic-assisted friction stir spot welding" (UAFSSW) technique. This process was employed to join dissimilar AZ31and AA6061alloy sheets. It was shown that ultrasonic vibration was significant for the upward flow of the bottom plate and to get a flawless joint. It was found also that the existence of ultrasonic vibration improved the stir zone width and led to finer grains in the stir zone.

In this research, a new method to increase the efficiency of FSSW is presented. The workpiece is vibrated normally to the tool plunge path while the tool is rotated. This new method is entitled friction stir spot vibration welding (FSSVW). Mechanical properties of friction stir spot vibration (FSSV) welded samples such as hardness, fracture surface, and shear strength are compared with those from friction stir spot (FSS) welded samples.

2. Materials and process

Workpieces with 25 mm width and 100 mm length are prepared from joining sheets. The chemical composition and mechanical properties of the studied sheet are presented in **Tables 1** and **2**, respectively. The workpieces in a lap position are installed on the machine, designed for doing the FSSW and FSSVW, while the thin one is on top. The schematic view of this machine is presented in **Figure 1**. The machine is fixed on the milling machine table.

According to **Figure 1**, the workpiece vibration is applied through a fixture. The motor shaft rotation is transformed into the linear and reciprocating movement of the fixture by a camshaft mechanism. The vibration amplitude is adjusted to be 0.5 mm. An AC motor with 0.5 kW power is used for FSSVW. Motor shaft speed which controls the vibration frequency is adjusted using a driver. A non-consumable tool (**Figure 2**), consisting of the pin, from carbide tungsten, and shoulder, from M2 heat-treated steel, is used for welding processes. The welding tool was rotated in the clockwise direction during the welding process.

Si	Mn	Cr	Cu	Mg	Fe	Al
0.5	0.4	0.25	0.1	4.8	0.4	Bal

Table 1.
Chemical composition of base metal (wt%).

Ys (MPa)	UTS (MPa)	Elongation %	E (GPa)	Hardness (V)
184	307	15	71	94

Table 2.
Mechanical properties of base metal.

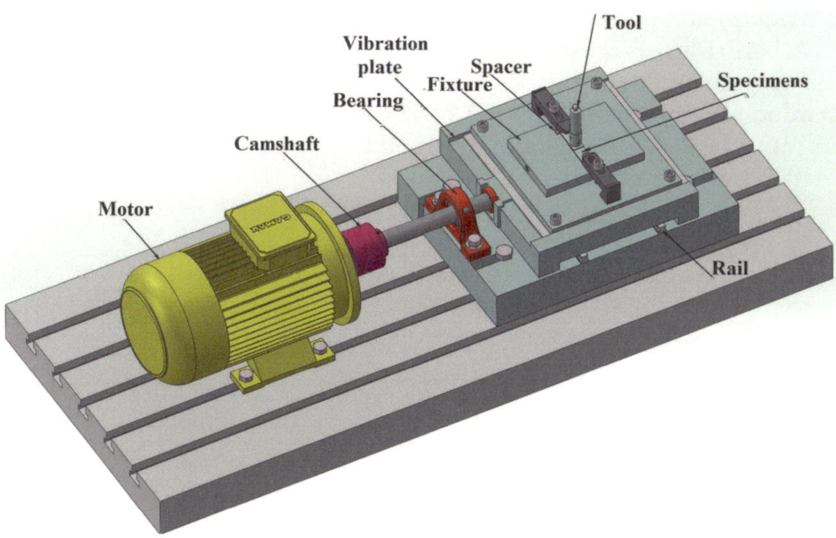

Figure 1.
Schematic view of the machine designed and manufactured for FSSVW.

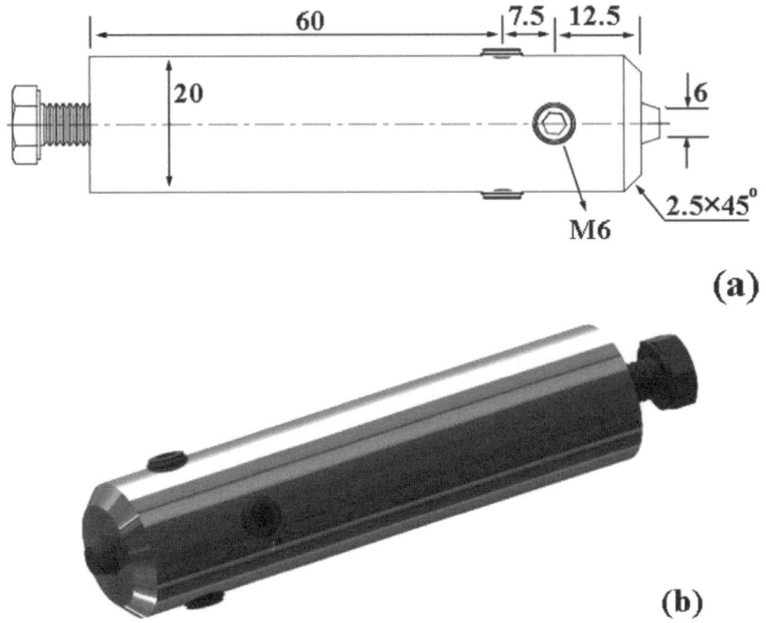

Figure 2.
a) the geometry and, b) the design of pin and shoulder utilized for FSSW and FSSVW processes.

During the FSSVW process, different vibration frequencies are implemented to investigate the effect of vibration frequency on microstructure and mechanical characteristics. **Table 3** shows the welding condition for different welded samples.

Metallography techniques based on ASTM are used to reveal the microstructure of weld regions. Mounted samples were ground by rotating discs of abrasive paper of silicon carbide and then were polished and etched. The linear intercept method (ASTM-E112-13) is applied to measure the grain size. Shear tensile test according to ANSI/AWS/SAE/D8.9-97 is used to obtain lap shear strength-displacement curves of welded specimens. For each welding condition, four specimens are tried. Vickers micro-hardness method based on ASTM-E384 is applied to assess the hardness. For hardness tests, the load is 300 gf and the dwell time is 10 s. For each welding condition, 4 data are measured.

Parameter	Vibration	Vibration frequency (cycles/s)	Rotational speed (rpm)	plunge depth (mm)
Vibration	+ (state 1)	42	1500	2
	– (state 1)	–	1500	2
	+ (state 2)	42	1500	2.5
	– (state 2)	-	1500	2.5
Frequency	+	25	1150	2.5
	+	42	1150	2.5
	+	55	1150	2.5

Table 3.
Welding conditions for different welding trials (+ donates FSSVW and – donates FSSW).

3. Discussion and results

3.1 Microstructure

Figure 3 shows the microstructures of welded samples relating to different welding conditions, according to **Table 3**. According to **Figure 3**, different weld region zones, namely, stir zone (SZ), thermo-mechanically affected zone (TMAZ) and heat-affected zone (HAZ) are also observed for FSSV welded specimens, as well as FSS welded specimens. The microstructures of the stir zone for FSS and FSSV welded specimens are shown in **Figure 4**. It is obvious that the presence of vibration during welding reduces the grain size of the stir zone for both rotational speeds; additionally, grain sizes of samples welded samples with high plunge depth are lower than those welded by low plunge depth. These can be related to the effect of plastic deformation on dislocation production in metals.

High plunge depth or presence of vibration increases the plastic deformation. Studies have noted that dislocation density increases as plastic deformation increases. As dynamic recrystallization (DRX) is the main reason for grain refinement during FSW [21, 22], an increase of dislocation density leads to enhanced DRX and correspondingly, finer grains are developed.

Figure 5 shows the stir zone grain size values for different welding conditions. It is observed that the stir zone grain size for all welded specimens is lower than the base metal grain size. Additionally, **Figure 5** shows that FSSV welded specimens have lower grain sizes for FSS welded specimens. Based on to Kaibyshev [23], the microstructure modification during severe plastic deformation includes two consecutive processes: (i) the formation of three-dimensional arrays of low angle boundaries (LABs) and (ii) the gradual transformation of LABs into high angle boundaries (HABs) (≥15°). LABs with low misorientation (~1°) are constantly formed in pure aluminum and its alloys by dynamic recovery during deformation by rearrangement of accumulating lattice dislocations (**Figure 6a**). At high strain values, mobile dislocations migrate across sub-grains and are trapped by sub-boundaries increasing their misorientation. Extensive rotation of sub-grains leads to increasing misorientation of LABs with strain within sub-grains. These processes result in the formation

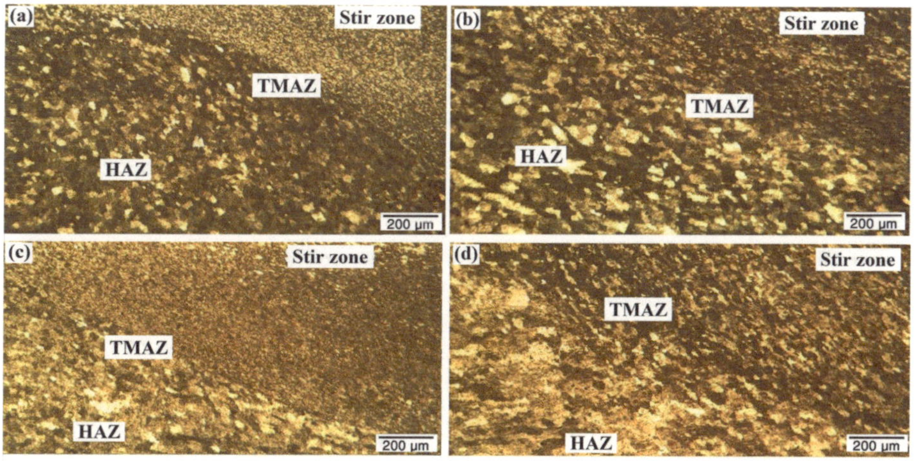

Figure 3.
Microstructures of TMAZ, HAZ, and SZ of different samples, (a, c) FSSVW and (b, d) FSSW (a and b relate to the welding situation 1 of ***Table 3****, c and d relate to the welding situation 2 of* ***Table 3****).*

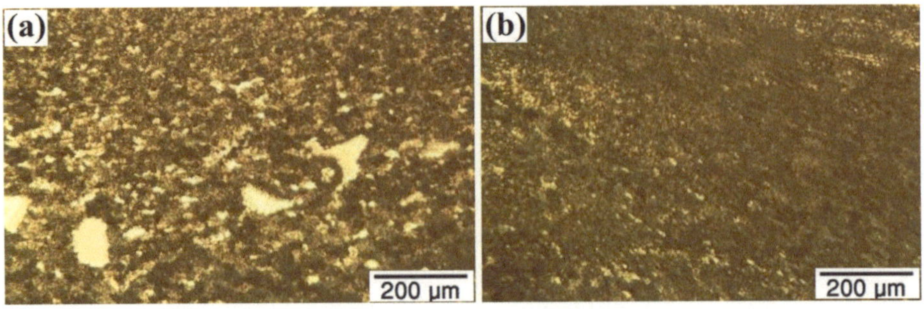

Figure 4.
SZ microstructure of FSS (a) and FSSV (b) welded samples (welding situations 2 of ***Table 3****).*

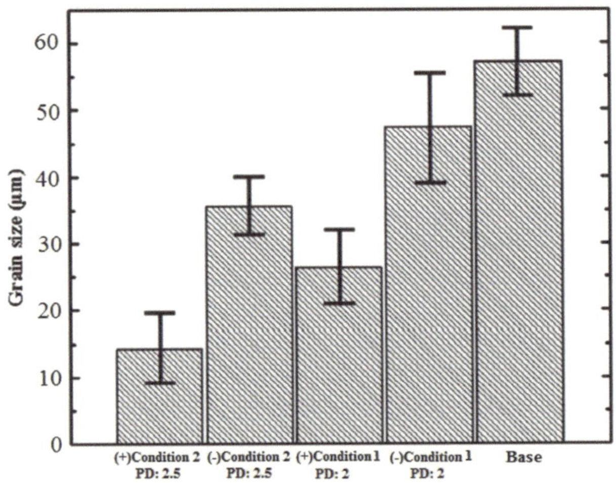

Figure 5.
Stir zone grain size results for different samples welded by different welding conditions (welding factors values were based on ***Table 3****; (−) and (+) signs indicate non-presence and presence of vibration, respectively).*

of individual segments of HABs, and this can be considered as proof for the occurrence of dynamic recrystallization (**Figure 6b**). The recrystallized grains persistently replace sub-grains evolved at small strains through the continuous transformation of their boundaries, and accordingly, grain size refinement occurs [24].

3.2 Mechanical characteristics

Shear strength curves of different welded specimens are presented in **Figure 7**. According to **Figure 7**, samples welded using the FSSVW method have higher strength compared to samples welded using the FSSW method, and additionally, maximum shear load increases as plunge depth increases. It was observed (**Figure 3**) that the presence of vibration, decreases the grain size. As grain size decreases, the volume fraction of grain boundaries increases, and the movement of dislocations decreases. According to the Hall–Petch equation ($\sigma = \sigma_0 + kD^{-1/2}$) [25], strength (σ) increases as grain size (d) decreases. Additionally, as plunge depth increases, more volume fraction of material enters within the stir zone and more mixing of up and down workpieces is carried out in the weld area and this leads to more strength of the weld.

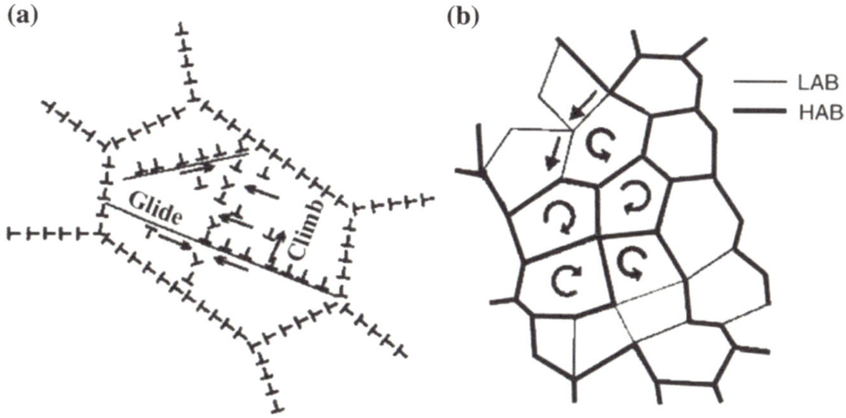

Figure 6.
Schematic illustration of dynamic recrystallization: A dynamic recovery and formation of LABs and b grain size refinement due to gradual transformation of LABs into HABs [23].

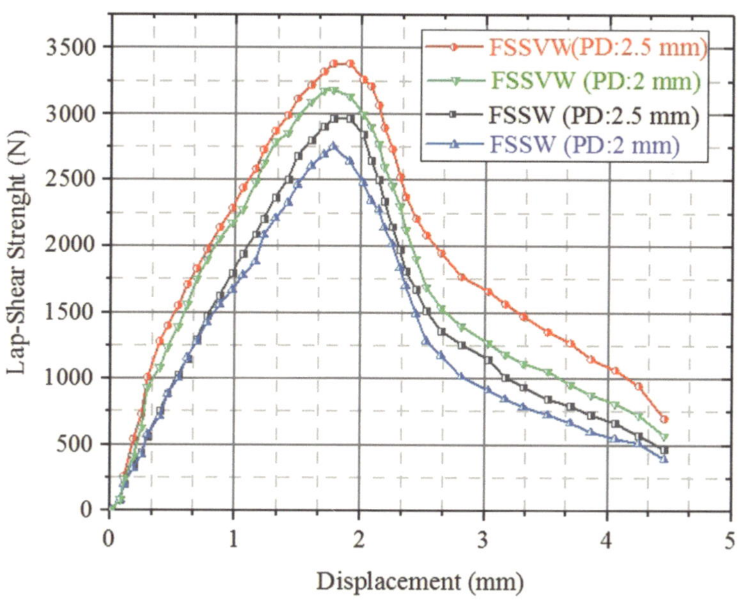

Figure 7.
Lap-shear strength curves of FSS and FSSV welded specimens.

Fracture surfaces of FSS and FSSV welded specimens, after the shear test, are seen in **Figure 8**. According to **Figure 8**, fracture surfaces of all specimens show dimples. The presence of dimples is characteristic of ductile fracture surfaces [26]. It is known that during the straining of ductile materials, voids form within the microstructure, and as straining proceeds, voids coalescence and grow. These voids are responsible for the constitution of dimples [27, 28]. It is observed in **Figure 8** that dimples for FSSV welded specimens are smaller than those observed in FSS welded specimens and dimples for specimens welded under high plunge depth are smaller than those constituted in specimens welded under low plunge depth.

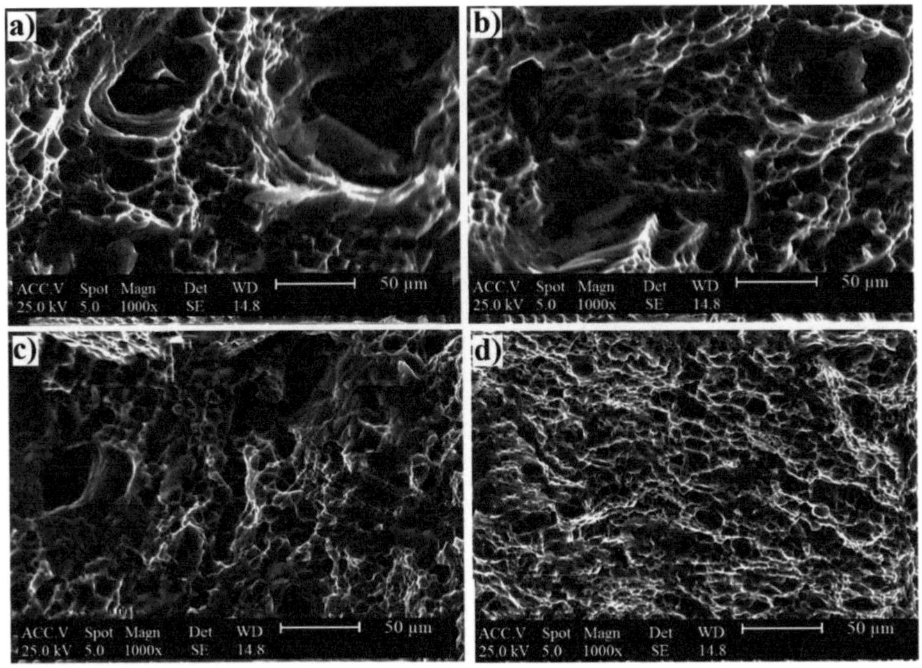

Figure 8.
SEM fracture surface of a) FSV welded sample with PD: 2 mm, b) FSVS welded sample with PD: 2 mm, c) FSS welded sample with PD: 2.5 mm and d) FSV welded sample with PD: 2.5 mm.

Generally, less ductile metals show dimples with larger sizes and fracture occurs at lower values of strain [26]. Correspondingly, more ductility is predicted for FSSV welded specimens compared to FSS welded specimens. Additionally, more ductility is anticipated for specimens welded with higher plunge depth compared to those welded with lower plunge depth. These predictions are in agreement with the results presented in **Figure 7**. It is obvious in **Figure 7** that displacement at maximum load for FSSV welded specimens are higher than that for FSS welded specimens and this variable increases as plunge depth increases.

Figure 9 shows the hardness values of different weld zones of FSS and FSSV welded specimens. Although, the average grain size in the SZ is smaller than the BM, the microhardness values of the SZ are lower than the BM. It can be explained by the existence of two competing phenomena. First, the reduction in the average grain size induced by DRX results from severe plastic deformation which contributes to the increase in the microhardness. Second, the dissolution of the iron-rich phases and the precipitates resulting from intense mixing under severe plastic deformation and the high temperature contributes to the softening of the material. These two competing mechanisms have a strong influence on the final mechanical properties of the different zones and the entire weld. Based on **Figure 9**, hardness values of SZ and TMAZ regions for FSSV welded specimens are higher than those relating to FSS welded specimens. Additionally, **Figure 9** shows that the hardness value increases as plunge depth increases. These can be related to the effect of grain size refinement as vibration is applied and plunge depth increases. It was observed (**Figure 3**) that the presence of vibration and increase of plunge depth both result in more grain refinement. As grain size decreases, the impediment to dislocations movement enhances, and strength and hardness increase. Grain size refinement is known as a strengthening mechanism [29].

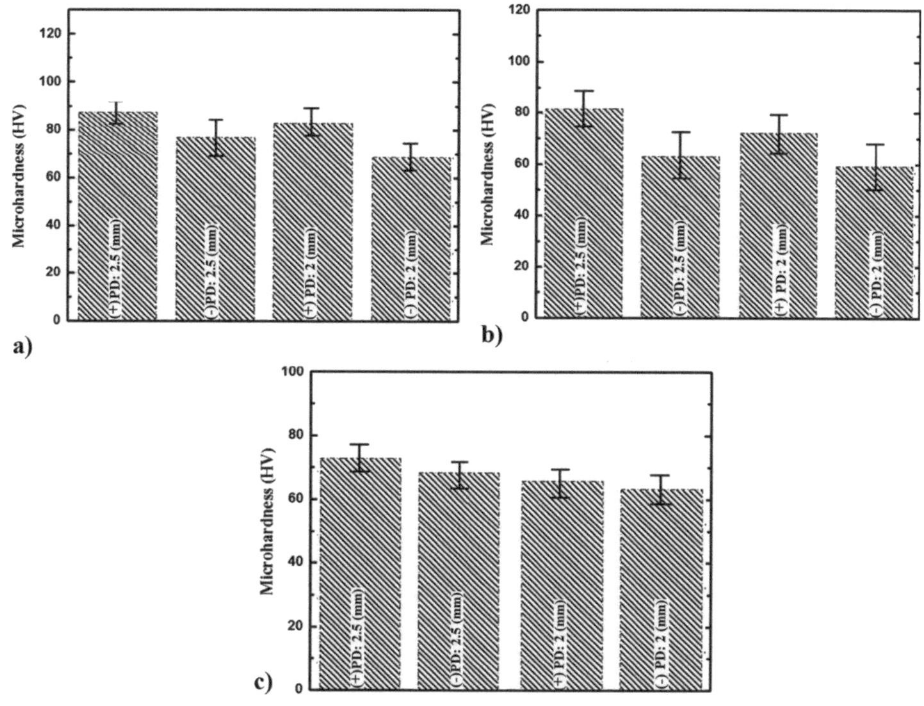

Figure 9.
Micro-hardness values of various zones of FSSW and FSSVW samples: a) SZ, b) TMAZ, and c) HAZ (– and + sign indicate without and with vibration, respectively).

3.3 Effect of vibration frequency

Figure 10 shows the shear strength curves of various FSSV welded specimens and **Figure 11** shows the SZ hardness values of these specimens. For all of these specimens, the welding conditions are the same but the vibration frequency is different. According to **Figure 10**, the maximum shear strength increases as vibration frequency increases. It should be mentioned that DRX is the main mechanism for grain refinement during FSSW [30]. As vibration frequency increases, more strain is applied to the material within the stir zone. It has been known that dislocation density increases as straining increases [31–33]. Higher dislocation density leads to more DRX and correspondingly, finer grains are developed and higher strength and hardness are obtained.

The fracture surfaces of the base metal and FSSV welded specimens with various vibration frequencies are presented in **Figure 12**. Dimples, which are characteristics of ductile fracture surfaces, are seen in fracture surfaces of all specimens. **Figure 12** shows that dimples for base material are the largest and dimple size decreases as vibration frequency increases. This can be related to the effect of vibration frequency on grain size refinement. Barooni et al. [34] found that SZ grain size decreases as vibration frequency increases. Voids, which are responsible for the constitution of dimples in the fracture surface of ductile materials, form around the second phase particles and inclusions as well as dislocation locks. As deformation proceeds, the voids grow and coalescence of them form large voids. Grain boundaries act as barriers to the growth of voids. As grain size decreases, more nuclei for void formation are constituted and on the other hand, the voids cannot grow large and the voids are

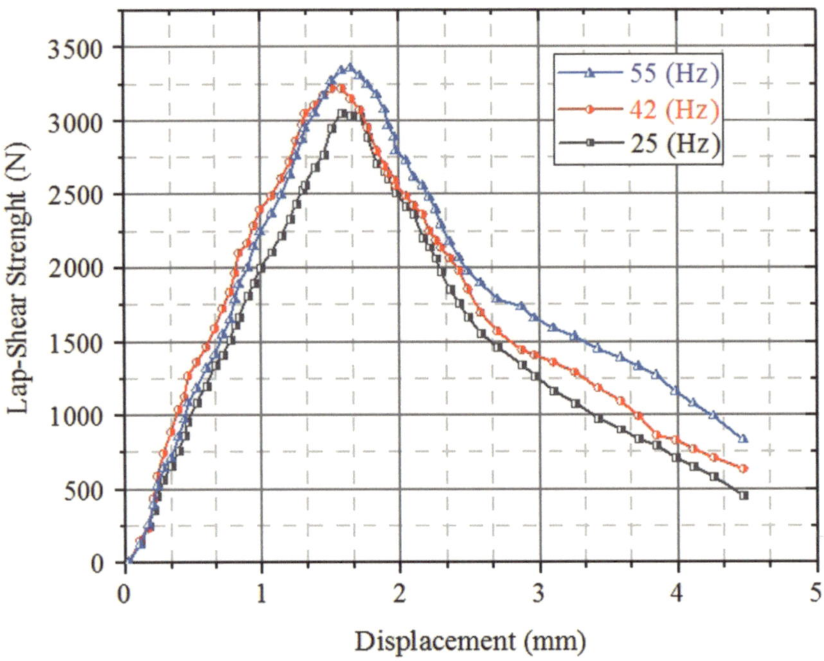

Figure 10.
Shear strength curves of FSSV welded samples with various vibration frequencies (welding factor values were based on ***Table 3****).*

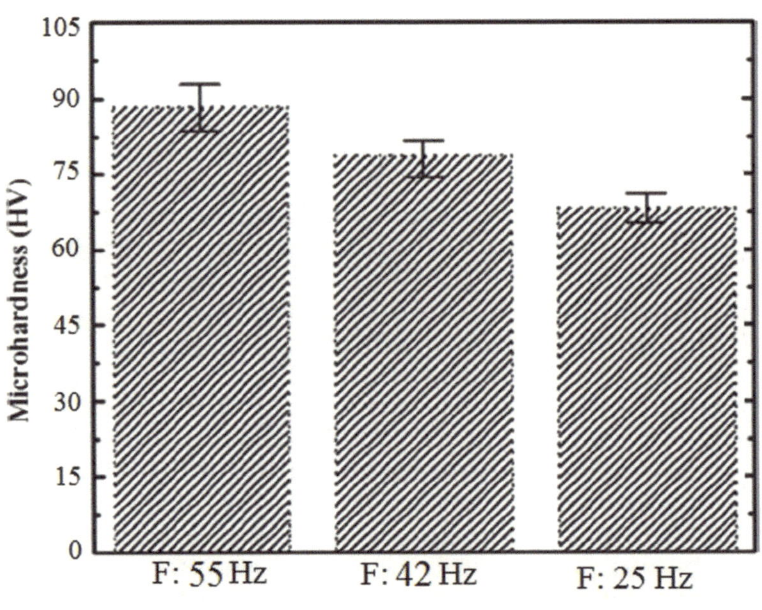

Figure 11.
Stir zone micro-hardness values of FSSV welded samples with various vibration frequencies (welding factor values were based on ***Table 3****).*

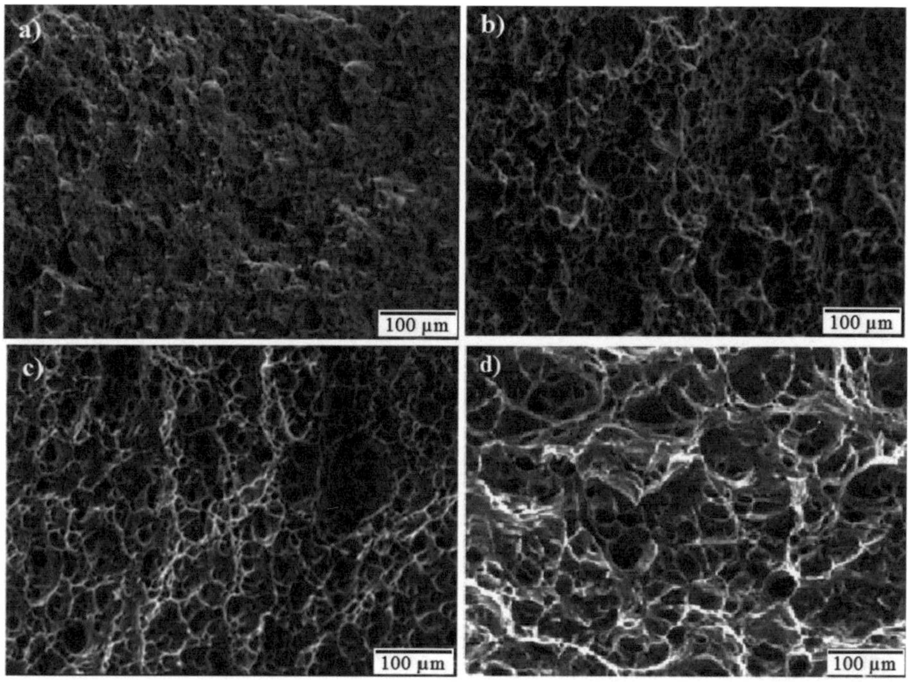

Figure 12.
SEM fracture surface of FSSV welded specimens with different vibration frequencies a) 55 Hz, b) 42 Hz, c) 25 Hz, and d) BM.

smaller. In **Figure 12**, the smallest voids are seen for specimens welded with the highest vibration frequency.

4. Conclusion

In this study, the microstructure and mechanical properties of the Al5083 joint fabricated by FSSW and FSSVW methods were compared. The results showed that the presence of vibration affected the microstructure and led to more grain refinement. The results also indicated that:

The grain size of the stir zone and thermo-mechanically affected zone decreased as the vibration was applied during FSSW.

The shear strength and ductility of FSSV welded specimens were higher than those relating to FSS welded specimens.

Fracture surfaces of all specimens showed that dimples for FSSV welded specimens with high plunge depth are smaller than those observed in FSS welded specimens with low plunge depth.

Shear strength and hardness values of the FSSV welded specimen increased as vibration frequency increased.

Author details

Behrouz Bagheri[1*], Mahmoud Abbasi[2] and Farzaneh Sharifi[3]

1 Department of Mining and Metallurgy, Amirkabir University of Technology, Tehran, Iran

2 Faculty of Engineering, University of Kashan, Kashan, Iran

3 School of Mechanical and Materials Engineering, Washington State University, Pullman, WA, USA

*Address all correspondence to: b.bagheri@aut.ac.ir

References

[1] Tutar M, Aydin H, Yuce C, Yavuz N, Bayram A. The optimization of process parameters for friction stir spotwelded AA3003-H12 aluminium alloy using Taguchi orthogonal array. Materials and Design. 2014;**63**:789-797. DOI: 10.1016%2Fj.matdes.2014.07.003

[2] Pathak N, Bandyopadhyay K, Sarangi M, Panda SK. Microstructure and mechanical performance of friction stir spot-welded aluminum-5754 sheets. Journal of Materials Engineering and Perform. 2013;**22**:131-144. DOI: 10.1007/s11665-012-0244-x

[3] Bozkurt Y, Bilici MK. Application of Taguchi approach to optimize of FSSW parameters on joint properties of dissimilar AA2024-T3 and AA5754-H22 aluminum alloys. Materials and Design. 2013;**51**:513-521. DOI: 10.1016%2Fj.matdes.2013.04.074

[4] Babu S, Sankar VS, Janaki GDR, Venkitakrishnan PV, Reddy GM, Rao KP. Microstructures and mechanical properties of friction stir spot welded aluminum alloy AA2014. Journal of Materials Engineering and Perform. 2013;**22**:71-84. DOI: 10.1007/s11665-012-0218-z

[5] Buffa G, Fanelli P, Fratini L, Vivio F. Influence of joint geometry on micro and macro mechanical properties of friction stir spot welded joints. Procedia Engineering. 2014;**81**:2086-2091. DOI: 10.1016/J.PROENG.2014.10.290

[6] Sun YF, Fujii H, Takaki N, Okitsu Y. Microstructure and mechanical properties of dissimilar Al alloy/steel joints prepared by a flat spot friction stir welding technique. Materials and Design. 2013;**47**:350-357. DOI: 10.1016%2Fj.matdes.2012.12.007

[7] Malafia AMS, Milan MT, Oliveira MF, Spinelli D. Fatique behavior of friction stir spot welding and riveted joints in an Al alloy. Procedia Engineering. 2010;**2**:1815-1821. DOI: 10.1016/j.proeng.2010.03.195

[8] Shen Z, Yang X, Zhang Z, Cui L, Li T. Microstructure and failure mechanisms of refill friction stir spot welded 7075-T6 aluminum alloy joints. Materials and Design. 2013;**44**:476-486. DOI: 10.1016%2Fj.matdes.2012.08.026

[9] Czechowski M. Effect of anodic polarization on stress corrosion cracking of some aluminium alloys. Advances in Materials Science. 2007;**7**(11):13-20 http://polona.pl/item/34289887

[10] Awang M, Mucino VH. Energy generation during friction stir spot welding (FSSW) of Al 6061-T6 plates. Materials and Manufacturing Processes. 2010;**25**:167-174. DOI: 10.1080/10426910903206758

[11] Gharavi F, Matoria KKA, Yunus R, Othman NK. Investigation of the nugget zone corrosion behavior in friction stir welded lap joints of 6061-T6 Aluminum alloy. Materials Research. 2014;**17**(6):1563-1574. DOI: 10.1590/1516-1439.275914

[12] Piccini JM, Svoboda HG. Effect of pin length on friction stir spot welding (FSSW) of dissimilar Aluminum-steel joints. Processing and Materials Science. 2015;**9**:504-513. DOI: 10.1016/j.mspro.2015.05.023

[13] Aydin H, Tuncel O, Tutar M, Bayram A. Effect of tool pin profile on the hook geometry and mechanical properties of a friction stir spot welded AA6082-T6 aluminum alloy. Transactions

of the Canadian Society for Mechanical Engineering. 2021;**45**(2):114-118. DOI: 10.1139/tcsme-2020-0035

[14] Jedrasiak P, Shercliff HR, Reilly A, McShane GJ, Chen YC, Wang L, et al. Thermal Modeling of Al-Al and Al-steel friction stir spot welding. Journal Materials Engineering and Performance. 2016;**25**(9):4089-4098. DOI: 10.1007/s11665-016-2225-y

[15] Capar N, Kumru U, Baser TA, Tekin G, Saray O. Friction stir spot welding for automotive applications. International Journal of Advanced Automotive Technology. 2017;**1**(3):114-118 https://dx.doi.org/10.15659/ijaat.17.07.542

[16] Chavan A, Shete MT. Optimization of friction stir spot welding process using artificial neural network. International Journal of Science Technology and Engineering. 2015;**1**(10):353-358 https://ijste.org/Article.php?manuscript=IJSTEV1I10114

[17] Astarita A, Bitondo C, Squillace A, Armentani E. Bellucci F: Stress corrosion cracking behavior of conventional and innovative aluminium alloys for aeronautic applications. Surface International Analysis. 2013;**45**(10):1610-1618. DOI: 10.1002/sia.5234

[18] Shekhawat RS, Nadakuduru VN. Impact of post *weld* heat treatment on mechanical and microstructural properties of underwater *friction stir spot welded* 6061 aluminium alloy. Materials Today: Proceedings (In Press). 2021. DOI: 10.1016/j.matpr.2021.09.207

[19] Rostamiyan Y, Seidanloo A, Sohrabpoor H, Teimouri R. Experimental studies on ultrasonically assisted friction stir spot welding of AA6061. Archive of Civil and Mechanical Engineering. 2015;**15**:335-346. DOI: 10.1016/j.acme.2014.06.005

[20] Ji SD, Li ZW, Yue YM, Gao SS. Investigation of ultrasonic assited friction stir spot welding of magnesium alloy to aluminum alloy. Strenght of Materials. 2016;**48**:2-7. DOI: 10.1007%2Fs11223-016-9730-y

[21] Woo W, Ungar T, Feng Z, Kenik E, Clausen B. X-ray and neutron diffraction measurements of dislocation density and subgrain size in a friction-stir-welded aluminum alloy. Metallurgical and Materials Transactions A. 2010;**41**:1210-1216. DOI: 10.1007/s11661-009-9963-5

[22] Abbasi M, Abdollahzadeh A, Omidvar H, Bagheri B, Rezaei M. Incorporation of SiC particles in FS welded zone of AZ31 Mg alloy to improve the mechanical properties and corrosion resistance. International Journal of Materials and Research. 2014;**107**(6):566-572. DOI: 10.3139/146.111369

[23] Kaibyshev R, Shipilova K, Musin F, Motohashi Y. Continous dynamic recrystallization in an Al-Li-Mg-Sc alloy during equal-channel angular extrusion. Materials Science Engineering: A. 2005;**396**:341-351. DOI: 10.1016/j.msea.2005.01.053

[24] Jonas JJ, Quelennec X, Jiang L, Martin E. The avrami kinetics of dynamic recrystallization. Acta Materialia. 2009;**57**:2748-2756. DOI: 10.1016%2Fj.actamat.2009.02.033

[25] Abbasi M, Bagheri B, Keivani R. Thermal analysis of friction stir welding process and investigation into affective parameters using simulation. Journal of Mechanical science and Technology. 2015;**29**(2):861-866. DOI: 10.1007/s12206-015-0149-3

[26] Dieter G. Mechanical Metallurgy, SI Metric Edition. New York, NY, USA: McGraw-Hill; 2005

[27] Bagheri B, Abbasi M, Abdollahzadeh A, Omidvar H. An advanced approach to modify friction stir spot welding process. Metals and Materials International. 2020;**26**:1562-1573. DOI: 10.1007/s12540-019-00416-x

[28] Bagheri B, Rizi AAM, Abbasi M, Givi M. Friction stir spot vibration welding: Improving the microstructure and mechanical properties of Al5083 joint. Metallography, Microstructure, and Analysis Journal. 2019;**8**(5):713-725. DOI: 10.1007/s13632-019-00563-y

[29] Callister WD, Rethwisch DG. Materials Science and Engineering. New York, NY, USA: John Wiley & Sons; 2011

[30] Bagheri B, Abbasi M, Givi M. Effects of vibration on microstructure and thermal properties of friction stir spot welded (FSSW) Aluminum alloy (Al5083). International Journal of Precision Engineering and Manufacturing. 2019;**20**(7):1219-1227. DOI: 10.1007/s12541-019-00134-9

[31] Hull D, Bacon DJ. Introduction to Dislocations. Vol. 37. Amsterdam, The Netherlands: Elsevier; 2011

[32] Abdollahzadeh A, Bagheri B, Abbasi M, Sharifi F, Mirsalehi SE, Moghaddam AO. A modified version of friction stir welding process of Aluminum alloys: Analyzing the thermal treatment and Wear behavior. Proceedings of the Institution of Mechanical Engineers Part L Journal of Materials Design and Applications. 2021;**235**(10):2291-2309. DOI: 10.1177/14644207211023987

[33] Abbasi M, Givi M, Bagheri B. Application of vibration to enhance the efficiency of friction stir processing. Transactions of Nonferrous Metals Society of China. 2019;**29**:1393-1400. DOI: 10.1016/S1003-6326(19)65046-6

[34] Barooni O, Abbasi M, Givi M, Bagheri B. New method to improve the microstructure and mechanical properties of joint obtained using FSW. International Journal Advanced Manufacturing and Technology. 2017;**93**:4371-4378. DOI: 10.1007/s00170-017-0810-3

Chapter 7

Analysis of the Effect of Heat Treatment Conditions of a Ferritic Stainless Steel on Residual Stresses and Tribological Behavior

João Pedro de Castro Valente Lenz Ferreira, Marcelo de Matos Macedo, Jorge Humberto Luna-Domínguez, Vikas Verma, Wilian da Silva Labiapari and Ronaldo Câmara Cozza

Abstract

In recent years, the scientific community has shown a great interest in the study of the wear performance of metallic materials under different test conditions, together with the measurement of residual stresses. Thus, the objective of the present work was to analyze the effect of heat treatment on residual stresses and tribological behavior of P410D ferritic stainless steel. The results showed that, with the increase in hardness of the material—derived from different heat treatment conditions, the resistance to micro-abrasive wear of P410D ferritic stainless steel increased, characterized by a decrease in wear volume. The residual stresses reported were "*tractive*"; additionally, it was observed that the lowest residual stresses values were related to the lower wear volumes values.

Keywords: stainless steel, heat treatment, residual stress, X-ray diffraction, micro-abrasive wear, tribological behavior

1. Introduction

Stainless steel is an iron-chromium-carbon alloy that has high resistance to oxidation and corrosion in varying environments. In addition to the presence of nickel (Ni), the main alloying element that ensures the chemical resistance to oxidation and corrosion is chromium (Cr), with 11% concentration by mass; such resistances are associated with the formation of a mixed oxides layer and its dissolution in the medium in which it is exposed. Based on the predominant constituent phase in its microstructure, stainless steels are classified mainly into three categories: *austenitic*, *ferritic*, and *martensitic* [1]. Thus, in Fe-Cr phase diagram (**Figure 1** [2]), the "*ferritic stainless steels*" are at the right of the austenitic field [3] and are classified as "*ferritic*

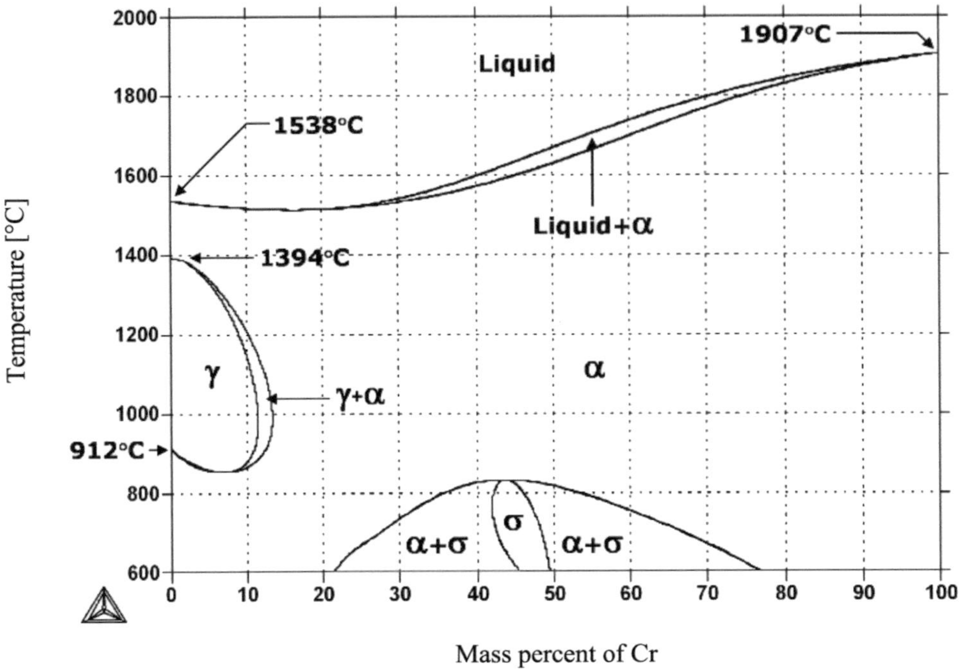

Figure 1.
Fe-Cr phase diagram [2].

stainless steels" the materials with a percentage greater than 12% Cr—by mass [3], possessing body-centered cubic crystalline structure.

In the ferritic microstructure, chromium (Cr) is configured as a substitutional atom, influencing hardening by solid solution [4], where "*hardness*" is a direct indicator of the wear resistance of ferritic stainless steels influenced by the type of heat treatment conducted. Additionally, some ferritic stainless steels can be austenitized at high temperature and cooled in a quench process, to obtain martensitic microstructure, which presents a high density of discordances.

In another scientific segment, the "*X-ray diffraction*" technique is applied for measuring residual stresses in metallic materials. It is conceptualized in the deviation of the propagation trajectory suffered by X-ray waves and is compared with the positions of the atoms of the analyzed material due to the changes caused by residual stresses themselves of the network parameters in the material [5] (**Figure 2**).

In X-ray diffractometer, there is a coordinated movement between the Coolidge tube—generator of the X-rays—and the specimen. The rotational motion of the Coolidge tube defines the "*angle of incidence of the X-rays—θ*" on the surface of the material and the rotational movement performed by material analyzed defines the "*angle of rotation of the specimen—ψ*" (**Figure 3**).

As the specimen is rotated by a defined angle ψ, the angle of incidence of the X-rays (θ) is altered accordingly. The angle of rotation of the specimen (ψ) is measured in relation to a normal axis to the atomic planes, defined by "*Miller indexes—hkl*", parallel to the surface of the specimen and the angle of incidence of the X-rays (θ) is referenced in relation to the surface of the material.

From there, a graph of $\theta = f(\text{sen}^2\psi)$ is raised, which results in a 1st Degree Equation (Eq. (1)), by which the residual stress is calculated. This technique is called "*sen²ψ*".

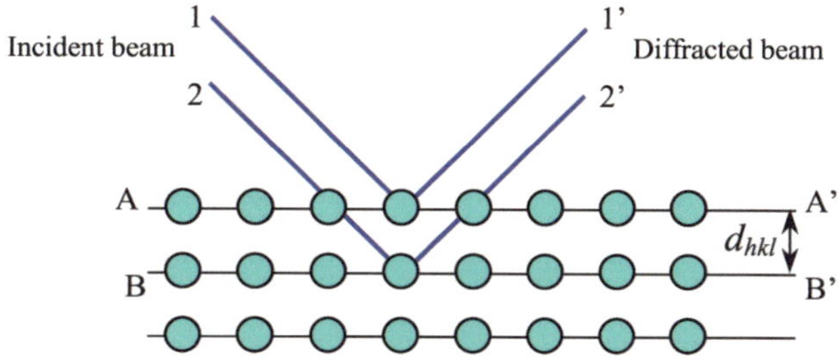

Figure 2.
Physical principle of the "X-ray diffraction" *technique.*

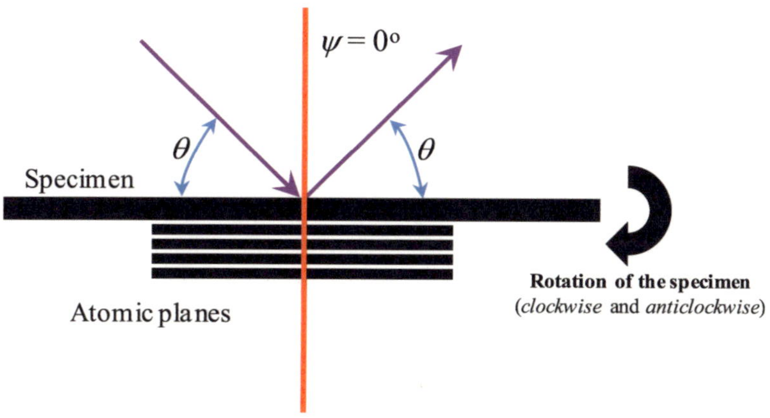

Figure 3.
"X-ray diffraction": *Definition of the angles "θ" and "ψ"*—"angle of incidence of the X-rays" *and* "angle of rotation of the specimen", *respectively.*

$$\sigma_{res} = m\frac{E}{2(1+\nu)}\, cot\,(\theta_0) \tag{1}$$

In Eq. (1), "m" is the angular coefficient of the generated straight line, "E" and "ν" are the longitudinal elasticity modulus and the Poisson coefficient of the analyzed material, respectively, and "θ_0" is obtained for ψ = 0°. For $m < 0$, the residual stress acting on the material will be "*compressive*" and, for $m > 0$, the residual stress will be "*tractive*".

On the tribological side, "*wear*" can be defined as "*damage on the solid surface, involving progressive loss of mass, due to the relative movement between the surfaces and contact with other material or materials*" [6]. Together with the given general definition, each type of wear has a specific setting—"*Abrasive wear*", as discussed in this work, is due to hard particles, or hard protuberances, forced against and moving along a solid surface [7].

In industrial sectors where wear causes downtime, there occur decrease in production and involve high maintenance costs, so it is not enough to acquire knowledge only in mechanical and/or metallurgical manufacturing of materials and processes—it is also important to research and to understand the wear processes that act in specific working conditions.

Through analyses of wear craters generated during micro-abrasive wear tests by rotating ball, it is possible to predict or, at least, estimate the abrasive wear behavior of a material or any mechanical component in real working conditions. Additionally, such analyses can be expanded and better understood along with residual stress measurements conducted by the "*X-ray diffraction*" technique.

Thus, the objective of this work was to analyze the effect of heat treatment on residual stresses and tribological behavior of a dual-phase stainless steel, under conditions of micro-abrasive wear.

2. Materials, equipment, and scientific methodology

2.1 Materials

Six P410D ferritic stainless steel specimens were used—named "*Specimen 1*", "*Specimen 2*", ..., "*Specimen 5*" and "*Specimen 6*". The chemical composition of P410D ferritic stainless steel is specified in **Table 1**.

Each specimen had dimensions of, approximately, 27*x*10*x*5 [mm], being conditioned under different parameters of heat treatment of quench, in a combustol oven, whose application temperatures were monitored by thermocouples. **Table 2** shows the temperature values adopted for the heat treatments of the specimens.

After heat treatments of quench, the specimens were subjected to metallographic analysis, being, initially, hot embedded, sanded, polished and chemically etched with CATELA—2 g of picric acid, 6 ml of acetic acid, 3 ml of hydrochloric acid, and 100 ml of ethyl alcohol. After, microstructural images were acquired by optical microscopy.

As counter-body, an AISI 52100 steel bearing (quenched and tempered) was used, with diameter D = 25.4 mm (D = 1″—*standard size*).

Figure 4 presents an image of the AISI 52100 steel bearing microstructure and its chemical composition. Its microstructure is composed only of two phases: the tempered martensitic matrix (with the characteristic shading contrast) and the small M_3C carbide precipitates homogeneously distributed. The darker and lighter areas in the matrix are typical of tempered martensite and show the gradients of etching depending on the orientation of the martensite lenses and the density of the carbide precipitation in different regions (due to small differences in chemical composition).

The abrasive slurry was prepared with black silicon carbide (SiC)—average abrasive particle size of a_p = 3 μm—and distilled water. **Figure 5** shows an image of the SiC

Chemical element	Quantity—% mass
Cr	11.1334
Mn	0.5332
Si	0.5260
Ni	0.3046
P	0.0271
C	0.0107
S	0.0005
Fe	Balance
N [ppm]	128

Table 1.
Chemical composition of P410D ferritic stainless steel—% mass.

Specimen	Socking temperature (°C)
1	822
2	854.1
3	879.5
4	895.4
5	953.3
6	973

Table 2.
Temperature values defined for the heat treatments of P410D ferritic stainless steel specimens.

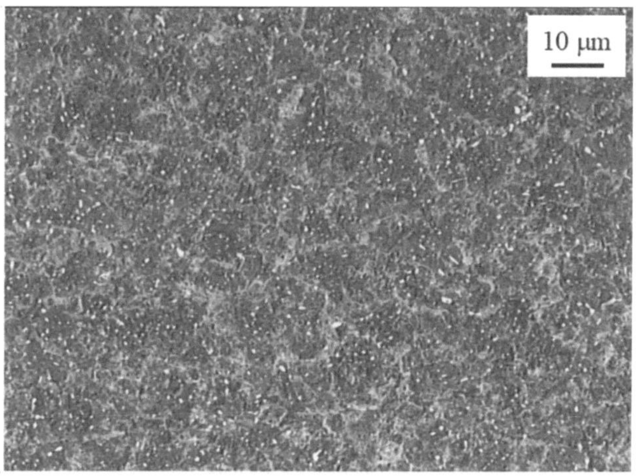

Chemical Element	Quantity – % mass
C	1.04
Mn	0.35
Si	0.25
Cr	1.45
Fe	Balance

Figure 4.
Microstructure and chemical composition of the AISI 52100 bearing steel sphere—Metallurgical state: Quenched in oil at 860°C and stress-relieved at 200°C for 1 h.

abrasive particles (**Figure 5a**), which was obtained by scanning electron microscopy (SEM), and its abrasive particle size distribution (**Figure 5b**).

Table 3 presents the hardness values of the materials used in this work (specimens, test ball, and black silicon carbide). The numbers of the specimens were established in ascending order, along with the respective hardness values.

2.2 Tribometer

Figure 6 shows the ball-cratering tribometer used in this work. Having a "*fixed-ball*" mechanical configuration, the test shaft was divided into two distinct parts,

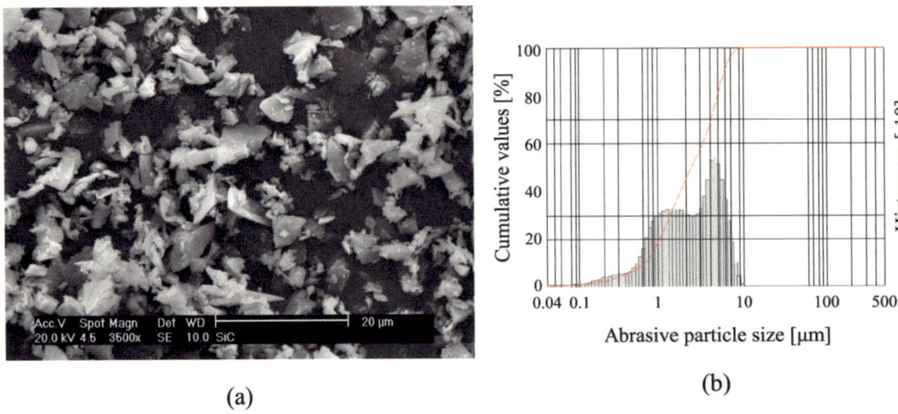

Figure 5.
Abrasive particles of black silicon carbide (SiC): (a) image obtained by scanning electron microscopy (SEM) and (b) particle size distribution.

	Material	Hardness—[HV]
Specimen 1	P410D ferritic stainless steel	177
Specimen 2	P410D ferritic stainless steel	200
Specimen 3	P410D ferritic stainless steel	222
Specimen 4	P410D ferritic stainless steel	238
Specimen 5	P410D ferritic stainless steel	297
Specimen 6	P410D ferritic stainless steel	304
Test ball	AISI 52100 steel bearing	856
Abrasive material	Black silicon carbide (SiC)	2380–2630

Table 3.
Hardness of the materials used in this work.

called "*motor test shaft*" and "*moving test shaft*" (**Figure 6a**). In turn, each of these parts has a face with a concave radius of R = 12.7 mm (R = ½"), thus, enabling the accommodation of a test sphere of diameter D = 25.4 mm (D = 1"). For the application of the normal force—N, was adopted a "*dead weight*" mechanical system (**Figure 6b**).

The "*motor test shaft*" is driven by a direct current electric motor of power P = 30 W (**Figure 6a**), under a rotating speed of n = 56 rpm.

Finally, the fixation of the specimen is performed by device shown in **Figure 7**. **Figure 8** shows one of the specimens before the tribological tests.

2.3 Research methodology

Table 4 shows the test conditions established for the micro-abrasive wear experiments.

A normal force value was defined for the tribological tests, N = 2 N, together with an abrasive slurry concentration of C = 25% SiC + 75% distilled water—in volume.

The test time, for all wear experiments, was established at the value of t = 20 min. With the test ball diameter of D = 25.4 mm and a ball rotating speed of n = 56 rpm, a sliding distance of $S \approx 90$ m was calculated.

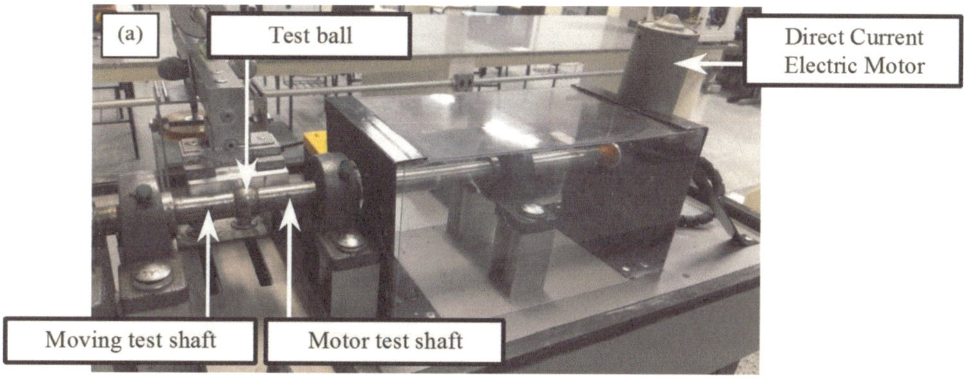

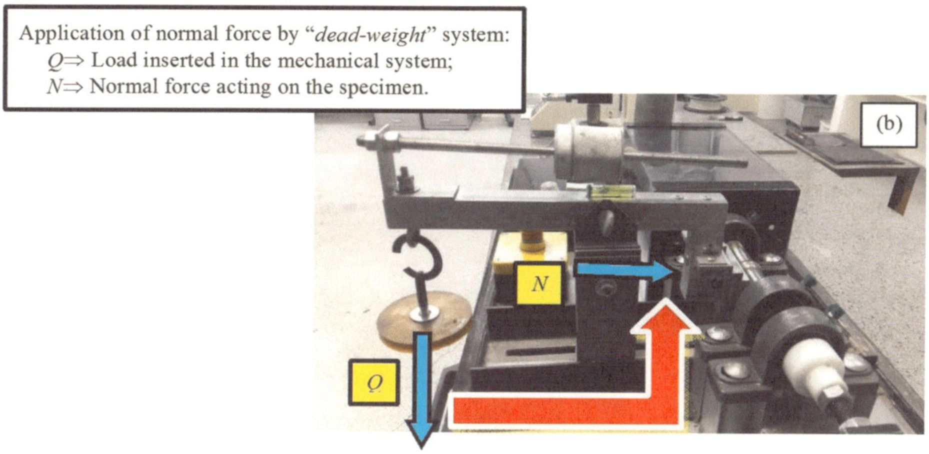

Figure 6.
Tribometer "ball-cratering" *of* "ball-fixed" *mechanical configuration: (a)* "motor test shaft", "test ball" *and* "moving test shaft" *mounted on the tribometer; (b) application of normal force by* "dead-weight" *mechanical system.*

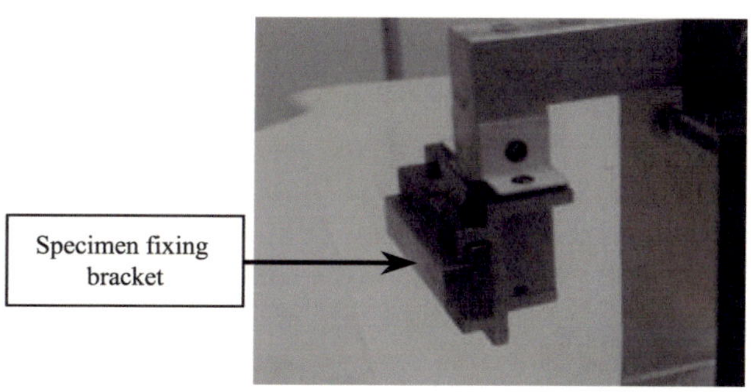

Figure 7.
Device for fixing specimen.

For each specimen, three ball-cratering micro-abrasive wear tests were conducted and, during the experiments, the abrasive slurry was continuously dripped between the specimen and the test ball.

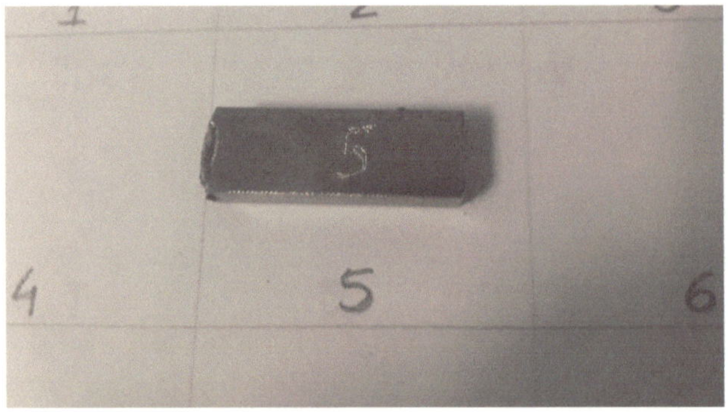

Figure 8.
Image of "Specimen 5", before the ball-cratering wear tests.

Test conditions	
Normal force—N [N]	2
Abrasive slurry concentration—C [% vol.]	25% SiC +75% distilled water
Test time—t [min]	20
Rotation of the test ball—n [rpm]	56
Sliding distance—S [m]	≈90
Number of repetitions	3

Table 4.
Test conditions defined for the micro-abrasive wear tests by fixed rotating ball.

All wear craters were generated without removing the specimens from the clamping device available in the equipment since it has the "horizontal" and "vertical" positioning displacements feature.

Diameters of the wear craters (b) developed during micro-abrasive wear tests were measured by optical microscopy. Subsequently, the values of the wear volume (V) of the respective wear craters were calculated using Eq. (2):

$$V \cong \frac{\pi . b^4}{64.R} \text{ for } b \ll R \tag{2}$$

Where "R" is the radius of the test ball.

Finally, the effect of the heat treatment conditions on P410D ferritic stainless steel was validated based on the statistical analysis of the wear craters volumes and on the behavior of the wear volume as a function of the hardness of each specimen—$V = f(H)$, respectively.

3. Results and discussion

3.1 Microstructural analysis

Figure 9 presents the metallurgical microstructures of P410D ferritic stainless steel specimens heat-treated under different socking temperatures.

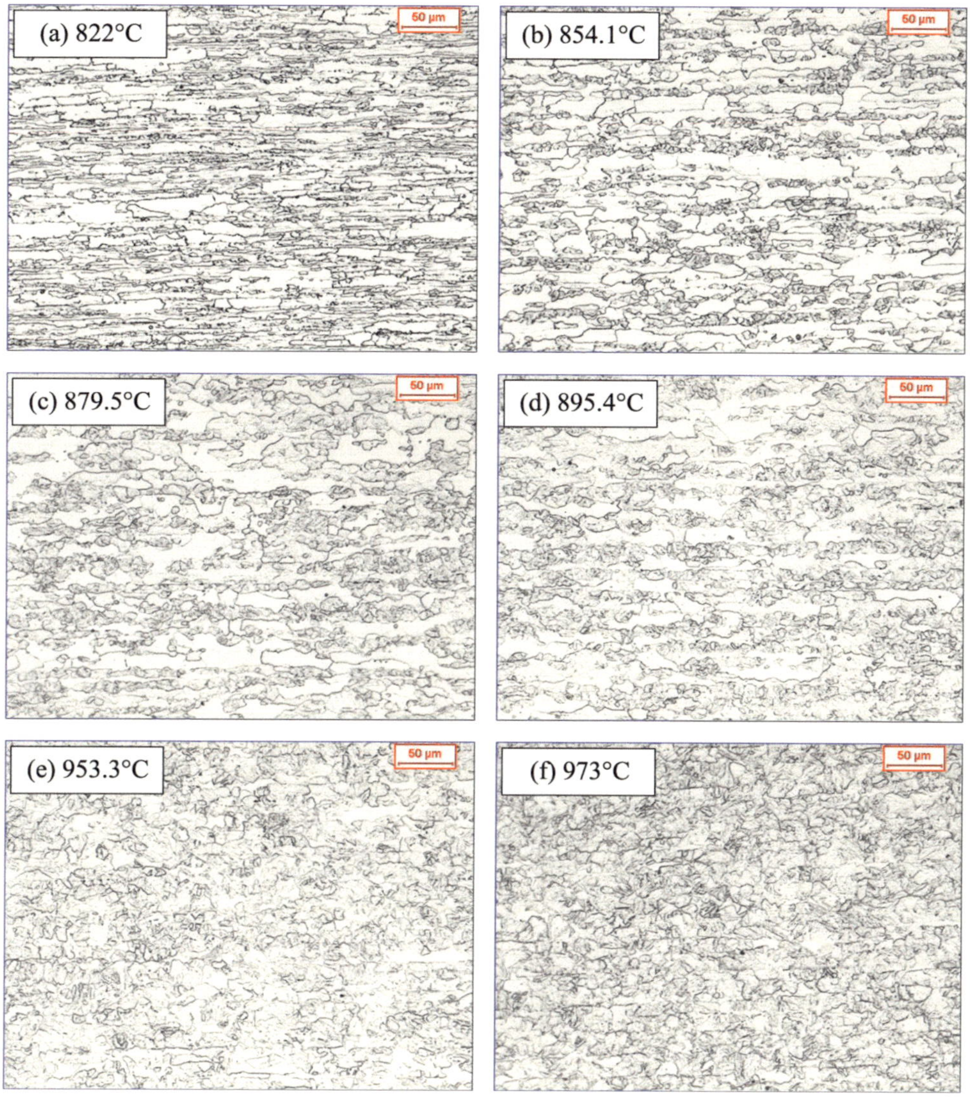

Figure 9.
Metallurgical microstructures of the P410D ferritic stainless steel specimens after different heat treatment conditions. (a) "Specimen 1"—822°C, (b) "Specimen 2"—854.1°C, (c) "Specimen 3"—879.5°C, (d) "Specimen 4"—895.4°C, (e) "Specimen 5"—953.3°C, and (f) "Specimen 6"—973°C.

It was observed in *Specimen 1* (**Figure 9a**) that the socking temperature was not high enough for the recrystallization of the ferritic matrix, due to the presence of deformed grains. The presence of recrystallized grains, deformed grains, pores, and martensitic phase can be observed in the *Specimen 2* (**Figure 9b**). In *Specimen 3*—**Figure 9c**, there is a greater presence of recrystallized grains with an increase in the martensitic phase. The *Specimen 4* (**Figure 9d**) presented a higher amount of martensitic phase and grains apparently smaller than the previous samples, with the presence of some deformed grains. In *Specimen 5* (**Figure 9e**), deformed grains are hardly noticed and an increase in the amount of martensite phase was observed. Finally, in *Specimen 6* (**Figure 9f**) occurred the maximum point of martensitic structure formation, with no more deformed grains.

3.2 Residual stresses

Table 5 and **Figure 10** presents the residual stress values reported for the P410D ferritic stainless steel specimens, together with the respective hardness values. All residual stress values measured were "*tractive*" and inversely proportional to the hardness of each specimen.

3.3 Tribological behavior

Figure 11 presents the specimens used for the analysis of results, already with all micro-abrasive wear tests by rotating ball conducted on their surface, and **Figure 12** displays images of wear craters produced.

Table 6 shows the arithmetic-mean of the wear volumes (V), as well as the respective values of the standard-deviations, for each one of the specimens; it is noted that, for all specimens, the standard-deviation, in reference to the arithmetic-mean of the values of wear volumes (V), was below 10%.

Figure 13 displays the behavior of the wear volume (V) as a function of the hardness (H) of the specimen—$V = f(H)$.

Specimen	**Hardness—[HV]**	**Residual stress—[MPa]**
1	177	+165
2	200	+156
3	222	+128
4	238	+102
5	297	+95
6	304	+87

Table 5.
Values of residual stress measured in the specimens of P410D ferritic stainless steel.

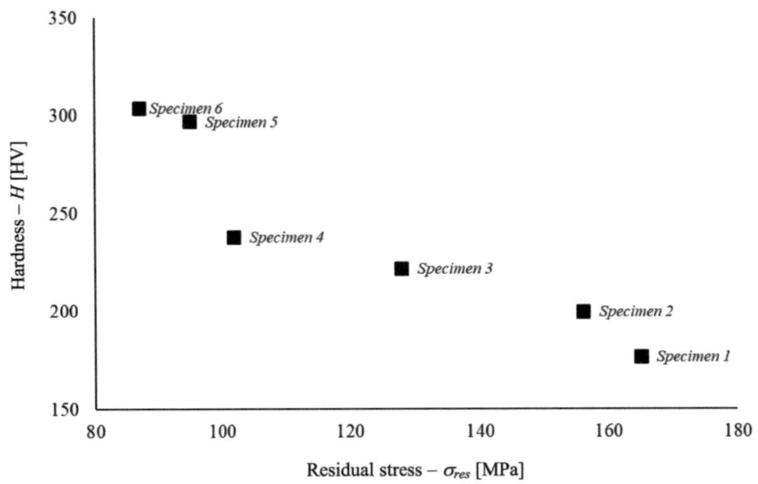

Figure 10.
Residual stress value acting on each specimen of P410D ferritic stainless steel, related to its hardness.

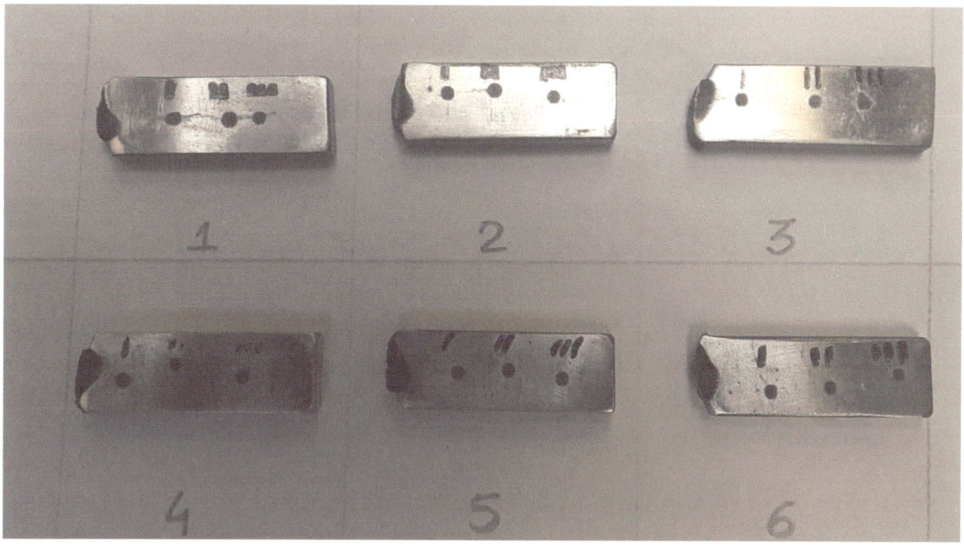

Figure 11.
Wear craters generated on P410D ferritic stainless steel specimens.

It was observed that, with the increase in hardness of the material, the volume of worn material—the volume of the wear crater—decreased, following, in qualitative agreement, the *Archard Equation* (Eq. (3)):

$$\xi = K_W.K_F \int \frac{P.v}{H_{CP}(T)} dT \tag{3}$$

By directing the quantities pertinent of the *Archard Equation* to the "*ball-cratering*" wear test, they can be defined as:

- ξ is a quantitative quantity, where the higher its value the greater is the severity of the micro-abrasive wear process—or, the greater the volume of worn material;
- K_W and K_F are constant quantities related to materials belonging to the tribological system—specimen, abrasive particles, and test ball—during the micro-abrasive wear process;
- P is the contact pressure reported in the tribological system "*specimen—abrasive particles—test ball*", defined by Eq. (4):

$$P = \frac{N}{A} \tag{4}$$

Where A the projected area of the wear crater.

- v is the tangential velocity of the test ball, defined by Eq. (5):

$$v = \pi.D.n \tag{5}$$

- H is the hardness of the specimen, as a function of the temperature (T)—$H = f(T)$.

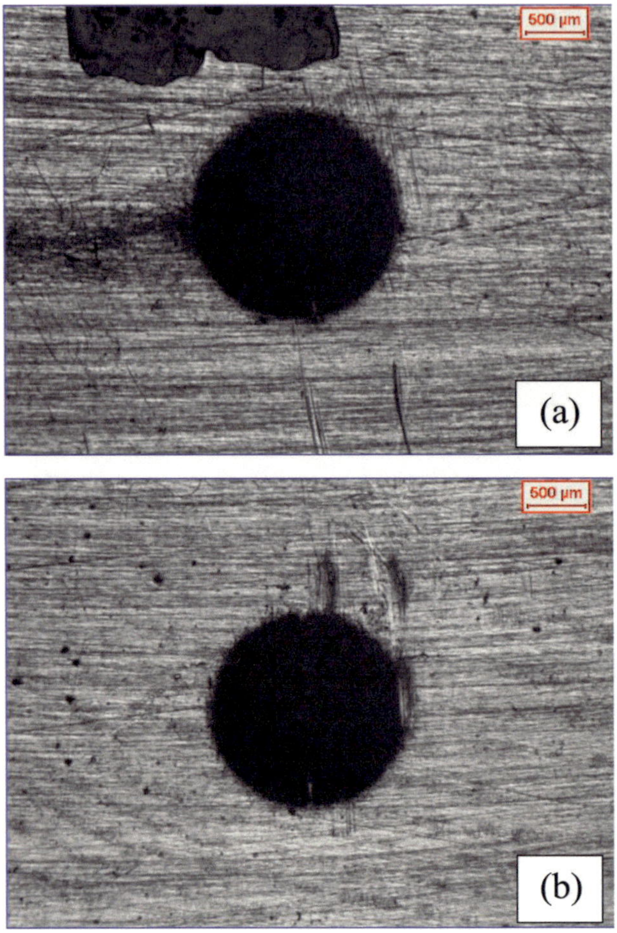

Figure 12.
Wear craters obtained from "ball-cratering" wear tests conducted in this work: (a) Specimen 2 *and (b)* Specimen 4.

Specimen	**Hardness—*H* [HV]**	**Wear volume—*V* arithmetic-mean [mm^3]**	**Wear volume—*V* standard-deviation [mm^3]**
1	177	0.0139	0.0009
2	200	0.0119	0.0010
3	222	0.0118	0.0002
4	238	0.0115	0.0010
5	297	0.0111	0.0006
6	304	0.0102	0.0005

Table 6.
Arithmetic-mean of the values of wear volumes (V) *with the respective values of standard-deviations.*

Analyzing the behavior of the physical parameters of the *Archard Equation* during the micro-abrasive wear process, it is noted that the tangential velocity value of the test ball was the same for all test conditions—n = 56 rpm; with this, the tangential velocity of the test ball remained constant for all specimens.

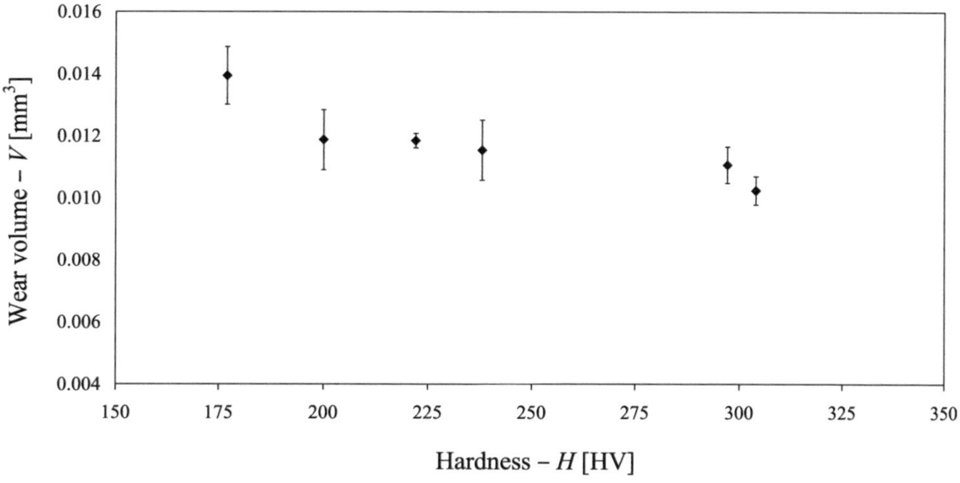

Figure 13.
Behavior of the wear volume (V) *as a function of the hardness* (H) *of the specimen*—V = f(H).

Additionally, the temperature of each specimen remained constant at room temperature, resulting in the hardness (*H*) of the materials analyzed remain unchanged during all ball-cratering wear tests.

However, following the *Archard Equation*, ξ and H are inversely proportional; therefore, the increase in H caused a decrease in ξ, characterizing, consequently, lower severity of wear related to a lower volume of wear (V) generated.

Finally, the only physical quantity that varied in a decreasing way during the ball-cratering wear tests was the contact pressure (P)—since the normal force (N) remained constant, the projected area (A) of each wear crater increased with the progressive increase in the sliding distance.

In fact, in all wear tests, the contact pressure (P) followed, under the qualitative agreement, the approach detailed by Cozza [8, 9] in previous works, where P decreased as a function of the projected area of the wear crater for each specimen. For each hardness value, different values of ξ were obtained, related to the calculated values of wear volumes (V).

Finally, based on **Figure 13**—which exhibited practical results of the variation of wear volume (V) as a function of the hardness of the specimen (H)—and by theoretical complementation departing from *Archard Equation* (Eq. (3)), it can be said that the results generated are within the technical-scientific agreement of micro-abrasive wear.

4. Conclusions

The results obtained in this scientific work showed that the wear volume decreased with an increase in the hardness of P410D ferritic stainless steel specimens—due to the different heat treatment conditions.

Increase in oven temperature led to an increase in solution treatment temperature and hardness of P410D ferritic stainless steel along with a decrease in residual stress.

Author details

João Pedro de Castro Valente Lenz Ferreira[1], Marcelo de Matos Macedo[2], Jorge Humberto Luna-Domínguez[3], Vikas Verma[4], Wilian da Silva Labiapari[5] and Ronaldo Câmara Cozza[1,6*]

1 Department of Mechanical Engineering, University Center of FEI—Educational Foundation of Ignatius "Priest Sabóia de Medeiros", São Bernardo do Campo, São Paulo, Brazil

2 Department of Materials Science, Federal University of ABC (UFABC), Santo André, São Paulo, Brazil

3 Universidad Autónoma de Tamaulipas, Tampico, Tamaulipas, Mexico

4 Research and Development Cell, R V Institute of Technology (Bijnor), Dr. A.P.J. Abdul Kalam Technical University, Lucknow, Uttar Pradesh, India

5 Aperam South America Research Center, Timóteo, Minas Gerais, Brazil

6 Faculty of Technology of Mauá (FATEC-Mauá), Department of Mechanical Manufacturing, "Paula Souza" State Center of Technological Education (CEETEPS), Mauá, São Paulo, Brazil

*Address all correspondence to: rcamara@fei.edu.br; ronaldo.cozza@fatec.sp.gov.br

References

[1] Callister WD Jr, Rethwisch DG. Materials Science and Engineering: An Introduction. 9th ed. John Wiley & Sons; 2013. ISBN-10: 1118324579. ISBN-13: 978-1118324578

[2] https://en.wikiversity.org/wiki/File:Fe-Cr_Phase_Diagram.gif

[3] da Costa e Silva ALV, Mei PR. Aços e Ligas Especiais. 3ª Edição ed. Editora Blucher; 2010. ISBN-10: 852120518X. ISBN-13: 978-8521205180

[4] McGuire MF. Stainless Steels for Design Engineers. ASM International; 2008. ISBN: 978-0-87170-717-8

[5] Perry AJ, Sue JA, Martin PJ. Practical measurement of the residual stress in coatings. Surface and Coatings Technology. 1996;**81**:17-28. DOI: 10.1016/0257-8972(95)02531-6

[6] ASTM G77-91: Standard Test Method for Ranking Resistance of Materials to Sliding Wear Using Block-on-Ring Wear Test

[7] ASTM G40-96: Standard Terminology Relating to Wear and Erosion

[8] Cozza RC. Effect of pressure on abrasive wear mode transitions in micro-abrasive wear tests of WC-Co P20. Tribology International. 2013;**57**:266-271. DOI: 10.1016/j.triboint.2012.06.028

[9] Macedo MM, Luna-Domínguez JH, Verma V, Schön CG, Cozza RC. Assessment of micro-abrasive wear tribological properties of H10 tool-steel under conditions of "*constant normal force ⇒ variable pressure*" and "*constant pressure ⇒ variable normal force*". Wear. 2021;**476**:203664. DOI: 10.1016/j.wear.2021.203664

Chapter 8

Non-Destructive Evaluation of Residual Stresses in Welding

Santanu Saha

Abstract

During welding, due to the highly localized transient heat input, considerable residual stresses and deformations may occur. Welding residual stresses and welding distortion may greatly impair manufacturing and strength. Residual stresses are internal forces without external forces acting. The total residual stresses superimpose on the stresses from the external load, i.e., the load stresses. Residual stresses are the result of microstructural deformations, e.g., dislocations etc. and can be divided as follows: i) volumetric (or "dilatoric") ii) distortional (or "deviatoric"). Volumetric strain generally is caused by sectioning by thermal expansion, chemical conversion, microstructural transformation or change in state; distortional strain is generally caused by time-independent plastic or time-dependent visco-plastic deformation. There are several methods available to measure the residual stresses or strains non-destructively. One of them is X-ray diffraction. X-rays are diffracted crystal lattices and produce interference phenomena, from which it is possible to draw conclusions relating to the interplanar spacing of the lattice. Other methods are the neutron diffraction method, ultrasonic method and the magnetostriction method. In the magnetostriction or Barkhausen noise method, the stress state is deduced from the value of the local magnetization restraint. The magnetic flux density in a ferromagnetic material subjected to a time-varying magnetic field does not change in a strictly continuous way, but rather by small, abrupt, discontinuous increments called Barkhausen jumps. The jumps are due primarily to discontinuous movements of boundaries between small magnetically saturated regions called magnetic domains in the material. This chapter describes the causes and measurement of residual stress induced during welding.

Keywords: arc welding, residual stress, Barkhausen noise, magnetically induced velocity changes, stress dependence

1. Introduction

Residual stresses in welded components develop during cooling due to expansion/contraction hysteresis across the weld and the heat-affected zone. There are several ways to reduce or minimize the residual stresses. By proper pre-welding management and taking precautions, welding residual stresses can be eliminated or minimized to an acceptable level. This can be achieved by several techniques including but not limited to the following:

a. Pre heating;

b. Proper welding procedure;

c. Welding sequence;

d. Post weld heat treatment;

e. Natural aging;

f. Peening;

g. Proper design;

h. Selection of suitable material for construction.

Not all the above variables apply to all welding processes and materials, but controlling the applicable parameters and use of proper welding process and sequence can effectively reduce welding residual stresses to acceptable levels.

Residual stresses can be present in the component even though proper actions were taken during welding to minimize the stress levels and can add to the applied stress, which causes the failure of the component. Although apparently, the applied stress is much less than the yield stress or design stress of the component, the component fails due to hidden residual stress which may add to the applied stress and can eventually lead to failure under normal operating stresses. Most of the earlier development to measure the amount of residual stress are destructive in nature which means cutting the component at several strategic locations thus making the component unusable for the intended purpose. This is a more complicated onsite fabrication process where there is no provision to cut the component to determine the residual stresses because the component will then become unusable. Sometimes the large size of the structures precludes the destructive way of determining the residual stresses. Hence the development of non-destructive measurement of residual stresses comes into considerations. Most commonly, non-destructive measurement of residual stresses is typically done using X-ray diffraction (XRD) and/or neutron diffraction (ND). However, these methods are generally not portable, and XRD is limited to near-surface measurements because of the short penetration depth of X-rays in metals ($\sim$101 μm). Therefore, there is a strong interest in exploiting alternative technologies for residual stress characterization in engineering components. Developments in ultrasonic, eddy current, and thermal-optic residual stress measurement techniques to determine their feasibility as field-deployable characterization tools makes the approach to detect and measure residual stresses in the welded components easier and reliable. One of the recent developments of non-destructive measurement of residual stresses is the Magnetic Berkhausen Noise technique (MBN). Here, we will discuss the principles and advances in non-destructive techniques, particularly the MBN technique.

2. Residual stresses in the welded joints

The main origin of residual stresses resulting from the welding process can be categorized into three factors:

a. Shrinkage during cooling, which is primarily of mechanical type;

b. Thermal stresses due to temperature gradient;

c. Microstructural changes cause structurally induced stresses.

In the case of arc welding, the workpiece is raised to the melting point within a very limited area, but the temperature drops sharply as the distance from the heat source is increased. This is due to:

i. The high thermal conductivity of the workpiece. Generally, metals have very high thermal conductivity;

ii. The subsequent rapid dissipation of heat because of high thermal conductivity;

iii. The insignificant volume of the heated metal because the heat input in welding is very localized in a small area.

The above conditions prevent the uniform expansion of the metal and may produce serious internal stresses, distortion and warping. **Figure 1** describes the influence of such a sharp thermal gradient in a simple butt-welded plate.

Types of welding distortion are classified as out-of-plane types (such as buckling, longitudinal bending (bowing), and angular change and in-plane types (such as transverse shrinkage, longitudinal shrinkage and rotational distortion). Several methods for controlling welding distortion are described; some can be applied during welding, and others after welding are completed. As some methods may reduce one distortion mode but increase another, it is imperative to identify the distortion mode of a particular structure before selecting the appropriate distortion mitigation method. Welding-induced buckling is eliminated by ensuring that the compressive longitudinal residual stress is lower than the critical buckling stress of the plate, either by increasing the critical buckling stress of the plate, reducing the welding residual stress, or modifying the residual stress after welding. Angular distortion is usually controlled by the use of presetting, restraints, or back-side heating. Bowing (also referred to as camber) distortion is controlled by either reducing the welding heat input or balancing the welding residual stress over the cross-section of a structure to minimize the bending moment [1].

Residual stress and distortion affect the fracture behavior of materials by contributing to buckling and brittle fractures at low applied-stress levels. When residual stress and the accompanying distortion are present, buckling may occur at lower compressive loads than would otherwise be predicted. In tension, residual stress may lead to high local

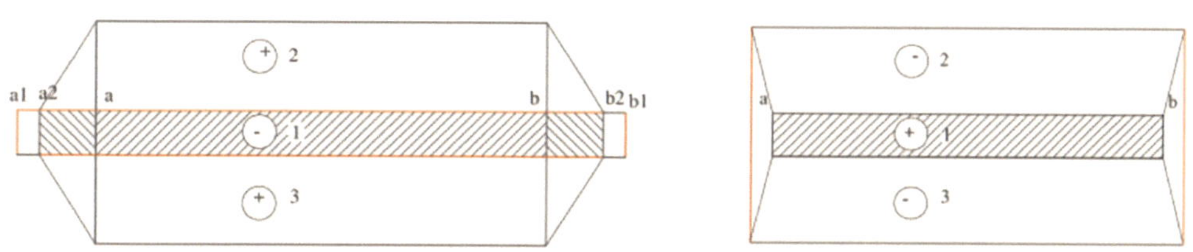

Figure 1.
Stresses set up in steel but welded plate (a) on heating (b) on cooling.

stress in weld regions of low notch toughness. This local stress may initiate brittle cracks that are propagated by any low overall stress that is present. In addition, residual stress may contribute to fatigue or corrosion crack growth and failures [1].

While residual stresses play a major factor in the analysis of fracture response, residual stresses can often be neglected when materials exhibit very high toughness.

The distribution of longitudinal residual stresses can be approximated by the following Equation:

$$\sigma_x\,(y) = \sigma_m\left[1 - (y/b)^2\right] e^{-(1/2)\,(y/b)^2} \tag{1}$$

where,

σ_m = maximum residual stress which may be as high as the yield stress of weld metal.

b = width of the tension zone of $\sigma_{x:}$

σ_x; Residual stress in X direction (along the axis of the weld).

σ_y: Residual stress in Y direction (perpendicular to the axis of the weld).

X: Direction along the weld axis.

Y: Direction perpendicular to the weld axis.

3. Technique for measurement of residual stresses

Many different methods for measuring the residual stresses have been developed during the past several years. Techniques to measure residual stresses may be classified as either destructive or semi destructive or non-destructive. **Figure 2** below classifies the techniques in a detailed manner. The destructive and semi destructive techniques, called also the mechanical method, are dependent on inferring the original stress from the displacement incurred by completely or partially relieving the stress by removing material. Several destructive and semi destructive methods are available out of which the following are most common:

i. Sectioning;

ii. Hole drilling;

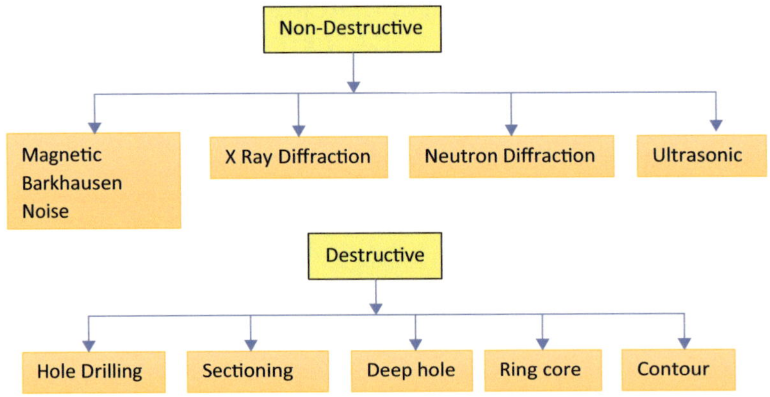

Figure 2.
Various stress measurement methods.

iii. Ring core method;

iv. Deep hole method.

For the non-destructive evaluation of residual stresses, the following methods are most common:

i. X-Ray diffraction;

ii. Neutron Diffraction;

iii. Magnetic Barkhausen method;

iv. Ultrasonic velocity comparison method.

3.1 Destructive methods

All destructive methods of residual stress measurement are dependent on the stress relaxation behavior of metals. When a small portion of material from the stressed part is removed, the locked-up stresses are relaxed and results in micro deformation (strain). By accurately measuring the relaxations, residual stresses can be calculated by one or more of the following techniques:

3.1.1 Hole-drilling technique

The hole drilling method is widely used and a very simple method. Generally, it is used to measure biaxial stresses. Residual stress is determined by measuring the relived radial strain by drilling a small hole in the welded plate. Measurement locations are outside the drilled holes. Measurement can be performed by measuring balls or by strain gauges. The primary requirement of the hole drilling method is that the material should be isotropic. It can be significant in the case of anisotropic materials. It is relatively reliable and accurate. Moreover, there is very little wastage or damage caused to the material due to the application of this technique. Generally, strain gauges are placed where the residual stresses are to be measured. Appropriate calibration standards are used for the measurement of stresses. Relaxation of stresses causes local strains which may be measured by the strain gauges and thus can be measured by comparing with the calibration standards.

There may be issues for accuracy of the results if the hole is drilled very near to the component edge or near to another hole. The calibration standard is generally based on infinite size or large plates. Moreover, there might be plastic deformation due to drilled holes which may affect the actual measurement of stress. If the residual stress is high (60% of yield stress or more) or if the drilling is not proper, this error may be as high as 15–16% depending on the hole diameter.

Generally, in most cases, the residual stresses are not uniform with the depth of the material from the surface. In such cases, the Incremental hole drilling method may be used. This method is based on the same principle as the basic hole drilling method stated above, but in this case, drilling is performed in a series of small steps. Drilling is performed by a high-speed pneumatic drill (more than 200,000 rpm). This high-speed drilling introduces very little stress into the material due to the drilling process, thus enabling the measurement more accurate.

3.1.2 Deep hole method

The deep hole method is a hole trepanning method instead of direct drilling hole. At the first stage, the hole is drilled out. Then the diameter of the hole is measured accurately. After that, a trepanning tool is used to remove the metal from the inside surface of the hole. This process relaxes the residual stresses in the core. Again, the diameter of the hole is measured after trepanning out. The residual stress is calculated from the change in diameter of the hole. One of the great features of this technique is that it can measure the residual stresses in the interior of the material.

3.2 Non-destructive methods

3.2.1 X-ray diffraction method

X-ray diffraction is one of the non-destructive techniques for the measurement of residual stresses on the surface of materials. Under the action of stress, either applied or residual, the resulting elastic strains cause the atomic planes in the metallic crystal structure to change their spacings. This change of the inter-planar atomic spacings can be detected and measured by X-ray diffraction. Thus, the total stress on the metal can then be obtained. ASTM E 2860 describes residual stress measurement technique by X ray diffraction for bearing steels.

Most metal components of practical concern consist of many tiny crystallites (grains), randomly oriented for their crystalline arrangement. This method applies to materials that are crystalline in nature. Most metallic components are crystalline and produce diffraction of definite pattern depending on X ray beam direction. Because of residual stresses present within the material, elastic strains are produced within the crystal lattice within the elastic limit of the material. By Hooke's law, this strain is proportional to the applied stress. Estimation of stress can be performed based on accurately measuring the distance between the crystallographic planes. Portable diffractometers that can be taken out into the field for measurements of structures such as pipelines, welds, and bridges are now available. Sometimes the layer-removal technique is used to remove a thin layer from the surface and stress profile again measured with X ray diffraction, but then the method becomes destructive and such cutting must be done with care to ensure the stress state is not unduly altered. The speed of measurement depends on several factors, including the type of material being examined, the X-ray source, and the degree of accuracy required [2].

Third-generation synchrotron sources provide access to high X-ray energies. They produce very high-energy X rays, which can penetrate greater depths within the material. Moreover, High energy X-ray beams are very collimated and X ray intensity is concentrated mainly in the central axis of the beam. They provide very high spatial resolution and very short data acquisition time. Because of these advantages, this technique is more useful to collect detailed data of strain fields in two dimensions as well as three dimensions. Even very minor phase transformation may be detected. There are also some serious drawbacks in the application of the synchrotron method. Mean that the sampling gauge is usually very elongated. The spatial variation is very different in different directions due to low scattering angles. Since the scattering angle is very short, it is good for simple geometries like plates, pipes, etc. but for large or geometrically complex samples it can be difficult to achieve short path lengths for all measurement directions [3].

3.2.2 Neutron diffraction method

The neutron diffractions method is also based on a similar principle of crystallographic planes spacing measurements as with X ray diffraction. Instead of X ray beam, a neutron beam is used. The advantage of the neutron diffraction methods in comparison with the X-ray technique is its larger penetration depth. X-ray diffraction technique has limits in measuring residual stresses through the thickness of a welded structure. Neutron beams can penetrate a few centimeters into the material. Hence, internal residual stresses can also be measured in some cases. Since X ray cannot penetrate deep into the high atomic number of materials because of high absorption coefficient, neutron diffraction can be used in such materials. Studies have revealed that the neutron beam can penetrate up to 30 mm in steel and up to 300 mm in light alloys like Aluminum. Because of high spatial resolution provided by neutron beams, this technique can be used to evaluate and measure stresses in three dimensions. Complete three-dimensional maps of the residual stresses in the material can be provided. However, compared to other diffraction techniques such as X-ray diffraction, the relative cost of application of neutron diffraction method, is much higher, mainly because of the equipment cost. It is too expensive to be used for routine process quality control in engineering applications [4].

3.2.3 Magnetic Barkhausen noise method

One of the recent developments of residual stress measurement by non-destructive methods is the magnetic Barkhausen Noise technique, popularly known as the MBN technique. The magnetic Barkhausen noise was discovered by Heinrich Barkhausen (1919). Ferromagnetic materials are composed of tiny magnetic zones called magnetic domains. In the non-magnetic state, all these magnetic domains are randomly oriented so that the resulting magnetic field is zero in and out of the material. When the ferromagnetic material is placed under a strong magnetic field, the tiny magnetic domains try to align themselves along the direction of the externally applied magnetic field. As discussed, the MBN technique is mostly applicable to ferromagnetic materials. Although, the MBN technique can be used to determine several other material parameters, the most common use to determine the residual stresses.

Magnetic domains are called as bloch walls also. During the magnetic hysteresis cycle, the bloch walls try to align themselves along the direction of externally applied magnetic fields and overall magnetization can be described in the form of hysteresis curves/cycles [5]. A close view of the hysteresis curve reveals that the curve is not smooth as can be seen from the cycle, rather it consists of small steps or jumps which can be seen in a zoomed view of the curve (**Figure 3**).

The MBN signals can only be seen in an enlarged cut-out of the hysteresis curve (indicated by a circle, see **Figure 3**). Barkhausen noise signal can have a range of frequencies. The effective depth of signal penetration is between 0.01 mm and 1 mm. The Barkhausen noise signal is damped due to the skin effect which is caused by the opposing eddy currents induced by the changing magnetic field. An estimation of the penetration depth of the BN signal can be calculated using the following formula- Eq. (2):

$$\delta = 1/\sqrt{\pi\mu\sigma f} \tag{2}$$

where δ denotes the penetration depth, μ represents the magnetic permeability, σ means the electrical conductivity and f denotes the frequency of the alternating magnetic field [5].

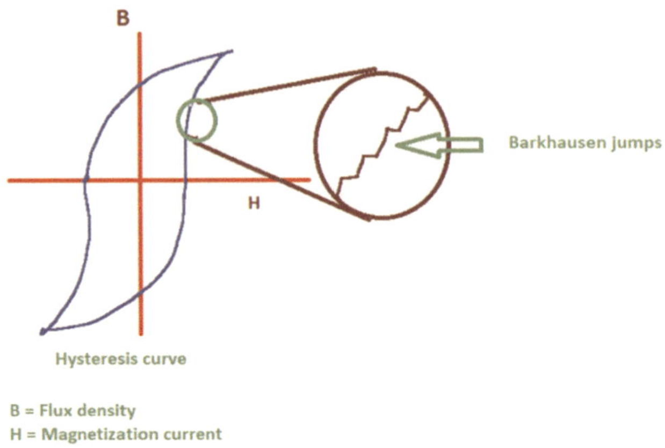

Figure 3.
Magnetic Barkhausen jumps.

In general, microscopic Grain boundaries, lattice dislocations within and around the grains, second phase materials (e.g., carbides in iron) and impurities in the ferromagnetic material prevent the movement of domain walls. When a high amplitude magnetization force is applied, the above restraining resistance forces are overcome by the external magnetization forces and the domains try to align in the field direction. This movement is not continuous rather by small microscopic jumps. When an inductive coil is placed near the specimen being magnetized, the coil can sense the jumps and the magnitude because the change in magnetization induces an electrical pulse in the coil. When all electrical pulses produced by all domain movements are added together a noise-like signal called Barkhausen Noise is generated [6]. **Figure 4** shows the setup for the detection of MBN signals, where an electromagnet is placed on the object and produces an alternating magnetic field, which can be chosen between 0.1 and several 100 Hz depending on the testing problem. The inductive sensor is located between the magnetic poles along with a Hall probe to measure the tangential magnetic field strength (**Figure 4**). The voltage pulses induced in the sensor are amplified at various stages between 60 and 100 dB. They are then filtered and rectified and their envelope called the inductive Barkhausen noise amplitude, M is

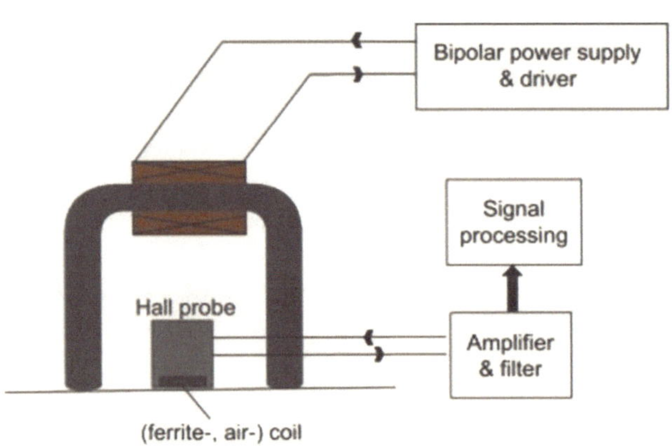

Figure 4.
Barkhausen noise measurement set up.

recorded and plotted against the tangential magnetic field strength H_t (**Figure 5**). The inductive sensor may be pick up coil or ferrite core coil or any other device. The frequency content of the Barkhausen events extends from a few hundred Hz up to the MHz range. Barkhausen events excited at greater distances from the surface will result in voltage pulses with lower frequency content than near-surface events. By changing the analyzing frequency of a bandpass filter from low to high, a weighted characterization of the Barkhausen noise from near-surface regions can be obtained. **Figure 5** shows a typical Barkhausen noise profile curve [5]. Typical measuring quantities derived from the magnetic Barkhausen noise are the maximum of the profile curve M_{MAX}, the H-field position of this, H_{CM}, and the half widths ΔH by 75%, 50%, and 25% of the maximum value (**Figure 5**).

It was observed that as we move away from the weld, the MBN noise response depends on the stress state of the region of the material. Near the edge of the weld the amplitude of and the type of the MBN signals changes abruptly suggesting a high accumulation of stress at the weld toe. MBN is sensitive to changes in applied stress. The interaction of elastic strain with the magnetic domains is called "magneto-elastic

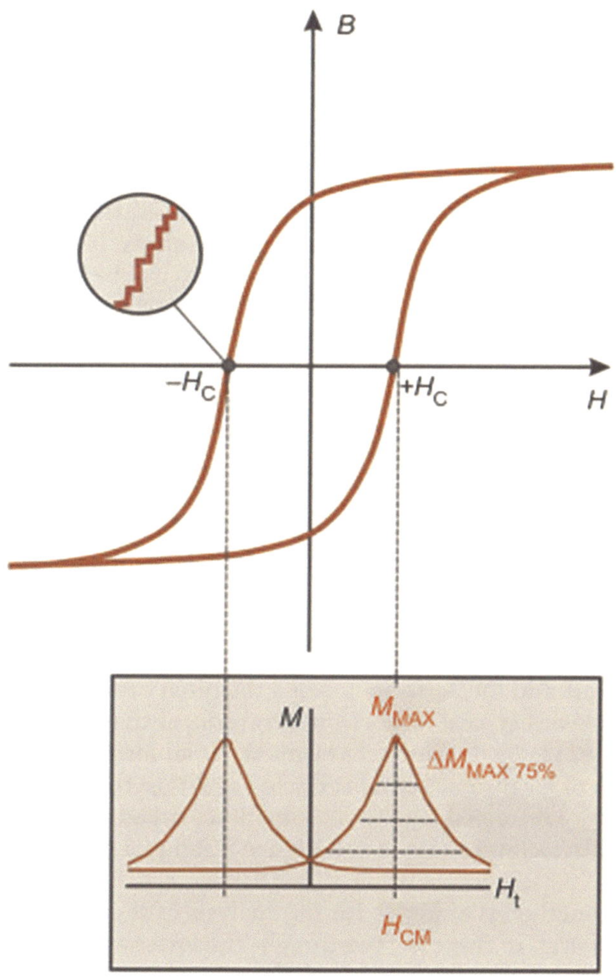

Figure 5.
Magnetic Barkhausen noise curve and derived measuring quantities: compare with magnetic hysteresis curve.

interaction". Accordingly, in ferromagnetic materials like iron, most steels and cobalt, compressive stresses will decrease the intensity of Barkhausen noise while tensile stresses increase it. This fact can be exploited so that by measuring the intensity of Barkhausen noise the amount of residual stresses can be determined. Barkhausen noise is also affected by the microstructural state of the material. Hence, the signal amplitude and the characteristics of MBN signals will not be the same for various materials. Therefore, for MBN to be effective in determining residual or applied stresses, different materials must be calibrated individually. Hence, the calibration procedure is very important for accurate and reliable results. Each zone having remarkably different microstructure should be separately considered for calibration.

For obvious reasons, it is known that the amplitude of the MBN signal decreases with the reduction in grain size. Also, the signal response increases with increasing misorientation angles at the grain boundary. Because of that, it is generally difficult to use the MBN to assess residual stresses in weldments containing heat-affected zones (HAZ), since HAZ has very rapid microstructural gradients. To reflect the micro-structural variations in the heat-affected zone, calibration samples should be similar to the actual component to be tested. That means calibration should contain similar welding parameters, process, heat input, etc. the signals from the calibration samples (weld, HAZ and base metals) should be compared with the actual test object to validate the test results.

3.2.4 Ultrasonic method

Another promising non- destructive technique for residual stresses measurement is the Ultrasonic determination of residual stresses. The ultrasonic method, also called as critically refracted longitudinal (L_{CR}) wave techniques can be utilized for residual stresses measurements on thick samples. Residual stress states can be determined by measuring L_{CR} velocities in the material. Since the velocity of ultrasonic waves is dependent on the elastic constant and density of the material, the change in elastic stress state can be determined by this acoustic-elastic effect. The relationships between the changes of the velocities of longitudinal ultrasonic waves and shear waves with orthogonal polarization under the action of tensile and compressive loads in steel and aluminum alloys are established in several publications. The intensity and character of these changes can be different, depending on the material properties. Different configurations of ultrasonic equipment can be used for residual stresses measurements. A piezo electric transducer can be used as a transmitter of ultrasound and once the sound is passed through the material/zone of interest, it can be received by the same transducer or another dedicated transducer. The first one is called the pulse echo technique and the 2nd one is called the pitch catch or transmission technique. **Figure 6** shows a typical piezoelectric transducer that generate L_{CR} waves at steel from a Perspex wedged transducer element. At an incident angle of 27.5°, the longitudinal mode of Ultrasonic sound refracted at 90° to the normal to the interface, producing a critically refracted L_{CR} wave, sometimes called a creeping wave. **Figure 7** shows a typical pulse echo technique and **Figure 8** shows a typical pitch catch technique.

The ultrasonic method is effective for the analysis of residual stresses in the interior of the material. In this case, the trough- thickness average of the residual stresses is measured. Comparing to X-ray diffraction technique, the depth of measurement is higher and this technique is free from radiation hazard (X-ray technique uses X-ray radiation which is harmful to human body).

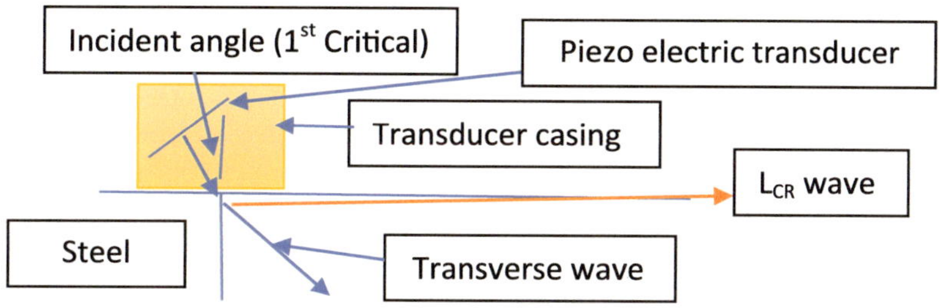

Figure 6.
Typical piezoelectric transducer.

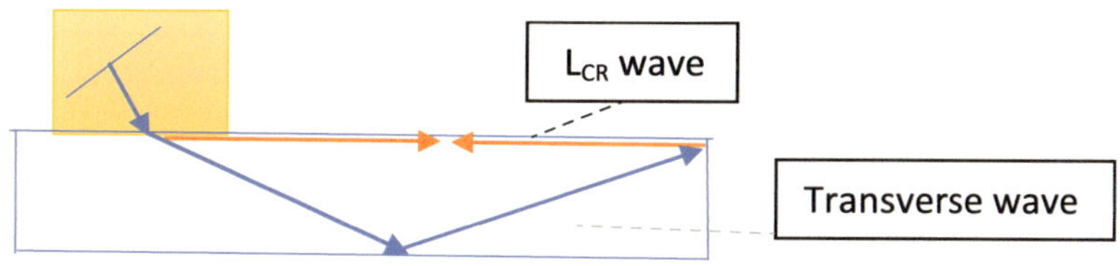

Figure 7.
Pulse echo technique.

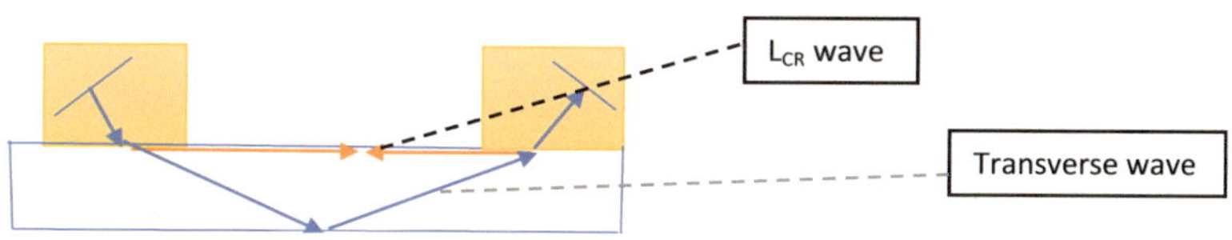

Figure 8.
Pitch catch technique.

In the above arrangements, velocities of longitudinal ultrasonic waves and shear waves with orthogonal polarization are measured at a considered point to determine the uniaxial and biaxial residual stresses. The bulk waves in this approach are used to determine the stresses averaged over the thickness of the investigated elements. In general, the change in the ultrasonic wave velocity in structural materials under mechanical stress amounts to only tenths of a percentage point. Therefore, the equipment for practical application of the ultrasonic technique for residual stress measurement should be of high resolution, reliable, and fully computerized. Currently, Aluminum alloys are widely used in the automotive, aerospace and other industries because of their high strength/weight ratio. The L_{CR} technique offers greater advantages over other techniques in the case of Aluminum welded structures.

Using calibration specimens, the acoustic-elastic constants are measured and those constants are used to determine the stresses. The greatest sensitivity is obtained when the wave propagates in the same direction as the stress. The stress can be calculated according to:

$$\left(V_{pp} - V_L{}^0\right)/V_L{}^0 = k_1\sigma_p + k_2\left(\sigma_q + \sigma_s\right) \tag{3}$$

$$\left(V_{pq} - V_T{}^0\right)/V_T{}^0 = k_3\sigma_p + k_4\sigma_q + k_5\sigma_s \tag{4}$$

where $V_L{}^0$ = Longitudinal Ultrasonic Velocity in the material.
$V_T{}^0$ = Transverse Ultrasonic Velocity in the material.

V_{ij} is the velocity of a wave traveling in the direction i polarized in direction j,
s = principal stress direction.
V_{pp} = Wave propagation direction parallel to the principal stress direction.
V_{pp} = Wave propagation direction parallel to principal stress direction with polarized in direction q.
K_I = coupling constants.

The ultrasonic stress measurement method provides a measure of the macro stresses over large volumes of material. Since, ultrasonic wave velocities also depend on microstructural inhomogeneities, it is sometimes difficult to differentiate the effect of stress from the microstructural inhomogeneities. Nevertheless, being portable and cheap to undertake, the method is well suited to routine inspection procedures and industrial studies of large components.

4. Conclusion

There are several methods of determining the residual stresses in a component. However, different techniques have their own advantages and limitations. Although, the destructive methods are much more accurate and reliable, they are very difficult to apply to the production component, where most of the problems of residual stresses exist. Due to damage of the component, it is not also possible to apply the destructive testing technique in most of the cases. On the other hand, non-destructive techniques are the most opted technique because it does not cause any damage to the component under examination. But they are less accurate and depend on several parameters which affect the repeatability of the results. Advances in computer software make results more accurate, reliable in non-destructive techniques also. Out of the common non-destructive technique, the MBN technique is most promising and can change the way of residual stress measurement which also helps in more reliable failure analysis and fitness for service determination studies.

Author details

Santanu Saha
GPTI Group of Companies, Sharjah, United Arab Emirates

*Address all correspondence to: santanurini@gmail.com

References

[1] Parmar RS. Welding Engineering & Technology. Darya Ganj, New Delhi: Khanna Publishers; pp. 213-235

[2] Krzywosz K. Non-Destructive Evaluation and Measurement of Residual Stress. EPRI Interim report. Charlotte, NC, USA: Electric Power Research Institute; 2009. pp. 46-55

[3] Osman A. Residual stress measurement by X ray diffraction [Master of Science Thesis]. Texas. USA: Texas A & M University, College Station; 2004. pp. 40-54

[4] Withers PJ, Bhadesia HKDH. Residual stress–Measurement techniques. Vol. 17. University of Cambridge, Cambridge, UK: Material Science & Technology; 2001. pp. 355-365

[5] Wilcox M, Mysak T. An introduction to MBN noise and its applications—A paper. Herefordshire, UK: Insight NDT Equipment Ltd; 2004

[6] Suvi-Santa-aho, Laitinen A, Sorsa A, Vippola M. Barkhausen noise probes & modelling: A review. Journal of Nondestructive Evaluation. 2019;**38**(4): 2-4

Chapter 9

Residual Stresses Distribution Posterior to Welding and Cutting Processes

Asma Manai

Abstract

Welding is a joining process that leads to considerable change in the local material and the formation of welding residual stresses (RS). Welding residual stresses can be compressive (beneficial for the fatigue life) or tensile (harmful for the fatigue life). In this chapter, a probabilistic analysis of residual stresses distribution posterior to welding processes is carried out. Several researchers stated that the type of the introduced stresses either compressive or tensile depends on several factors. Some of these factors are listed in this chapter. Welding of mega-structures is carried out in the workshops, then a cutting process takes place to construct the exact size of the structural components. This cutting process has a significant effect on the weld residual stresses re-distribution. A study of the re-distribution of the weld residual stress after cutting was performed. It was found that independent of the weld seam length, the residual stresses re-distributed up to 60 % of the weld seam length.

Keywords: weld parameters, tensile residual stresses, compressive residual stresses, as-welded, cutting processes

1. Introduction

Residual stresses have a significant effect on the fatigue life of structures. Surface tensile residual stresses (TRS) can cause harm to structures, components or specimens. However, surface compressive residual stresses (CRS) can improve the fatigue life of the structures, components or specimens. In most cases, welding introduces TRS at the surface. Few cases reported that CRS can be found at the surface of welded structures, components or specimens.

Rossini et al. [1] define residual stresses as the stresses that remain within the Structure in the case of absence of external load or thermal gradients after manufacture and material processing (refer to **Figure 1**). The equilibrium of the self-balanced stress can be translated to equilibrium in x-direction gives Eq.(1).

$$\int_{-h/2}^{h/2} \sigma_{RS,x} \, dy = 0 \tag{1}$$

Where, h is the plate thickness and ($\sigma_{RS,x}$) is the residual stresses in the x-direction.

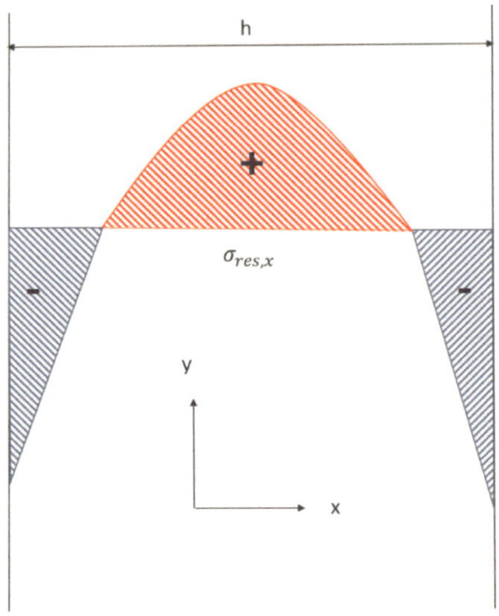

Figure 1.
Schematic presentation of residual stresses distribution.

An external load applied to a structure, component or specimen will lead to a stress distribution. If the structure (component or specimen) has residual stresses and its behavior is still elastic, the material will respond to the sum of the stress distribution of the external load and the residual stresses. Eq.(2) expresses the relation between stress distribution in the material (σ), the externally applied stress (σ_{Ex}) and the residual stresses (σ_{RS}).

$$\sigma = \sigma_{Ex} + \sigma_{RS} \tag{2}$$

In case where the structure, component or specimen is submitted to external cyclic loading (σ_{Ex}), the residual stresses do not affect the stress amplitude ($\sigma_{a,Ex}$), as it is permanently present in the material, therefore, (Eq.(3)). However, it affects the mean stress ($\sigma_{m,Ex}$) (refer to Eq.(4))

$$\sigma_a = \sigma_{a,Ex} \tag{3}$$

$$\sigma_m = \sigma_{m,Ex} + \sigma_{RS} \tag{4}$$

There are several sources that introduce residual stresses, such as production process, heat treatment, welding process, post-weld treatments, etc.

Residual stresses can be classified into two scales namely macro and micro residual stress. RS that occur over long distances within the material are characterized as macro RS. In Withers et *al.* [2], mentioned that the origins of macro stress are peening, welding, shot-peening and Tungsten Inert Gas (TIG) dressing. While, RS that exists either between grains or inside a grain due to coherence at interfaces, crystalline defects, and dislocation stress fields (Withers et *al* [2] and Donato et *al.* [3]) is named as micro residual stresses.

The scale of the residual stress, whether it is micro (intergranular) or macro scale, determines the measurement technique. There is no unique technique that is qualified for measuring all the stress types (micro and macro). Within one specimen or

Technique	Type	Residual stress type
Mechanical	Destructive	Macro-residual stress
X-ray diffraction	Non-destructive (surface method)	Macro and micro residual stress
Neutron diffraction	Non-destructive	Macro and micro residual stress
Ultrasonic	Non-destructive	Macro and micro residual stress
Magnetic	Non-destructive	Macro and micro residual stress

Table 1.
Techniques of measuring residual stresses.

component, measurement of residual stresses using two different techniques, give completely different results. Therefore, for reliable results, it is recommended to select a suitable method for each case. Nasri et *al.* [4], reported that the choice of the measurement technique depends on the scale of the RS.

There are many techniques for residual stresses measurements. These techniques can be grouped into three types namely, nondestructive, semi destructive, and destructive. The following bullets enumerate examples of these techniques:

- Nondestructive: X-ray, neutron and synchrotron diffraction, ultrasonic method, and Barkhausen noise method.

- Semi destructive: Hole-drilling, ring-core, and deep-hole methods.

- Destructive: Sectioning compliance techniques, and contour method.

Table 1 lists the measurements techniques with their corresponding type and residual stress scale that is aimed to measured.

For welding residual stresses, the most used techniques are x-ray diffraction (Monin et *al* [5]) and neutron diffraction (Paddea et *al.* [6]).

In this chapter, Section 2 is meant to determine the mean factors that determine the type of welding residual stresses at the surface and to provide a probabilistic analysis of the shape and type of welding residual stress at the surface. Section 3 gives an overview of the effect of residual stresses on fatigue life and lists the different causes, and reasons for residual stresses relaxation. A re-distribution of welding residual stresses after the cutting process was studied in section 4.

2. Weld residual stresses

The welding process is associated with intensive heating and cooling. This process leads to weld effects at the weld toe and root. Some of these effects are residual stresses, micro-cracks, high-stress concentration, and local change in the material properties. These effects have a significant influence on the fatigue life of the welded structure, component, or speciemen. Radaj [7] found that the residual stresses and the geometrical change at the weld toe are the most critical parameters that are determinantal for fatigue. In Manai et *al* [8], Manai [9] and Schijve [10], it was stated that CRS is beneficial for fatigue life, while TRS is harmful and reduces the fatigue life.

2.1 Factors affecting RS distribution

There are many factors that affect the residual stresses distribution in welded structure, component, or specimen. The main factors that determine what residual stresses are present in a welded structure (tensile or compressive at the surface) are listed in the following points:

- The existance of residual stresses in the plates that is resulting from the manufacturing process prior to welding, (before welding takes place).
- The material properties (micro-structures, thermal and mechanical properties) of the weld and base materials.
- The geometry and the shape of the plates being welded.
- The welding procedure (the welding conditions, and the pass sequence in multipass welds).

2.2 Change on the residual stresses distribution

Several factors might modify the residual stresses after welding. These factors can be either during manufacturing process or during the service life of the as-welded structure. Some - but not limited to - of these factors are:

- Surface treatments (peening, TIG dressing, Grinding, etc.), which might cause redistribution of residual stresses due to material removal.
- Cutting process.
- Mechanical loading, such as proof testing or vibration during transportation.
- Thermal treatments.
- Mechanical treatments such as vibrational stress relief.
- In-service repair.
- Crack initiation or loss of the material due to corrosion.

It is highly recommended to consider these factors while assessing the as-welded structure, component, or specimen.

2.3 Type of weld residual stresses

A literature study and a probabilistic analysis of welding residual stresses distribution were performed by Manai et *al* in [8]. They concluded that the probability of occurrence of TRS at the surface is 0.89, substantially, the probability of occurrence of CRS at the surface equal to 0.11. In addition, Manai et *al*. [8] developed a method that determines the shape of welding residual stresses distribution through the thickness direction by knowing only the magnitude of surface welding residual stresses. It was assumed that the residual stresses in the thickness direction at the weld toe has the shape showed in **Figure 2**. Three parameters were used to define this shape which are the magnitude of the surface residual stresses (σ_{RSA}), the maximum magnitude of the subsurface residual stresses (σ_{RSB}), and the depth of the maximum sub-surface residual stresses (D_B).

As welding residual stresses distribution depend on the material properties and the geometry of the plates (thickness of the welded plate), a normalization of the abovementioned parameters that define residual stresses shape is introduced. The magnitude of the residual stresses (σ_{RS}) was normalized by the yield strength (σ_{fy}) of the material, ($\frac{\sigma_{RS}}{\sigma_{fy}}$). The depth of the residual stress(D_B) was normalized by the plate thickness (T), ($\frac{D_B}{T}$). In the section below, a summary of the residual stresses distributions through the thickness direction at the weld toe is stated.

2.3.1 Tensile welding residual stresses at the surface

Based on the probabilistic analysis in Manai et *al.* [8], in the case where welding introduces TRS at the surface, the following conclusions are extracted (regardless of the material and the thickness of the welded plate):

- A linear regression line connecting the magnitude of the surface RS (σ_{RSA}) and the maximum magnitude of the sub-surface RS (σ_{RSB}) is investigated.
- The mean of the depth of the maximum sub-surface residual stresses (see **Figure 2**), D_B, is 35% of the plate thickness with a standard deviation of 24%.
- The surface magnitude residual stresses (σ_{RSA}) follows a log-normal distribution with a mean value of $-0.35\ \sigma_{fy}$ and standard deviation of $0.27\ \sigma_{fy}$.
- the maximum sub-surface residual stresses (σ_{RSB}) follows a log-normal distribution with a mean value of $0.2\ \sigma_{fy}$ and standard deviation of $0.25\ \sigma_{fy}$.

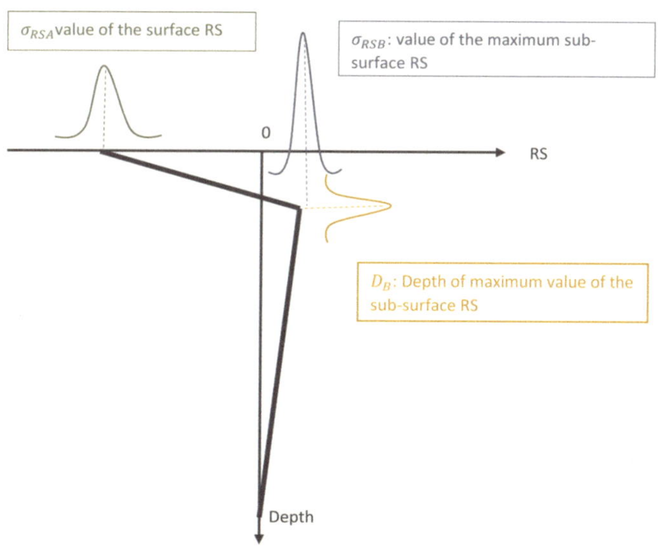

Figure 2.
Schematic presentation of residual shape of RS through the thickness direction.

2.3.2 Compressive welding residual stresses at the surface

In case where welding inroduces CRS at the surface and independent of the material and the thickness of the welded plate, the following points were concluded in Manai et al [8]:

- The residual stresses at the surface (σ_{RSA}) have a mean value of 0.57 σ_{fy} and a standard deviation of 0.12 σ_{fy}.

- The sub-surface residual stresses (σ_{RSB}) have a mean value of 0.59 σ_{fy} and a standard deviation of 0.39 σ_{fy}.

- The normalized depth of the maximum sub-surface residual stresses ($\frac{D_B}{T}$) follows a normal distribution with a mean value of 10% of the normalized thickness varying with a standard deviation of 7%.

3. Effect of residual stresses on fatigue

Welding residual stresses modify the mean stress experienced by a welded joint under the fatigue loading. In case where high TRS is presented at the welded area, it is assumed that cyclic stresses are fully damaging. Therefore, the effect of welding residual stresses must be taken into account when dealing with welded joints. This effect appears in the calculation of crack growth. As $\frac{da}{dN}$ (a is the crack depth and N is the number of cycles) and ΔK_{th} (stress intensity factors range threshold) depend through the stress ratio ($R = \frac{\sigma_{min}}{\sigma_{max}}$, where σ_{min} is the minimum stress and σ_{max} is the maximum stress) on the mean stress (σ_m). TRS increases the mean stress, therefore accelerates crack propagation. Ultimatly, CRS decreases the mean stress, therefore, leads to the retardation of crack propagation.

In the case where TRS is introdcued at the welded area, crack propagation occurs even when the structure, component, or specimen is subjected to external compressive stress cycles.

In Manai [9], simulations of the fatigue life of as-welded structures in the case of the presence of TRS at the surface and in the case of the presence of CRS at the surface were carried out. It was stated that the fatigue life increases with a factor of 4.5 times in the case of CRS occurred at the surface after the welding process in comparison to the case where TRS occurred at the surface (after welding).

4. Weld residual stress relaxation after cutting processes

In order to install mega-welded strucutres such as bridges, off-shores and so on, welding is carried out in the workshop and cutting processes are usually applied. This cutting process is determined depending on the exigences for edge detail and the application Barzoum et *al*. [11] and Cicero et *al*. [12]. The most techniques used for cutting are machine cutting and thermal cutting processes. Moreover, there are additional cutting techniques that has recentrly recognized such as plasma, laser and waterjet. For welded structures, component or specimen a cutting process will introduce relaxation of residual stresses. A few studies emphasized the re-distribution of the welding residual stresses caused by the cutting process [13, 14, 15]. An analysis of data presented in Liang et *al*. [15] was performed. The used material is Q355B with yield strength $\sigma_{fy} = 359$ MPa. The RS was measured using sectioning relaxation strain gauges.

In order to measure the RS in different weld seam lengths, step-by-step sectioning with measurement of the relaxation stress was performed. Residual stresses in specimens with widths vary between 30 mm and 160 mm were measured. In order to better analyze the data, the magnitude of the residual stresses is normalized with the yield strength of the material (σ_{fy}), and the length was normalized by the plate width (w). The following conclusions were extracted:

For longitudinal residual stresses:

- Regardless of the weld seam length, typical distributions of longitudinal welding residual stresses were found after cutting, (See **Figure 3**).
- After cutting, high tensile residual stresses were measured at the middle of the plate width and low tensile at both edges of the specimen.
- The maximal tensile residual stresses (measured at the middle of the welded plated), gradually decrease with the decrease of the width of the cutted specimen.

For transversal residual stresses:

- Regardless of the weld seam length, typical distributions of transversal welding residual stresses were found after cutting.
- After cutting process, low tensile residual stresses were measured at the middle of the weld seam length and a high compressive residual stresses were measured at both specimen edges.
- After the cutting process, 30% of the width of the welded plate contains tensile residual stresses and 60% contains compressive residual stresses (See **Figure 4**).

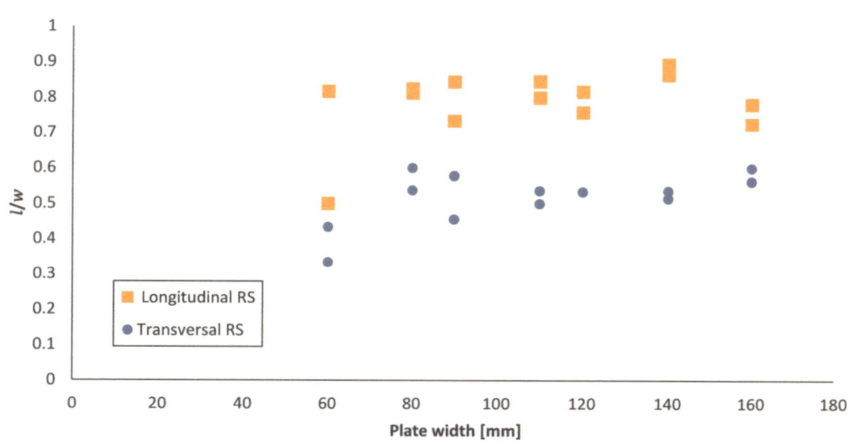

Figure 3.
Plate width as a function of the normalized measured length of CRS after cutting.

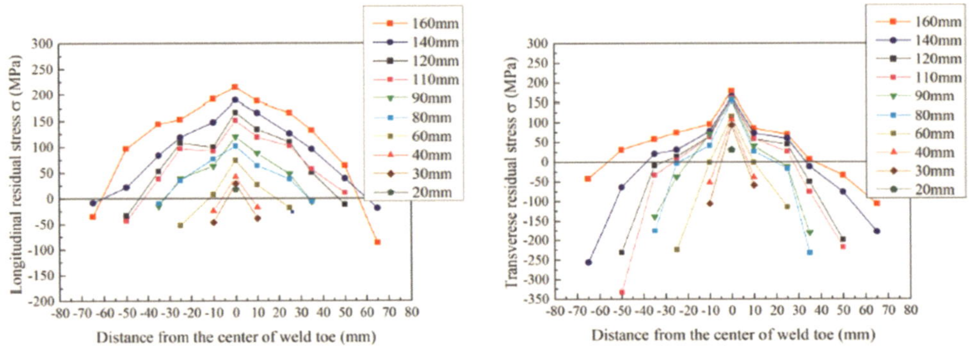

Figure 4.
Measured RS along the weld seam length after cutting, Liang et al. [15].

- Similar to the longitudinal RS, the magnitude of transversal RS decreases gradually with the decrease of the weld seam length (after cutting took place), (See **Figure 3**). In **Figure 3**, l is the length where CRS was measured.

Nomenclature

a	crack depth
D_B	depth of the maximum sub-surface residual stresses
h	plate thickness
l	the length where CRS was measured
N	number of cycles
R	stress ratio
σ_{RSA}	surface residual stresses
σ_{RSB}	maximum sub-surface residual stresses
T	plate thickness
w	plate width
σ_a	stress amplitude
$\sigma_{a,Ex}$	amplitude of the external load
σ_{Ex}	external stress
σ_{fy}	yield strength of the material
σ_m	mean stress
$\sigma_{m,Ex}$	mean of the external load
σ_{max}	maximum stress
σ_{min}	minimum stress
σ_{RS}	residual stresses
$\sigma_{RS,x}$	residual stresses in the x-direction
ΔK_{th}	stress intensity factors range threshold

Abbreviations

RS	residual stresses
TIG dressing	Tungsten Inert Gas dressing
TRS	tensile residual stresses
CRS	compressive residual stresses

Author details

Asma Manai
Chalmers University of Technology, Gothenburg, Sweden

*Address all correspondence to: asma.manai@chalmers.se

References

[1] Rossini N S, Dassisti M, Benyounis K Y, Olabi A G: Methods of measuring residual stresses in components. Materials and Design. 35: 572-588. [Accessed: 2012]

[2] Withers P J, Bhadeshia H K D H: Residual stress: Part 2-Nature and origins. Materials Science and Technology. 17(4): 366-375. [Accessed: 2001].

[3] Donato G H B, Magnabosco R: Modeling and characterization of residual stresses in material processing. Comprehensive Materials Processing: Materials Modeling and Characterization. 2: 219-233. [Accessed: 2014]

[4] Nasir N S M, Razab M K L A, Mamat S, Iqbal M: review on welding residual stress, Journal of Engineering and Applied Sciences. [Accessed: 9 May 2016]

[5] Monin V I, Gurova T, Castello X, Estefen S F: Analysis of residual stress state in welded steel plates by X-ray diffraction method. Reviews on Advanced Materials Science. 19: 172-175. [Accessed: 2009]

[6] Paddea S, Francis J A, Paradowska A M, Bouchard P J, Shibli I A: Residual stress distributions in a P91 steel-pipe girth weld before and after post weld heat treatment. Materials Science and Engineering: A. 534: 663-672.[Accessed: 2012]

[7] Radaj D: Welding residual stress and distortion: calculation and measurement. ISBN 3-87155-791-9 DVS-Verlag GmbH, Düsseldrof, pp. 332–350 [Accessed 2003]

[8] Manai A, Polach R F Al-Emrani M: A probabilistic study of welding residaul stress distribtuion and theri contribution to the fatigue life, Engineering Failure Analysis. [Accessed: 13 November 2020].

[9] Manai A: Effect of weld residual stress in the fatigue strength, Euro Steel 2020 UK. [Accessed: August 2021]

[10] Schijve J: Fatigue of structures and materials, Springer, [Accessed: 2009]

[11] Barsoum Z, Stenberg T, Lindgren E: Fatigue properties of cut and welded high strength steels – Quality aspects in design and production. Procedia Eng 2018;213: 470–6. [Accessed: 20 December 2016]

[12] Cicero S, Garcia T, Alvarez J A, Meizoso AM, Bannister A: Definition of BS7608 fatigue classes for structural steels with thermally cut edges. International Journal of Steel Structures 2016;120:221–231. [Accessed: 2016]

[13] Dattoma V, Giorgi M, Nobile R: On the evolution of welding residual stress after milling and cutting machine. Computer Structures 2006;84:1965–76. [Accessed: 2006]

[14] Zhang J, Dong P: Residual stresses in welded moment frames and implications for structure performance. Journal of Structure Engineering 2000;126: 306–315.

[15] Liang H, Kan Y, Chen H, Zhan R, Liu X, Wang D: Effet of cutting process in the residual stress and fatigue life of welded joint treated by Ultrasonic impact treatement. [Accessed: 16 Jully 2020].

Chapter 10

Use of Hybrid Methods (Hole-Drilling and Ring-Core) for the Analysis of the RS on Welded Joints

Bernardo Zuccarello

Abstract

The hybrid methods (HMs) for the residual stress (RS) analysis, such as the well-known hole-drilling method (HDM) and the ring-core method (RCM), have been widely developed since 80'. They are mechanical methods based on the partial relaxation of the RSs that occur when a proper geometry variation of the analysed component is introduced by drilling a hole (HDM) or a proper annular groove (RCM). The RS computation is performed by measuring the strains relaxed on surface and then by combine properly such measured strains with the influence coefficients previously computed accurately by using a numerical codes that consider the geometry of the particular component to be examined. In such a manner, the HMs can be potentially applied to any RS distribution independently from the cause that have caused them. In more detail, the HMs can be used for the analysis of the RS on welded joint by using both classical welding methods, as MIG or TIG processes, or modern methods as friction stir welding, etc. In the present chapter, after a brief presentation of the theory of the HMs, their application to various cases of welding joints are treated, and the possible limitation are discussed.

Keywords: residual stresses, hybrid methods, hole-drilling method, ring-core method

1. Introduction

The Hole Drilling Method (HDM) and the Ring-Core Method (RCM) are semi-destructive mechanical method widely used for the residual stress (RS) evaluation in mechanical components [1–12]. Each of two methods have a proper application field; respect to the more known HDM, the RCM is characterized by a higher stress relaxation that allows the user to extend the RS analysis up to about 5÷7 mm from the surface of the analysed component, whereas the HDM allows to analyse the RS up to about 1.5–2 mm. Although several applications have been performed in the various industrial fields along with a lot of research works for both the methods, the RCM has never been considered by standardization organisms, whereas the HDM has been standardised by the ASTM since the 80′. In practice, for both the methods, the most

IntechOpen

accurate computational approach used for the evaluation of a generic non-uniform RS distribution, is the so called Integral Equation Method proposed by Schajer (HDM) [6, 7] and Ajovalasit et al. (RCM) [13]; such computational approaches has been included in the ASTM E837-13a standard [14]. However, the very different geometrical variation introduced by the RCM (annular groove) and the HDM (simple hole) involve the presence of different influence parameters on the the RS analysis. In the past the HDM has been used more widely than the RCM thank to its lower damaging and easier experimental procedure. However, the use of the RCM is increased in recent years, especially in Europe, also thank to new modern equipment commercially available [15]. The diffusion of both the methods is corroborated by the high number of special strain gauge rosettes sold in the world by the most famous strain gauge manufacturer as the American MM or the Germany HBM. Also, both the method attract again various research activities having the aim to improve the accuracy of such methods, see as an example the works devoted to the error and uncertainty analysis [16, 17], recently reported in literature.

2. Procedure for the RS evaluation by the HMs

In general the application of the HMs (HDM and RCM) consist on:

1. installation of a special rectangular rosette (see **Figures 1** and **2**) on the surface of the component to be examined;

2. execution of a proper geometry variation (annular groove or hole) by n successive depth increments Δz_i $(i = 1,2 \ldots .n)$;

3. measuring the three relaxed strains (ε_{ai}, ε_{bi}, ε_{ci}) after each groove depth increment (see an example **Figure 1**);

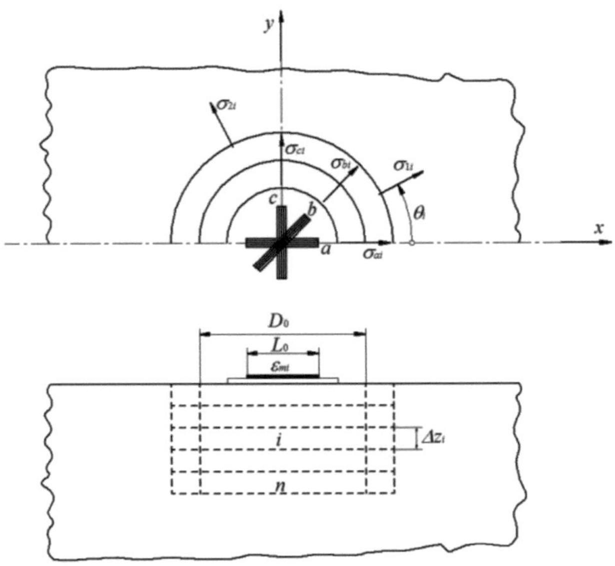

Figure 1.
Ring-Core Method: general notations and groove depth increments.

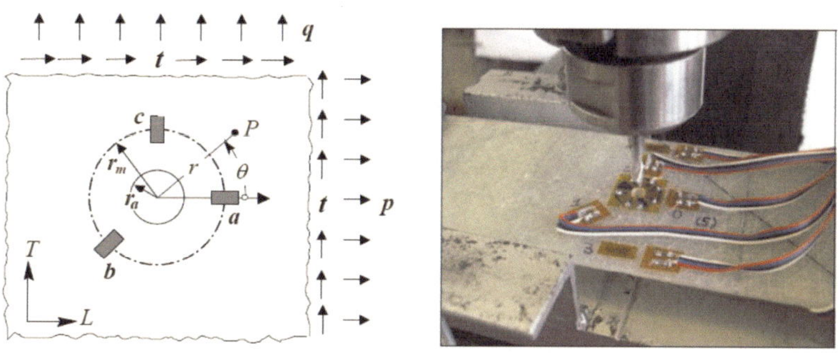

Figure 2.
Typical Experimental setup used by the HDM for the experimental RS analysis.

4. computing the three strain components p_i, q_i, t_i by the simple formulas:

$$p_i = \frac{\varepsilon_{ci} + \varepsilon_{ai}}{2}; \tag{1}$$

$$q_i = \frac{\varepsilon_{ci} - \varepsilon_{ai}}{2} \tag{2}$$

$$t_i = \frac{\varepsilon_{ci} + \varepsilon_{ai} - 2\varepsilon_{bi}}{2} \tag{3}$$

computing the corresponding three stress components P_i, Q_i, T_i (i = 1,2... n) by the iterative relationships (Eqs. 4, 5 and 6):

$$P_i = \frac{1}{a_{ii}}\left[p_i c_E c_\nu - \sum_{j=1}^{i-1} a_{ij} P_j\right] \tag{4}$$

$$Q_i = \frac{1}{b_{ii}}\left[q_i c_E c_\nu - \sum_{j=1}^{i-1} b_{ij} Q_j\right] \tag{5}$$

$$T_i = \frac{1}{b_{ii}}\left[t_i c_E c_\nu - \sum_{j=1}^{i-1} b_{ij} T_j\right] \tag{6}$$

where a_{ij} and b_{ij} (i = 1,2... n, j = 1,2 ... i) are the well known influence coefficients obtained by proper numerical simulations [6, 7, 14], whereas c_E and c_ν represent the elastic corrective coefficients for the actual material characteristics (E,v) of the examined component, that are equal to E/E_0 and $(1 + \nu_0)/(1 + \nu)$ respectively, being E_o and v_o the material characteristics used in the numerical simulations.

Computing the principal residual stresses $\sigma_{1i,2i}$ (i = 1,2... n) and the relative orientation θ_i (i = 1,2... n) as (Eqs. 7 and 8):

$$\sigma_{1i,2i} = P_i \pm \sqrt{Q_i^2 + T_i^2}; \tag{7}$$

$$\theta_i = \frac{1}{2}\arctan\left(\frac{-T_i}{Q_i}\right) \tag{8}$$

In the following **Figure 1** the experimental setup of the RCM is depicted, along with the relative general notations.

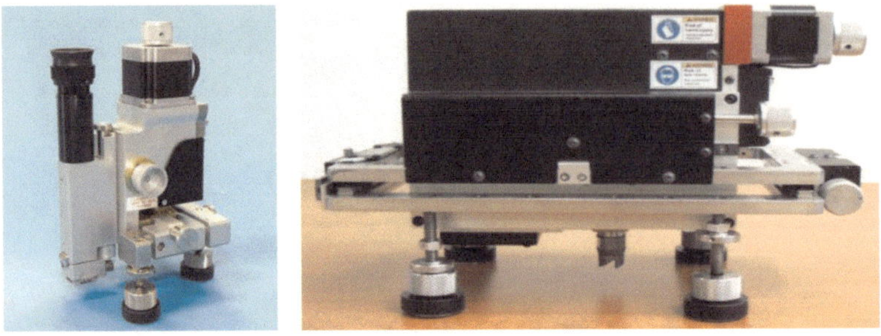

(a) (b)

Figure 3.
Modern equipment used for the RS analysis by the HDM (a) and RCM (b).

It is note that the experimental setup of the HDM differs from that of that of the RCM (compare **Figures 1** and **2**) simply for the different rosette used, a special rosette with separate grids that allows to drill a centered hole for a HDM, instead a stacked rosette that allows to minimize the dimension of the groove as well as to avoid the electrical disconnection of the rosette during the successive execution of the depth increments, performed by an annular cutter.

Recently the application of the HMs is performed by using automatic systems that use high speed drilling methods (obtained by proper miniaturized air turbines) and accurate centering systems that involve optical microscopes and step-by-step electric motors. As an example **Figure 3a** shows a diffuse system used for the practical application of the HDM, whereas **Figure 3b** shows a similar system used for the RCM.

It is important to note that the use of such modern systems allows the user to realize the hole or the groove by limiting significantly the further RS introduced by machining (thanks to the high speed machining), as well as center accurately the rosette thanks to the use of the optical microscope and step-by-step electric motors used to move accurately the cutter in the plane parallel to the component surface (see **Figure 3**).

3. Accurate evaluation of the influence coefficient by numerical methods

For the correct computation of the RS by using the HMs, an accurate evaluation of the influence coefficient a_{ij} and b_{ij} $(i = 1,2 \ldots n, j = 1,2 \ldots i)$ involved into Eq. 4–6 is necessary. As above mentioned, such coefficients are determined by using a numerical code (FEM, BEM etc) by considering the step-by step procedure for RS analysis. In detail, according to Eq. 4–6 the generic coefficient a_{ij} $(i = 1,2 \ldots n, j = 1,2 \ldots i)$ is determined by considering a numerical model that represent the component to be examined with a uniform hydrostatic stress distribution applied only to the *i-th* depth increment of the geometry variation (hole or annular groove) having *j* total depth increments. The b_{ij} $(i = 1,2 \ldots n, j = 1,2 \ldots i)$ coefficients, instead, can be determined numerically by considering a uniform shear stress distribution [13, 14]. Obviously, fixing the total depth, the accuracy of the in-depth RS profile determined by the HMs, increase with the number of depth increments used to reach the total depth although, as it has been demonstrated in [11, 14] if the depth increments are too small then the solution of the inverse problem represented by Eqs.(4–6), became ill-conditioned and also small errors in the strain measurements due to various spurious influence parameters lead to significant errors on the computed stresses. For this reason in general the use of about 6–8 non-uniform depth increments are advised by following the rule clear exposed in [11]. As an example, **Table 1** shows the optimum distribution of depth increments for the RCM by using a total steps included between 4 and 14 (from Ref. [14]); it is seen how the optimised depth increments are relatively larger at the first and specially at the last steps, whereas they are smaller at the intermediate steps. Similar optimized depth increments are provided in [11] for the HDM. Also, **Tables 2 and 3** show the relative optimized influence coefficients a_{ij} and b_{ij} for 8 total depth

Steps N.	Δz_1	Δz_2	Δz_3	Δz_4	Δz_5	Δz_6	Δz_7	Δz_8	Δz_9	Δz_{10}	Δz_{11}	Δz_{12}	Δz_{13}	Δz_{14}
4	1.00	0.90	1.00	2.10										
6	0.70	0.60	0.55	0.60	0.80	1.75								
8	0.60	0.45	0.40	0.40	0.45	0.50	0.70	1.50						
10	0.55	0.40	0.35	0.30	0.30	0.35	0.40	0.45	0.65	1.25				
12	0.50	0.30	0.30	0.25	0.25	0.25	0.25	0.30	0.40	0.45	0.60	1.15		
14	0.45	0.30	0.25	0.25	0.25	0.20	0.20	0.25	0.25	0.30	0.30	0.40	0.55	1.05

Table 1.
Optimum distribution of depth increment for RCM with total groove depth of 5 mm.

n=1	.0232							
n=2	.0424	.0284					$-a_{n1}$	
n=3	.0573	.0462	.0288					
n=4	.0686	.0579	.0449	.0265				
n=5	.0766	.0662	.0543	.0403	.0224			
n=6	.0821	.0719	.0604	.0476	.0337	.0176		
n=7	.0878	.0778	.0667	.0547	.0424	.0306	.0284	
n=8	.0897	.0798	.0688	.0571	.0452	.0341	.0396	.0109
	i=1	i=2	i=3	i=4	i=5	i=6	i=7	i=8
n=1	.0214							
n=2	.0385	.0224					$-b_{n1}$	
n=3	.0512	.0395	.0237					
n=4	.0658	.0554	.0414	.0217				
n=5	.0785	.0668	.0543	.0409	.0174			
n=6	.0892	.0777	.0652	.0505	.0358	.0124		
n=7	.1093	.0959	.0813	.0663	.0514	.0360	.0179	
n=8	.1198	.1066	.0920	.0760	.0605	.0451	.0478	.0106

Table 2.
Influence coefficients relative to a thick component (>100 mm) with standard dimension of the core-rosette assembly (Do=14 mm, Lo = 5 mm).

n=1	.0190							
n=2	.0372	.0246					$-a_{n1}$	
n=3	.0517	.0415	.0257					
n=4	.0632	.0534	.0411	.0238				
n=5	.0720	.0622	.0507	.0373	.0203			
n=6	.0782	.0685	.0574	.0449	.0315	.0161		
n=7	.0854	.0757	.0648	.0529	.0410	.0294	.0265	
n=8	.0883	.0785	.0677	.0560	.0443	.0333	.0383	.0102
	i=1	i=2	i=3	i=4	i=5	i=6	i=7	i=8
n=1	.0170							
n=2	.0335	.0186					$-b_{n1}$	
n=3	.0488	.0369	.0192					
n=4	.0620	.0523	.0403	.0175				
n=5	.0744	.0646	.0528	.0396	.0149			
n=6	.0862	.0755	.0633	.0495	.0348	.0104		
n=7	.1062	.0942	.0806	.0659	.0510	.0354	.0177	
n=8	.1190	.1059	.0913	.0755	.0598	.0449	.0475	.0094

Table 3.
Influence coefficients relative to a thin component (<15 mm) with standard dimension of the core-rosette assembly (Do =14 mm, Lo = 5 mm).

increments, with refer to a thick (**Table 2**) and a thin component analysed by the RCM (**Table 3**).

Although for standard dimensions of the geometry variation properly introduced by the HDM, the relative influence coefficients are provided by the same ASTM standard [13], and similarly accurate evaluation of the coefficients for the RCM are provided in [14], in particular practical cases where the analysed component cannot be considered as an infinite plate subjected to a plane stress field, the relative influence coefficient have to be determined by specific numerical simulations, that consider the exact geometry of the component to be examined. As occur in many inverse problems, the evaluation of the RS from the strains relaxed on surface after each depth increment is influenced by several influence parameters that can introduce into the evaluated RS a significant uncertainty, so that an accurate RS evaluation needs to a reliable estimation of the corresponding uncertainty. For this reason, interesting study are reported in literature that deal with the evaluation of the accuracy and of the uncertainty that affect the principal RS computed by using both the HDM [16] and the RCM [17]. Such interesting works contain the formulas that the user can use to an accurate estimation of the uncertainty of the RS after the analysis and the correction of the main influence parameters, as the thermal effects due to machining, the zero depth offset, the rosette eccentricity, the stresses induced by machining and so on. Synthetically, such studies have been demonstrated that using modern apparatus as that shown in **Figure 3**, in good experimental conditions the uncertainty of the principal RS computed by the RCM in general falls in the range 10–55 MPa [17]

DOI: http://dx.doi.org/10.5772/intechopen.102051

moving from the first to the last steps; similarly for the HDM the mean uncertainty is in general equal to about ±25 MPa [16].

4. Application of the HDM for the analysis of the RS in a welded joint

As above mentioned, the HMs can be advantageously applied for the RS analysis in welded joints between metal components. As an example, in the following the use of the HDM for the RS analysis on a I type welded joint between two stainless steel plates (AISI 304) is exposed in detail, and the results are compared with that computed numerically by using a FEM procedure accurately accomplished by using accurate thermal analysis of the welding process considered as a progressive process constituted by successive welding passes [18].

Figure 4 show the geometry of the joint considered and the relative FE model used for the RS analysis, accomplished by considering that the weld bead is formed by 5 successive passes that are simulated by changing the material properties according the temperature profile relieved experimentally by proper thermocouples installed near the weld bead.

The distribution of the RS along the y direction, computed by the numerical simulation process implemented, is reported in **Figure 5**. As expected, it is seen how

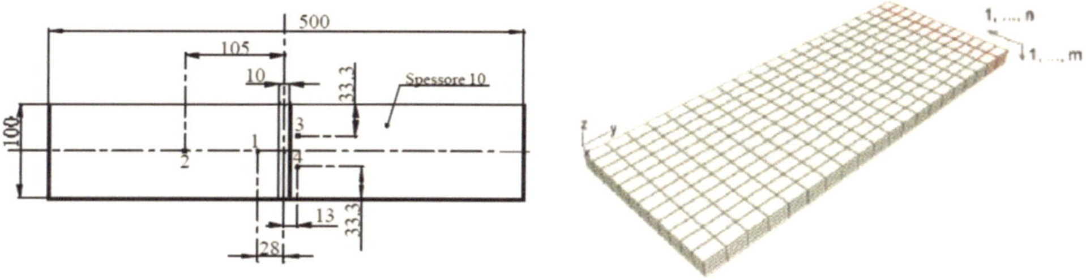

Figure 4.
Geometry of the welded butt joint considered (a) and relative FE model used for the numerical simulation of the welding process.

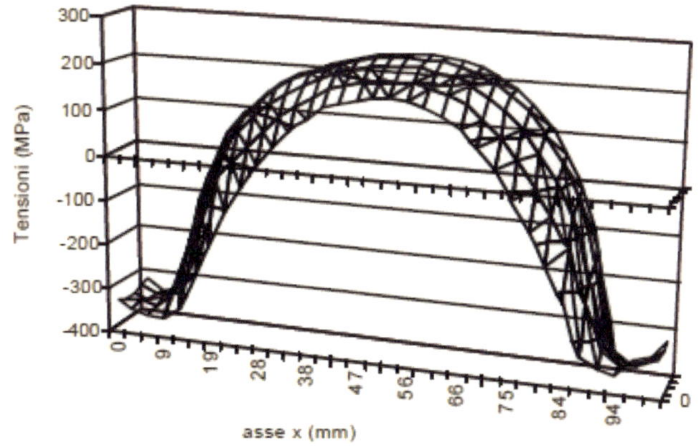

Figure 5.
Distribution of the RS along y direction (σy) computed by accurate FE simulations.

the maximum RS occur at the central zone of the weld bead, with values that fall in the range 200–205 MPa (about 60% the material yield stress).

From **Figure 5** it is seen how the RS vary significantly along the x direction, whereas they vary lightly in the y direction. For this reason the comparison between the RS computed numerically and that estimated experimentally by the HDM is carried out by considering the points 3 and 4 in **Figure 4**, i.e. by considering point near the weld bead in which the geometry of the surface is not significantly modified by the welding process. In detail, the following **Figures 6** and 7 show the comparison between the in-depth RS profile computed by numerical simulations and the RS distribution near the surface computed by using the HDM. The analysis of such figures show how the profile computed numerically is always inside the uncertainty range of the RS computed by means of the HDM, corroborating the accuracy of such an experimental method.

5. Analysis of through-thickness RS in aluminium FSW butt joints

The HMs can be also used for the analysis of the through-thickness RS in friction stir welding on metals. As an example in [19] the HDM have been advantageously used for the analysis of the RS in FSW butt joints between plate made by aluminium. In detail, using the HDM the in-depth RS distribution in the zone close to the tool shoulder border of the joint advancing side, has been accurately determined by considering also different aluminium alloys (AA1050 O, AA2024 T4, AA6982 T6, AA7075

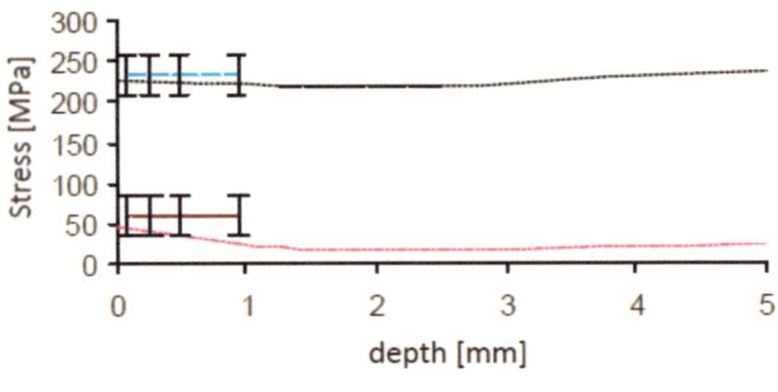

Figure 6.
*Comparison between the RS computed by the HDM in point 3 of **Figure 4** and the RS profile obtained by accurate numerical simulations.*

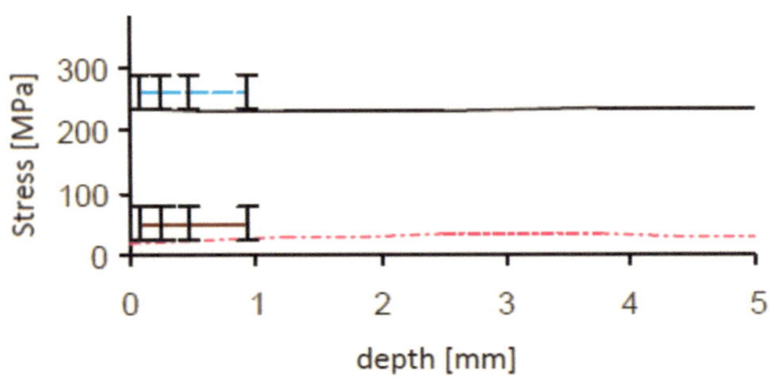

Figure 7.
*Comparison between the RS computed by the HDM in point 4 of **Figure 4** and the RS profile obtained by accurate numerical simulations.*

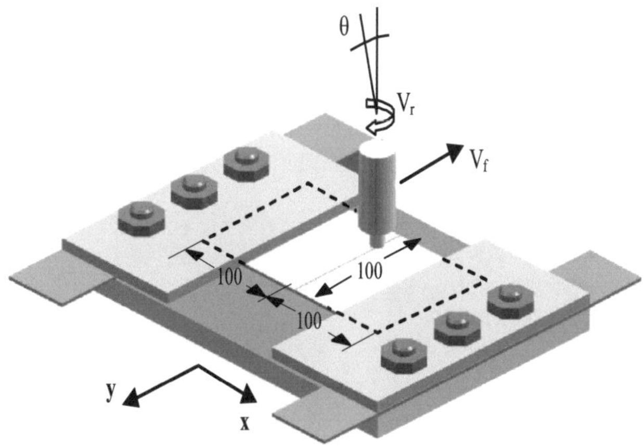

Figure 8.
Sketch of the FSW butt joint considered, during the welding process.

T6) and different specific thermal contribution (STC) that characterizes the velocity of the welding process. **Figure 8** shows the sketch of the FSW butt joint considered, during the welding process.

In detail, it is considered the case in which the FSW of a butt joint is obtained by inserting a specially designed pin, rotating with velocity V_r into the adjoining edges of the sheets to be welded, and then moving it all along the joint with velocity V_f (**Figure 8**). The realization of the weld bead leads to significant changes of the mechanical properties and of the microstructure of the material near the weld bead. In detail, as it shown in **Figure 9**, it is possible to distinguish the following areas:

a. Parent material, i.e. the remote material that is not subjected to the heat flux and conserves the original mechanical properties and microstructure;

b. Heat affect Zone (HAZ), i.e., the material subjected to a thermal cycle with a consequent microstructure and mechanical properties modification, although without any plastic deformations;

c. Thermo-Mechanical Affect Zone (TMAZ), i.e. the zone where the material is subjected to plastic deformation and significant heat flux with consequent variations of the mechanical properties and of the microstructure (no recrystallization phenomena);

d. Nugget, i.e. the recrystallized area in which the original grains are replaced with fine and equiaxed recrystallized grains with nominal dimensions of few micrometers.

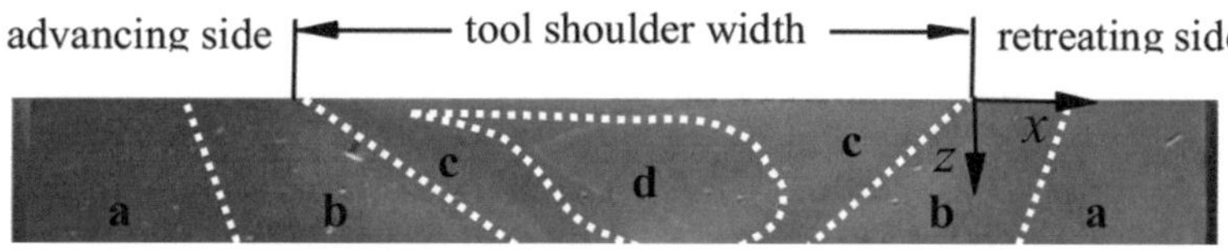

Figure 9.
Material microstructures in a typical transversal section of an aluminium AA6082-T6 FSW butt joint.

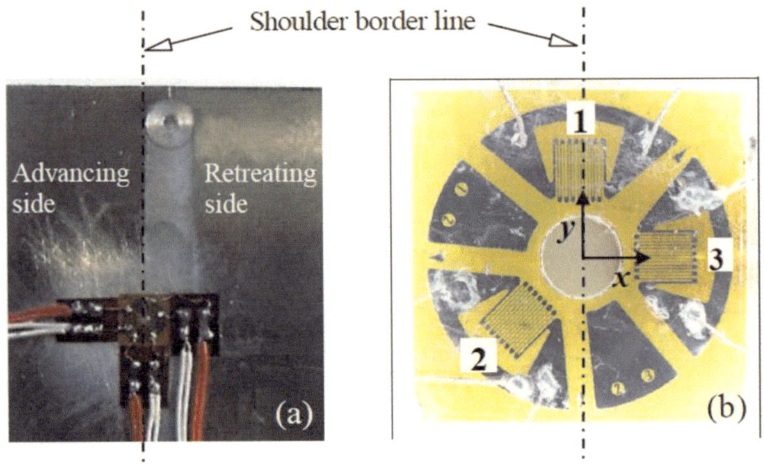

Figure 10.
(a) Rosette installation on a welding specimen and (b) details of the rosette.

In order to analyse the most relevant residual stress distribution that occur through the thickness of the welding joint, a special rectangular strain gauge rosette type MM EA-062RE-120, have been bonded into the welded specimens, with the 1 grid aligned with the tool shoulder border of the joint advancing side (see **Figure 10**), that is the zone of the welding seam where the maximum values of the RS are expected [19].

After the strain gauge rosette installation, it has been linked to a proper multichannel strain gauge monitoring machine (HBM UPM 100), and the three strains relaxed after each successive increase of the hole depth, have been collected by using a software [20] properly developed by the maker of the milling machine shown in **Figure 3**. In detail, such an advanced machine have an high speed air turbine and a special microscope, that permit respectively to limit the residual stresses due to machining as well as the rosette eccentricity. Also, the hole has been drilled by 25 successive steps of 0.1 mm (total hole depth of 2.5 mm), by using a tungsten carbide mill. In accordance with the ASTM standard [14], as well as with the good practice indications reported in Refs. [11, 12, 16], the measuring procedure have been performed in such a way to minimize the spurious residual stress induced by the hole drilling.

5.1 Residual Stress evaluation

The evaluations of the residual stresses through the thickness of the examined welded bead, have been carried out by using the calculation process exposed in chapter 2, and the discrete results have been fitted with simple polynomial functions. Then, it permits to perform the uniformity test prescribed by the ASTM standard [14] and then to compute uniform or non-uniform residual stresses by the Integral Method exposed in detail in the previous chapter 2. As an example, considering the case of the aluminum AA1050-O FSW joints with medium specific thermal contribution (MSTC), **Figure 11** shows the typical curves of the three relaxed strains (ε_1, ε_2, ε_3).

From **Figure 11** it is seen how, the surface relaxed strains take initially typical negative values due to the relaxation of positive RS; successively they increase in

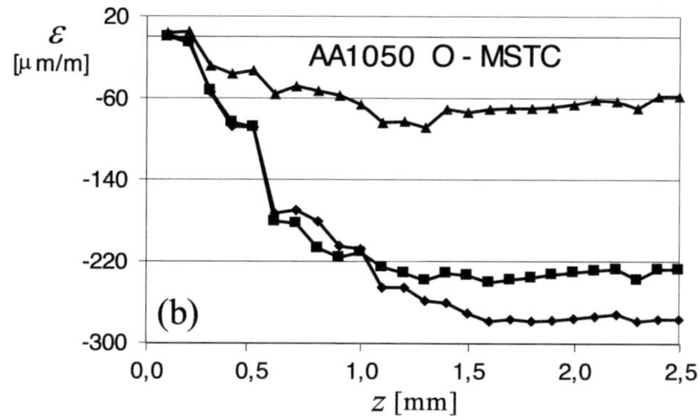

Figure 11.
Typical relaxed strain curves relative to AA1050-O aluminum joints, with MSTC.

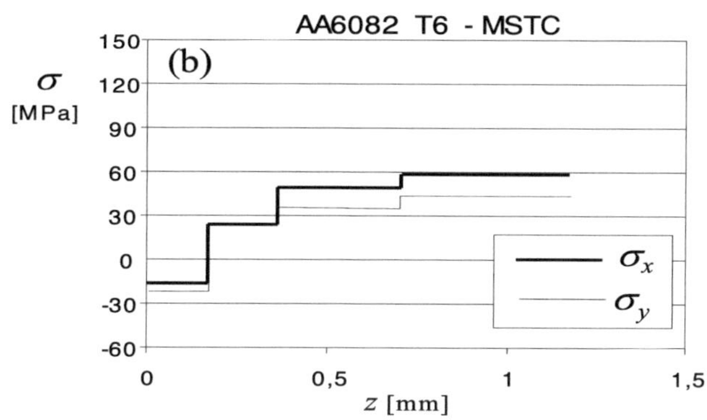

Figure 12.
Typical residual stresses relative to AA6082-T6 joints, with MSTC.

module (up to 140 to 280 μm/m) until hole depth of 1.5–2.0 mm, at which they take a flat trend (saturation of the surface relaxation phenomenon). Similar relaxation curves have been acquired for all the aluminium alloys considered and for all the manufacturing conditions (STC) studied.

As an example, **Figure 12** shows the RS computed by considering the case of AA6082 T6 FSW joints with medium specific thermal contribution (MSTC). Similar trends have been obtained for the other cases examined.

Figure 12 permits to observe that in all the three examined cases (AA6082 T6, AA1050-O, AA2024-T4), both the main RS components, acting on x and y directions respectively, take similar values and trends (increasing with depth). In detail, they range from negative value (on surface, from −8 MPa to −30 MPa), to positive values of about 60 MPa for MSTC (40 MPa and 90 MPa for LSTC and HSTC respectively) at depth of about 0.5 ÷ 1 mm. Therefore, for this aluminum alloy in the STC examined range (from LSTC to HSTC) the maximum stresses decrease if the STC value increase. Very interesting is the comparison of RS evaluated for the different materials and for the three different STC levels considered, which shows that unlike the traditional welded joints, in the examined FSW butt joints, the maximum RS do not occur on joint surface but inner to the weld at a depth from surface that varies from 0.5 to 1.25 mm. In more detail, for the aluminum alloys commonly used in the structural design (AA6082 T6 and AA2024-T4) the surface residual stress σ_s takes in practice

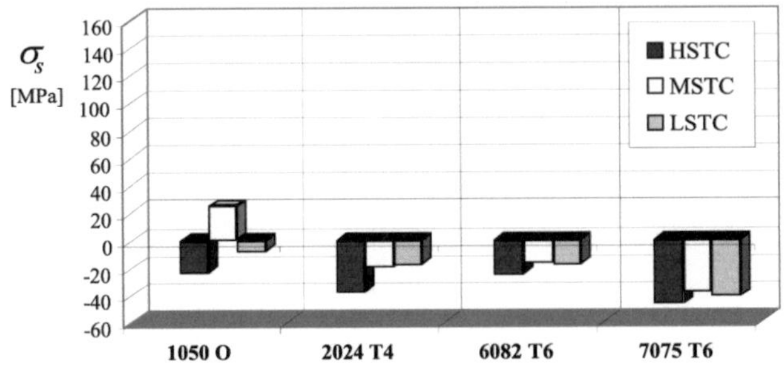

Figure 13.
Residual stresses measured on surface of the FSW joints.

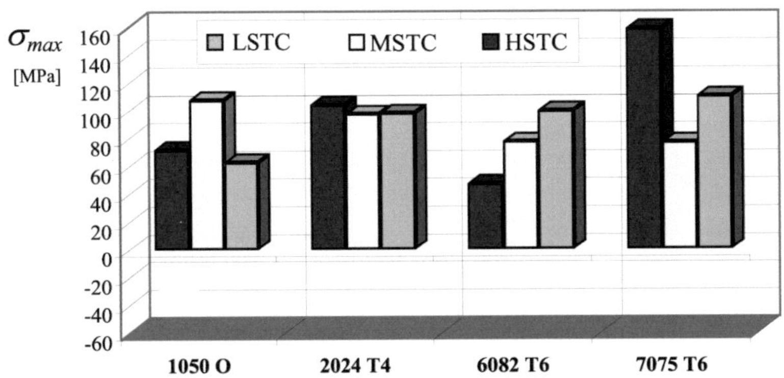

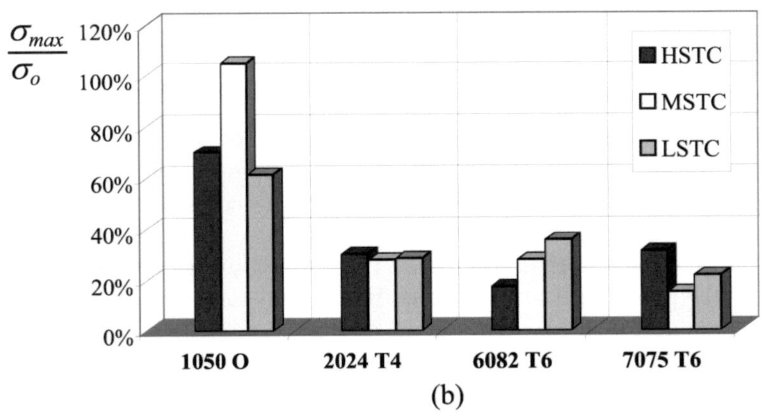

Figure 14.
Maximum residual stresses measured along the thickness of the joints: (a) absolute values and (b) percentage of the parent material yield stress value.

always negative values ranging from −20 to −40 MPa (see **Figure 13**), i.e. it takes always values less than 20% of the yielding stress σ_o.

Additionally, the HDM has allowed to highlight that the maximum residual stresses σ_{max} varies from about 50 to 150 MPa (see **Figure 14a**), i.e. from about 15–30% the yielding stress, depending to the particular material type and the STC level (see **Figure 14b**).

Finally, it is important to note that if the use of the HMs leads to significant plasticity effects at the bottom of the geometry variation (hole or groove) due to high RS levels, than values of the principal RS computed by the above exposed procedure should be corrected by using the procedure reported in literature [21, 22], to which the reader is addressed.

6. Conclusions

The above reported applications of the HDM show that in principle the HMs methods can be advantageously used for the experimental RS analysis in welded joints. As above mentioned the HDM allows to detect the RS until depth of about 1.5–2 mm under the component surface, whereas the RCM allows to reach depths until 4–5 mm. In detail, the accurate use of such methods need the use of modern systems equipped with a high speed machining cutter to minimize the RS due to machining, and a proper microscope to minimize the rosette eccentricity error. The main limitation of the HMs are related to the possible surface curvature of the welded bead and to the particular geometry of the joint. Obviously, in case of non-plane surface and non-plane welded components, the influence coefficients should be computed properly by specific and accurate numerical simulations. In general, to limit the experimental work it is convenient to apply the HMs directly into the zone where other independent considerations (as coarse numerical simulation or theoretical consideration s etc.)

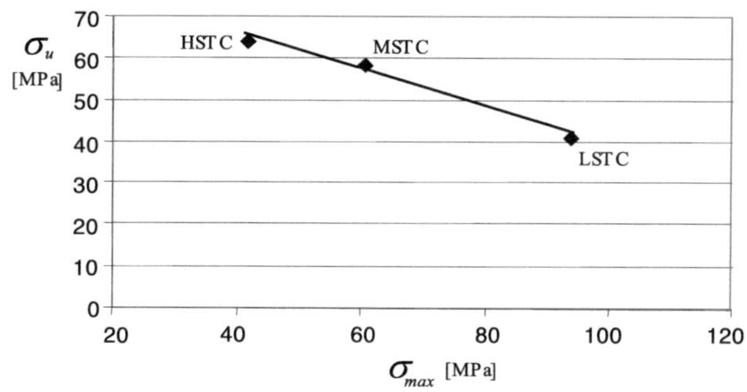

Figure 15.
Ultimate tensile stress of AA6082-T6 joints, with HSTC, MSTC and LSTC.

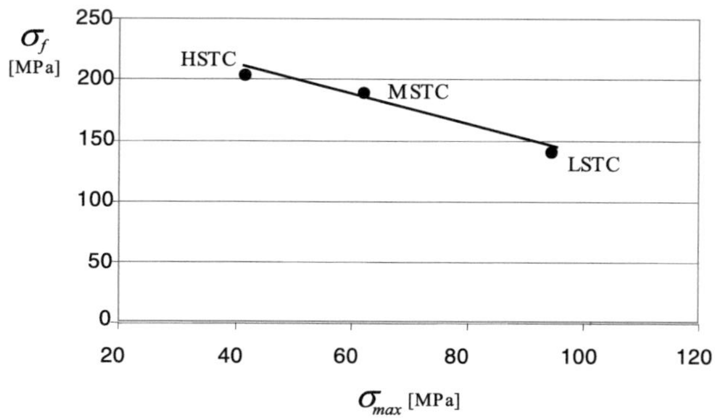

Figure 16.
Fatigue strength of AA6082-T6 joints, welded with HSTC, MSTC and LSTC.

indicate as the more stressed zone. As an example, as above clearly shown, in case of a butt joint the more stressed zone coincides with the zone near the central point, whereas in case of FSW butt joint the more stressed point is that close to the tool shoulder border in the advancing side of the joint. In such a manner the use of a unique experiment allows the user to evaluate the maximum residual stress that influence not only the static strength but also the fatigue strength of the welded joint. As an example **Figures 15 and 16** shows the correlation between the maximum RS (σ_{max}) computed by applying the HDM and the static (**Figure 15**) and the fatigue (**Figure 16**) strength of an AA6082 – T6 FSW butt joint, varying the STC.

It is seen a linear relationship between the maximum RS (σ_{max}) and the mechanical strength. Such a linear relationship is in accordance with the Goodman criterion commonly used in the mechanical design to estimate the mean stress effect on the metal fatigue resistance. Moreover, the generic good accordance relieved in the above exposed application (as in many other application here not mentioned for brevity sake), permits to establish that the HMs, widely employed in the industrial field due to its simplicity and low cost, allows the user an accurate estimation of the maximum residual stresses that occur in an a generic welded joint realized by traditional techniques (TIG, MIG etc.) or by innovative techniques (friction stir welding etc). Obviously, in general the HMs can be advantageously used to optimize the parameters that govern the welding process in order to minimize the maximum RSs and, consequently, the mechanical properties of the examined welded joint.

Author details

Bernardo Zuccarello
Dipartimento dell'Innovazione Industriale e Digitale (DIID), Università di Palermo, Palermo, Italy

*Address all correspondence to: bernardo.zuccarello@unipa.it

References

[1] Kelsey RA. Measuring non-uniform residual stresses using the hole-drilling method. Proceedings of the Society for Experimental Stress Analysis. 1956;**14**: 181-194

[2] Nickola WE. Practical subsurface residual stress evaluation by the hole-drilling method. In: Proceedings of the 1986 SEM Spring Conference. 1986. pp. 47-58

[3] Bijak-Zochoski MA. Semi destructive Method of Measuring Residual Stresses. VDI-Beritche. 1978;**313**:469-476

[4] Niku-Lari A, Lu J, Flavonet JF. Measurement of Residual Stress Variation with Depth by the Hole Drilling Method. Experimental Mechanics. 1985;**25**:175-185

[5] Flaman MT, Manning BH. Determination of the Residual Stress Variation whith Depth by the Hole Drilling Method. Experimental Mechanics. 1985;**25**:205-207

[6] Schajer GS. Measurement of non-uniform residual stresses using the hole-drilling method Part. I and II. Journal of Materials Engineering Technology. 1988; **110**(4):338-343

[7] Zuccarello B. Analisi delle Tensioni Residue con il Metodo della Cava Anulare. Cosenza: Proceedings of the XXIII Aias National Conference; 1994. pp. 87-91

[8] Zuccarello B. Optimization of Depth Increment Distribution in the Ring-Core Method. Journal of Strain Analysis for Engineering Design. 1996;**31**(4):251-258

[9] Schajer GS. Non-uniform Residual Stresses Measurement by the Hole-drilling Method. Strain. 1992;**28**(2):19-22

[10] Petrucci G, Zuccarello B. A New calculation procedure for non-uniform residual stress analysis by the hole drilling method. Journal of Strain Analysis. 1988;**33**(1):27-37

[11] Zuccarello B. Optimal calculation steps for the evaluation of residual stress by the incremental hole-drilling method. Experimental Mechanics. 1999;**39**(2): 117-124

[12] Schajer GS, Altus E. Stress calculation error analysis for incremental hole drilling residual method. Journal of Engineering Materials and Technology. 1996;**118**(1):120-126

[13] Ajovalasit A, Petrucci G, Zuccarello B. Determination of Non-uniform Residual Stresses Using the Ring-Core Method. Transactions of the ASME - Journal of Engineering Materials and Technology. 1996;**118**(2):224-228

[14] ASTM. Standard test method for determining residual stresses by the hole-drilling strain gauge method. In: International Designation E 837-13a, United States. 2013. p. 16

[15] VALENTINI E, BENINCASA A, BERTELLI L. An automatic system for measuring residual stresses by Ring-core method, Italian Stress Analysis Association, 40th National Congress. Palermo, Italy: University of Palermo; 2011

[16] Scafidi M, Valentini E, Zuccarello B. Error and Uncertainty Analysis of the Residual Stresses Computed by Using the Hole Drilling Method. Strain. 2011; **47**(4):301-312

[17] Zuccarello B. Error and uncertainty analysis of non-uniform residual stress evaluation by using the ring-core

method. Experimental Mechanics. 2016; **56**(9):20-35

[18] Cappello F, Carlisi G, Zuccarello B. Valutazione Numerica e Verifica Sperimentale dello Stato Tensionale in una Giunzione Saldata di Testa. In: Proceedings of the XXXI National Aias Conference. 2002. pp. 18-21

[19] Fratini L, Zuccarello B. An analysis of through-thickness residual stresses in aluminium FSW butt joints. International Journal of Machine Tools and Manufacture. 2006;**46**:611-619

[20] Valentini E. An automatic System for Measuring Non –Uniform Residual Stress by the Hole Drilling Method. Turin: XIII IMEKO World Congress; 1997. pp. 5-9

[21] Petrucci G, Zuccarello B. Effect of plasticity on the residual stress measurement using the groove method. Strain. 1996;**32**:97-104

[22] Beghini M, Bertini L, Raffaelli P. An account of the plasticity in the hole drilling method for residual stress measurement. Journal of Strain Analysis. 1994;**30**:227-233

Chapter 11

A Brief Study of Unconventional Variants of GMAW Welding: Parameters, Weld Bead, and Microstructures

Lino A.S. Rodrigues, Pedro P.G. Ribeiro, Ednelson da S. Costa, Tárcio dos S. Cabral and Eduardo de M. Braga

Abstract

The GMAW (Gas Metal Arc Welding) process is an electric arc welding technique widely used around the world due to its ease of use, low equipment cost and, mainly, due to the high deposition rate, the quality of the metal of solder, which makes it versatile and susceptible to modification. Thus, variants such as CW-GMAW (Cold Wire–Gas Metal Arc Welding), DCW-GMAW (Double Cold Wire–Gas Metal Arc Welding), and HW-GMAW (Hot Wire–Gas Metal Arc Welding) emerged from the conception of small adaptations to the original process that ended up generating better and more adjusted results than GMAW. Thus, variations of some parameters will be shown and their respective effects on the weld bead geometry, dilution, penetration, deposition rate, in addition to the effects on macro and microstructure. This provides the possibility of using the variants in different types of applications in the industry in general. Where the application in narrow 4 mm chamfer has already been observed, reduction of residual stresses, increase in fatigue resistance and coatings with special alloys.

Keywords: GMAW, variants, CW-GMAW, DCW-GMAW, HW-GMAW, welding metallurgy

1. Introduction

In a general context, the welding area currently develops on the conceptual and technological foundations of Industry 4.0, as well as the latter actively collaborates to advance the former. Welding joining processes have never presented so many changes in their techniques as in recent years. This is mainly due to the insertion of arc welding in additive manufacturing [1, 2], as robotic systems have sought to improve the manufacture of parts and components by deposition or coating of flat and tubular surfaces for any type of materials. Whether carbon steel [3] or special alloys [4, 5]. However, the difficulty of implementation and the cost of additive manufacturing favor more traditional processes such as the electric arc to remain for longer acting as the front line of the metalworking industry in the welding segment.

Thus, processes such as GMAW (Gas Metal Arc Welding) and FCAW-G (Flux Cored Welding with Gas Protection), among others, still remain firm, being used in the manufacturing and heavy assembly industries such as shipbuilding, oil industry, in addition to construction of structures and pipelines for the power generation industry, for example. Thus, it is known that the mentioned processes have high deposition rates, weld metal quality, in addition to versatility in their applications, which consolidates them in the market. However, due to the excessive need to increase productivity, many variants have emerged in order to assist in this procedure.

Currently, there are some aspects of the GMAW process, with the Cold Wire–Gas Metal Arc Welding (CW-GMAW), the Double Cold Wire–Gas Metal Arc Welding (DCW-GMAW), and the Hot Wire–Gas Metal Arc Welding (HW-GMAW). CW-GMAW and DCW-GMAW welding contributed significantly to increased productivity [6, 7]. CW-GMAW welding allowed for narrow bevel welding with 4 mm gap [8]. Furthermore, it is responsible for reducing the level of residual welding stresses [9] and increasing fatigue strength [10]. On the other hand, the HW variant presents as its main characteristic the increase in productivity, with the possibility of variation in penetration [11]. These variants have the possibility to achieve deposition rates ranging from small percentages from 10% to more than 100% of extra molten metal, also influencing the formation of microstructures such as acicular ferrite, which contribute to increase the properties, mainly mechanical, of the weld metal. In this context, some variants of the GMAW process will be presented below, some of which have already been tested, and others are still being tested for possible field applications, thus showing the importance of these welds for the academy and the manufacturing industry, assembly, and maintenance.

2. The GMAW process and its variants

The industry's anxiety to promote systems with high levels of productivity generated the need for adaptations in consolidated techniques such as welding with the GMAW process (also known as MIG—Metal Inert Gas and MAG—Metal Active Gas), which ended up being an important source from adaptations to consolidating alternatives that may well be better developed by companies and entering the market as proposals to increase melting and metal deposition rates. Therefore, there are many innovations for the advancement of welding in the industry, with a large number of different variations of the more traditional processes, such as the MIG/MAG process, where it can be said that the GMAW with double wire is the technical variant that most manufacturers have developed commercially. The authors [12] cite more variant techniques using this same conventional process (GMAW) as a basis, thus generating hybrid combinations such as MIG-Laser and Plasma-MIG systems. Thus, the use of GMAW welding is also being extensively studied in manufacturing by additive manufacturing, always aiming at the same objectives.

2.1 Cold wire (CW-GMAW) and double cold wire (DCW-GMAW)

The first variation of GMAW welding to be mentioned is the CW-GMAW, where the term CW – cold wire, consists of the addition of a non-energized wire that is inserted directly into the arc, the weld pool, or the transition zone, to increase the

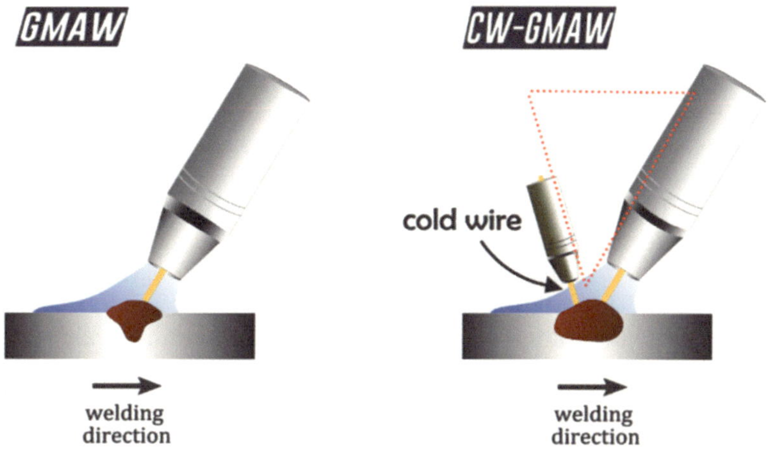

Figure 1.
Wire feeding scheme of the GMAW and CW-GMW processes.

rate of molten metal and subsequently the rate of material deposition, using only an electric arc (**Figure 1**). This helps to reduce the energy imposed on the part, in addition to the possibility of reducing the number of passes for filling the chamfer, decreasing dilution, decreasing the Heat Affected Zone (HAZ), and the application of coatings on surfaces, being able to be used in manual, semi, and fully modes automated. Thus, due to the similarity of the GMAW process with the FCAW (Flux Cored Arc Welding), the CW-FCAW (Cold Wire–Flux Cored Arc Welding) variant has already been tested for joining sheets in the naval industry and has been shown to have high potential for use in the assembly and manufacturing industry [13, 14]. Both were tested when welding parts in the flat position.

Preliminary work using CW-GMAW has been developed since the early 2000s. But the variant has been consolidated in the past decade with several applications that will be dealt with below. Primarily used to fill V-chamfers (with different opening angles), but with low cold wire feed rates. It should be noted that the feed rate of the cold wire is based on the feed rate of the electrode wire responsible for the electric arc. That is, the cold wire feed rate ratio is a percentage of the electrode feed rate and is called the electrodeless feed rate ratio (%), defined as follows:

$$R_W = \frac{W_s}{E} \tag{1}$$

where Ws is the cold wire feed rate in m/min, and E is the electrode feed rate in m/min. This parameter is used to decide the quantity of all the variants mentioned in this chapter, whether the variants with cold wires or with hot wires. Thus, the initial rates corresponded to low values from 10–60%, but currently some works such as [4] have already demonstrated the possibility of using rates of up to 140% and the use of CW-GMAW with pulsed current. Another important factor is the diameter of the cold wire, which also bases your choice on the electrode wire. In addition, it should be noted that it is possible to work with the possibility of mixtures of wires (electrode + cold) causing the formation of weld metals with the most varied chemical compositions and applications for joining and coating dissimilar materials, more precisely for wear-resistant coatings.

Otherwise, the equipment for the application of the CW-GMAW technique also has a relatively low cost in terms of the necessary adaptations to carry out the welding.

Bearing in mind that you only need an extra power head and a torch adapted to inject the cold wire at the desired location with coupling to the electrode wire welding torch.

Relevant works have been produced over the last few years, such as [15, 16], which studied several factors such as wire feed rate, pulsed current, energy efficiency, and their influence on electric arc stability, metallic transfer modes, bead geometry, as well as the possibility of applying this type of welding in narrow gap of 4 mm [8, 16]. Another study [17] evaluated the possibility of using the CW-GMAW by varying the electrode polarity in negative (DCEN—direct current electrode negative) and positive (DCEP—direct current electrode positive). In this way, these works help to consolidate the CW-GMAW process as a suitable process as an alternative for the implementation of high productivity with less energy to melt more metal.

This is confirmed by the various applications that have already been carried out, such as: [18] CW-GMAW welding was used to weld high-strength armor steel plates used for vehicle armoring; [19] performed steel welding for oil and oil pipelines (API X80) and [20] tested the process in automotive applications. On the other hand, the decrease in the penetration of the molten metal and the dilution that occur end up favoring the use of CW-GMAW in coatings, which is what was done by [7], who used this type of welding covering austenitic stainless steel plates for cobalt-chromium alloys (Stellite 21) and [21], which coated AISI-SAE 1020 carbon steel sheets with a nickel superalloy ER NiCrMo-4 (Hastelloy). On the other hand, other works compared the welded joint by GMAW and the CW-GMAW variant, the authors [10] joined sheets of naval steel and subjected the welded joint to fatigue cycles, noting the excellent resistance of the material, and the results of the variants are better than the original process, and also [9] performed similar work welding marine steel and observed that the CW-GMAW process helps to reduce residual stress peaks by up to 100 MPa, after measuring the sheets before and after welding, using X-ray diffraction and acoustic birefringence methods. Finally, many of the works mentioned also confirmed that due to the lower imposed heat transferred to the base metal, there is a decrease in the HAZ.

However, the DCW-GMAW variant uses the same idea as the CW-GMAW variant, using an energized electrode wire, but with the insertion of two non-energized wires (colds), hence the term DCW–double cold wire. The original idea started with [22], where he tested percentages of, up to 100%, of addition of cold wire in relation to the electrode wire, obtaining good results in terms of bead geometry (width and reinforcement), low dilution, and absence of discontinuities. The choice of insertion angle of the cold wires of the DCW-GMAW was based on the same proposal of the CW-GMAW; however, there is another option for the injection of these wires; this position was called coplanar, since they are inserted in the same plane as the welding torch [23] (**Figure 2**). In this same work, the authors concluded that there is a 15% loss in the hardness properties of the weld metal compared with the same GMAW metal using coplanar feeding. However, the author [24] verified that the DCW-GMAW process with angular feed and the CW-GMAW have better mechanical properties of hardness than the normal GMAW process.

2.2 Hot wire (HW-GMAW)

The HW-GMAW variant was designed using the idea of the CW-GMAW process; however, the additional wire that was cold and served only to add mass to the molten pool became a conductor of electric current, but that has no enough energy to strike an electric arc, and this function belongs only to the electrode wire. But now, in

DOI: http://dx.doi.org/10.5772/intechopen.104525

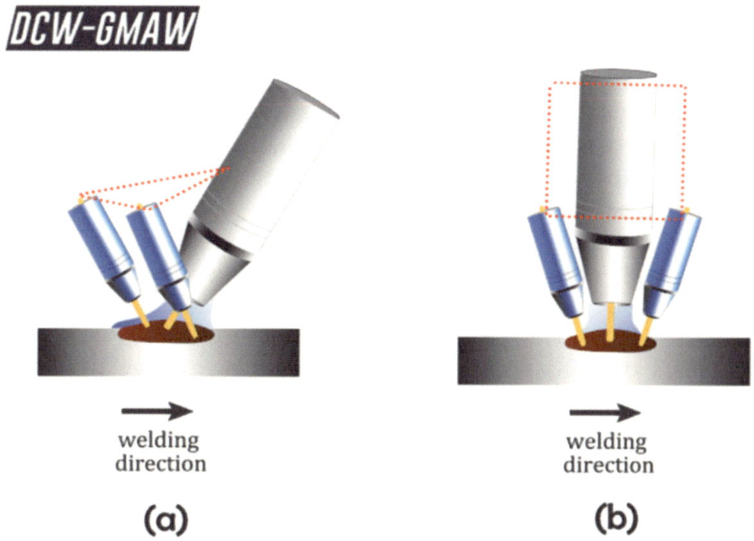

Figure 2.
DCW-GMAW variant wire feed scheme: (a) angular (b) coplanar.

addition to the extra power head, an auxiliary power supply is needed. One of the main motivations for hot wire (HW) is the preheating of additional wire, which reduces the amount of energy required to melt it, thus increasing the efficiency of the process.

Figure 3 schematically shows the operation of the variant. Thus the HW-GMAW was largely designed to increase the melt rate by Joule preheating the extra wire

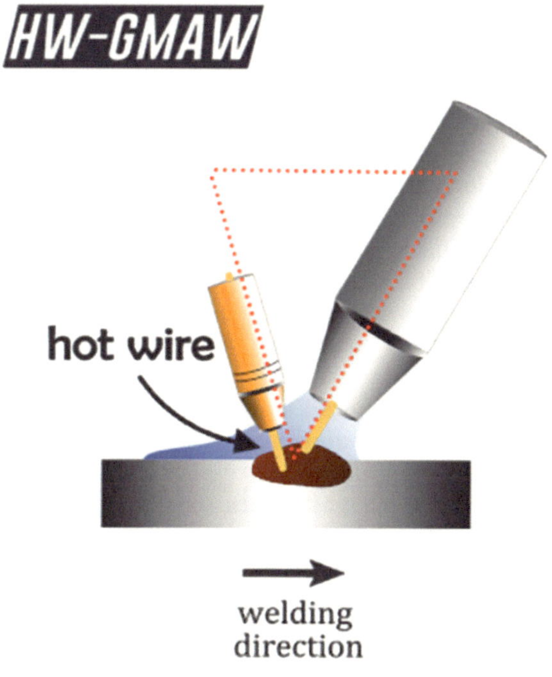

Figure 3.
HW-GMAW variant wire feed scheme.

without significantly increasing the total heat input to the substrate. The low dilution between the weld metal and the base metal makes this process suitable for the application of hardfacing using dissimilar materials [25–28].

The HW-GMAW process presents as advantages higher productivity due to its high melting and deposition rates, versatility in joint construction, together with low operating costs, due to the low system power required for melting the preheated wire when introduced into the weld pool.

According to studies by [11], the HW-GMAW process has shown promise for welding in narrow gap or in the deposit of corrosion-resistant coatings, and in this process, the preheated filler material reduces the main arc energy required for the electrode melting and test piece melting, in turn creating a cooler weld pool. These authors consider that the advance in the development of inverter welding sources, with advanced manipulation of waveforms, allows the increasing use of this variant.

It is worth mentioning that the intrinsic parameters of the welding processes each have their importance, and if changed together or individually, they promote different results, so that the effect of the parameters on the morphology of the weld beads can generate different results according to the methodology used for each work performed.

3. The geometry of weld bead and the metallurgical changes in the weld metal of the GMAW process variants

3.1 CW-GMAW

First, before dealing directly with metallurgy, there is a need to evaluate the geometry of the weld beads, considering that the first step to estimate whether a given process worked or not is to evaluate the geometric characteristics of the weld, and furthermore, to observe whether there is some apparent discontinuity. Thus, it will be possible to understand whether, in fact, the variant has validity and prospects for future development.

Several studies prove that the weld bead geometry is dependent on several factors such as current, voltage, polarity, welding speed, shielding gas, metal transfer mode, torch position, welding position, etc., that act directly on the weld pool and, consequently, provide a specific profile and dimensions of reinforcement, width, penetration, and dilution. This applies to the most varied types of arc welding, including GMAW, and it is still possible to predict this geometry according to the parameters used. Thus, the GMAW variants mentioned in this chapter also present their reinforcement convexity patterns and their respective characteristic dimensions of the cross sections, since the search for optimized welding parameters is always prioritized to propose a process stability standard used. Therefore, the authors [15] state that cold wires or hot wires must have smaller diameters than the electrode wire, since much larger cold wires can cause lack of fusion in the weld pool. Also, there is the possibility of simulating the bead profile using mathematical modeling of the dimensions in mm of penetration, dilution, height, and width of the bead, proposed by the same authors.

As an application of the CW-GMAW variant in welds for marine steel (ASTM A 131—different grades), the parameters of **Table 1** are generally used, all using wires of the AWS ER70S-6 class. The weld beads produced, most of the time, have a good surface finish and absence of discontinuities. Thus, the consolidation of the CW-GMAW variant provides a geometric pattern of the beads (**Figure 4**) when welded in the flat position in situations of simple deposition and V-bevels. GMAW

Parameter	Range
Current	280–380 A
Voltage	35–38 V
Shielded gas	Ar + 25% CO_2
Gas flow rate	1 a 1.5 m^3/h
Welding speed	50–70 mm/s
Wire feed speed	10–20 m/min
Wire diameter	0.8–1.2 mm
Contact-tip-to-workpiece distance (CTWD)	17 mm–22 mm

Table 1.
Standard welding parameter range for CW-GMAW.

process weld and the CW-GMAW variant for three electrode wire feed values and the percentages of cold wire added. It can be seen that the gradual increase in the melting rate and the deposition rate with the insertion of the cold wire provides an increase

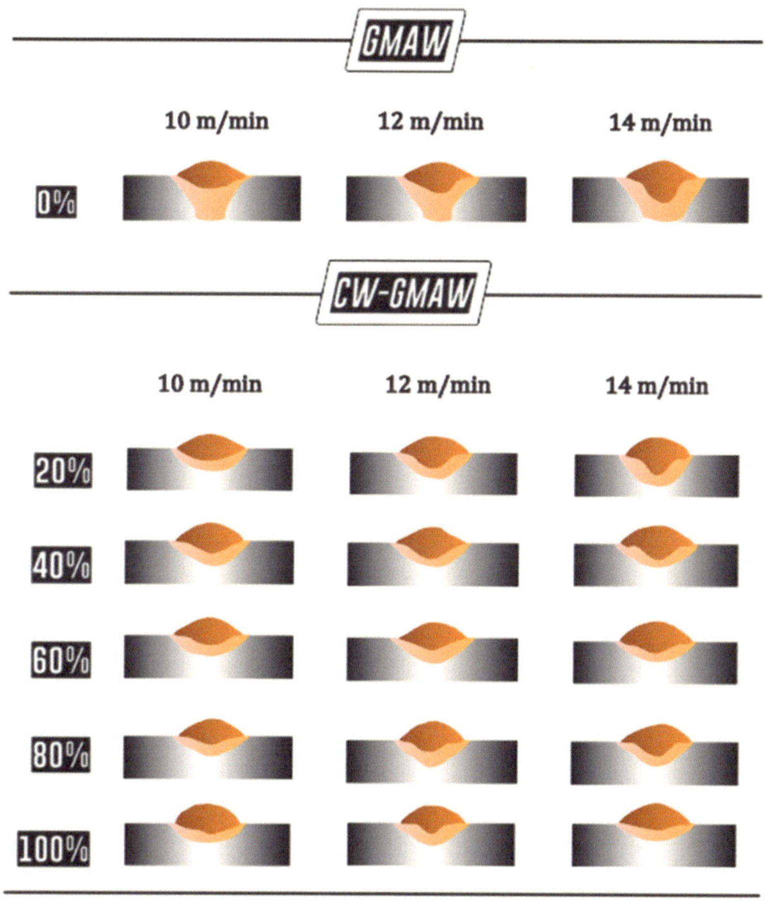

Figure 4.
Standard profile of welded beads with the GMAW and CW-GMAW processes with percentages of 20%, 40%, 60%, 80%, and 100% of cold wire. Based on the work of [22].

in convexity and a decrease in dilution, since the energy supplied to the part was converted to the melting of the wire additionally.

In this sense, the works of [9, 10, 22, 24] showed significant results regarding the CW-GMAW variants, the bead profiles are shown in **Figure 4**. All these authors welded steel sheets naval ASTM A 131 of different grades and dimensions of 9.5 mm X 150 mm X 300 V chamfer, 45° chamfer angle (bisel 22.5°), and the other parameters identical to **Table 1**, modifying the percentages used. Thus, using percentages of 20% and 40% with CW-GMAW, the authors [14] measured residual stress levels through X-ray diffraction and acoustic birefringence methods and using comparative analysis concluded that the variant helps to decrease the level of these stresses with the percentage increase of cold wire incorporated into the weld metal. **Figure 5** shows a specimen welded with CW-GMAW 40%, where it is observed that the sheet was clamped during welding to avoid distortion and measure residual stresses with restrictions. The geometric dimensions of the beads (in mm) were measured by determining the values of the width (w), penetration (p), height of the reinforcement (h), and the angle of wettability (α) of the weld metal in **Figure 6**. When comparing the measures presented, the conclusions can be reached: increase of the width of the bead and height of the reinforcement and the reduction of the penetration and of the HAZ.

Likewise, the results of [10] previously compared the fatigue strength of joints welded with GMAW and CW-GMAW, both in semiautomatic mode, showing the versatility in welding ASTM A131 grade A naval steel sheets when using wires with diameters varying from: 1.2 mm as electrode wire and cold wires of 0.8 mm and 1.0 mm, with cold wire feed rates of 50%. The dimensions of the sample body were the same (9.5 mm X 150 mm X 300 mm), requiring two passes for the total filling of the chamfer, the first pass was the one from scratch applied with GMAW to all parts, the second finishing pass, this being the comparative parameter between the GMAW process and the CW-GMAW. It was concluded that the fatigue behavior of the joints welded by the GMAW-CW process in both conditions is practically the same when compared with the conventional GMAW process. In addition, some metallurgical considerations were observed, such as the decrease of HAZ in the coarse-grained region and the formation of the fraction of primary ferrite and Widmansttäten, which influences the increase in hardness in the region, this occurs proportional to the diameter of the cold wire added, as the more cold wire, the greater the change.

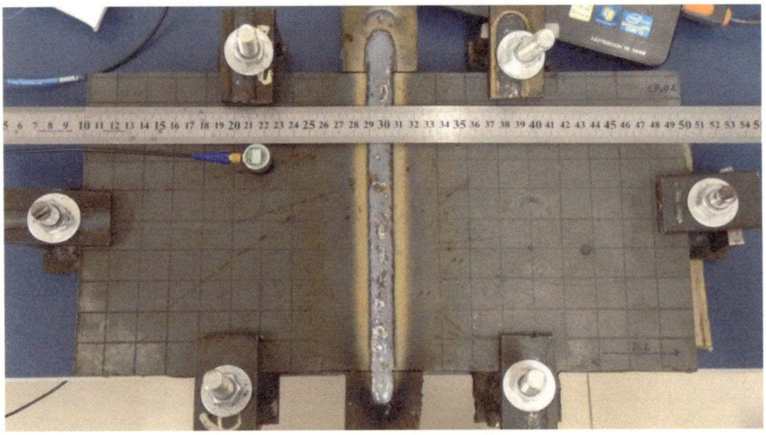

Figure 5.
Specimen welded with CW-GMAW 40% with clampers for measure residual stresses.

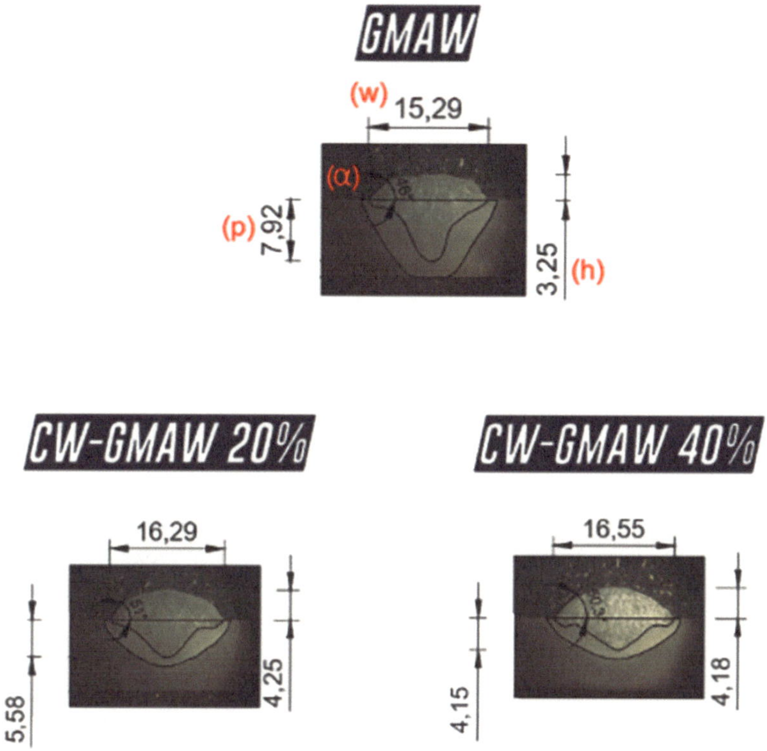

Figure 6.
The geometric dimensions of the beads of specimens welded with GMAW and CW-GMAW with clampers, 20% and 40% of cold wire.

This demonstrates that the addition of cold wire may be affecting the cooling rate due to the lower energy imposed on the weld pool.

However, works [22, 23] comparatively studied the GMAW process with the two variants, CW-GMAW and DCW-GMAW, with 03 wire feed speeds of 10 m/min, 12 m/min, and 14 m/min varying the percentages of 20%, 40%, 60%, 80%, and 100%. Thus, [22] studied the stability of both processes capturing the oscillograms, melting rates, deposition rates, and the geometry of the weld beads (**Figure 4**), where the following results were highlighted comparing the GMAW and the CW-GMAW:

1. Firstly, as a parameter and some possible analyses, the w/h ratio can be used, which helps to establish the ideal relationship for obtaining very convex beads that help to concentrate stresses and can be discarded, so it is estimated that values must be below 0.30 in order not to exceed this principle, studied by [29]. Thus, welds with percentages of cold wire ranging from 20–60% achieve ideal relationships with values, on average, of 0.25. Showing that in these cold wires, percentage ranges and all levels of electrode wire feed speeds, weld beads are the most suitable.

2. The linear penetration, on average, decreases by 31%, considering the 03 levels of feed speed and all percentages of cold wire, reaching 58%, with the level of 14 m/min and 100% of cold wire.

3. The dilution decreases, on average, by 32% for all percentages of cold wire, with the biggest drop being 48%, also with the level of 14 m/min and 100% of cold wire.

4. The highest deposition rate is the 14 m/min level and 100% cold wire with a value of 11.48 kg/h, corresponding to 61% higher than the GMAW deposition rate for the same level of wire feed speed.

5. While studying various metallurgical parameters exclusively for the wire feed speed of 12 m/min, [24] observed that there is an increase of up to 29% in the percentage of the average silicon content in the weld metal. On the other hand, the grain size decreases with the increase of the percentage of cold wire by up to 32%, which favors the mechanical properties by the Hall–Petch principle. This grain refinement can be seen in **Figure 7**, where images of weld metals etched with Nital solution (2%) from the GMAW, CW-GMAW, and DCW-GMAW processes are analyzed under an optical microscope.

Continuing, using the IIW (International Institute of Welding) C-Mn metal microstructure classification scheme [30], together with the measurement of the volumetric fraction of phases using images obtained by optical microscopy, both

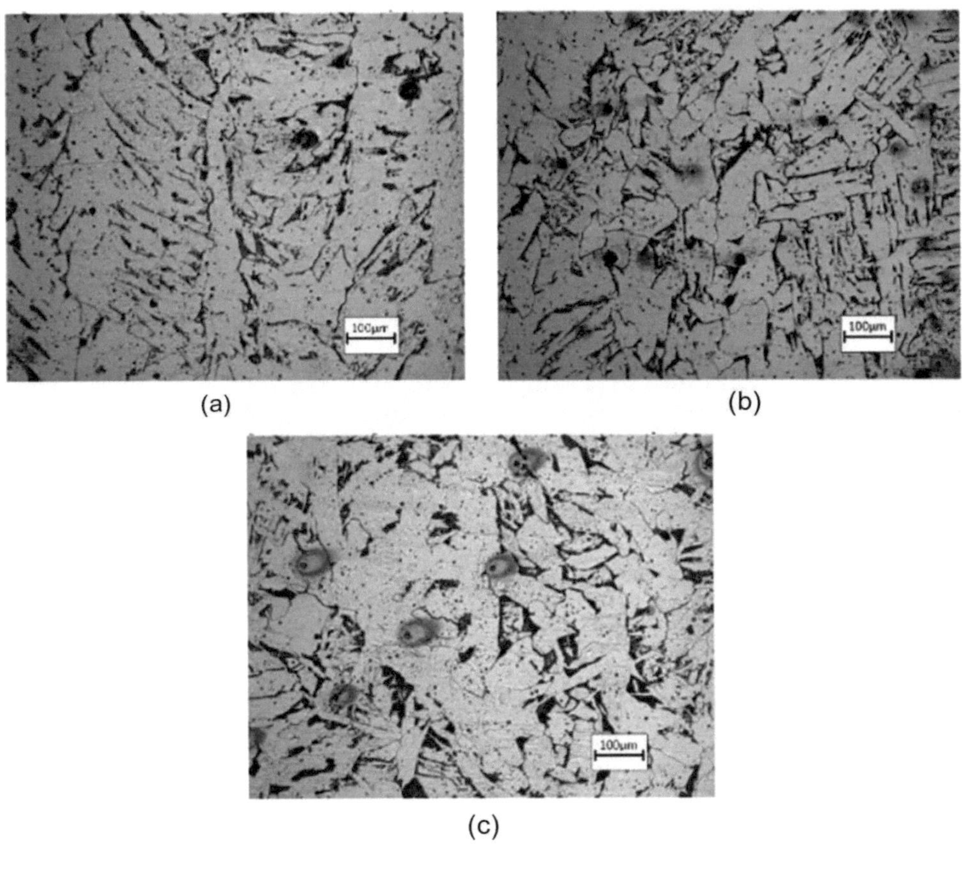

Figure 7.
Weld metal grain size: (a) GMAW, (b) CW-GMAW, and (c)DCW-GMAW. Nital solution (2%), optical microscope [17].

works mentioned above studied the influence of cold wire on the formation of the main microconstituents of the weld, mainly acicular ferrite (AF), since this microstructure has a desirable presence in the weld metal, due to its excellent mechanical properties. However, in addition to this, the presence of allotriomorphic ferrite and Widmansttäten ferrite is also part of the predominant phases in carbon steel welds as deposited [31]. Thus, [22, 24] found that in all parts the weld metal is formed by ferrite, in several different forms: primary ferrite (PF), grain boundary ferrite—PF(G), acicular ferrite (AF), intragranular polygonal ferrite—PF(I), non-aligned second-phase ferrite—FS(NA), and aligned second-phase ferrite—FS(A). PF, PF(G), and FS(NA) ferrites predominate in the composition of the microstructures present with almost 100% of the composition for the highest wire feed speeds.

However, for the three wire feed speeds, with up to 60% cold wire, there is an average acicular ferrite increase of around 24% compared with GMAW. For percentages of cold wire of 80% and 100%, an inverse behavior is observed with a decrease in the amount of AF, on average, of 36%. What can possibly be observed is that the microstructures are benefited or inhibited by the presence of certain chemical elements. As, for example, the presence of low and medium percentages of aluminum (Al) forms a TiO layer around the inclusions, where the circular ferrite is nucleated, thus favoring its growth. However, for high Al content, such formation does not occur [32]. Overall, the gradual increase in the insertion of cold wire improved the mechanical properties of hardness.

3.2 DCW-GMAW

The DCW-GMAW came from the idea of the other variant CW-GMAW to evaluate the ability of how much the GMAW process was able to increase the insertion of "cold" mass using only one electrode wire. One of the great challenges of this variant is the placement of the cold wires. Therefore, the profiles of the evaluated beads are based considering the entry of the cold wires in the angular position in relation to the welding torch, as shown in **Figure 2a**. Still not having significant results, however, [23] concluded that high percentages of cold wire, from 60%, cause a reduction of approximately 15% in the hardness properties of the weld metal. However, before that, it is necessary to deal with the geometry of the weld beads.

Based only on practical works, the DCW-GMAW variant was first tested and patented by [22] and soon after, also analyzed by [24], where in general it was observed that this process was capable of being applied in the industry in fact using the data obtained for this conclusion. Working with carbon steel, parameters similar to **Table 1**, percentages from 20–100%, with a variation of 20% and with wire feed speeds of 10 m/min, 12 m/min, and 14 m/min. **Figure 9** presents the standard profiles found for the weld beads based on [22], the summary below describes some significant results such as:

1. The w/h ratio, in general, was well below the failure limit, on average, of 0.23. Where for the lowest wire feed speeds and low cold wire contents (20%), we have the lowest values with 0.19. With the gradual increase in the rate of cold wire, the w/h ratio increases, with values of 0.28 for a rate of 100% of cold wire. But the value is still below the allowed 0.30.

2. Linear penetration decays similarly for both electrode wire feed speeds. For small cold wire rates, there is a drop from 10%. However, for rates of 80% and 100% of cold wire, this drop reaches values of 50%. That is, generating a weld of very low penetration. This is visible in **Figure 9**.

3. In the case of dilution, normal values of GMAW are around 50%; however, previously following the results of linear penetration of DCW-GMAW, there is a decrease in dilution up to 20.44% for cold wire rate of 100% at the lowest wire feed speed (10 m/min). This dilution represents an approximate 60% drop compared with normal values. As a rule, the dilution attenuates as the percentage of cold wire increases.

4. The deposition rate depends directly on the wire feed speed, so each of these speeds has a proportional characteristic value, so the lowest percentages start with a speed of 10 m/min with 20% cold wire with a slight increase of 17% until reaching 79% for addition of 100% cold wire. Overall, the value of 102% more stands out for the speed of 14 m/min with a rate of 14.41 kg/h compared with the value of 7.13 kg/h for the normal GMAW.

5. Considering only the wire feed speed of 12 m/min, in the work of [24], it was found that the silicon content for samples from 20–80% of cold wire has almost the same values, an average of 1.3%. Only, for the 100% cold wire sample, this increase is 0.1% over the average content. As for the grain size, the samples of 20% and 40% have, respectively, 74.78 μm and 73.06 μm, values close to the comparative GMAW (76.40 μm). However, considering the DCW-GMAW-100%, there is a significant decrease in grain size by 34%, an image of the grains can be seen in **Figure 7c**. This effect could probably have been caused by the increase of chemical elements refining the austenitic grains of the steel or by the disturbance of the weld pool by the volume of added metal.

In the case of the predominant microstructures, the phases in the forms of PF, PF(G) and FS(NA) ferrites constitute 98% of the composition of the microstructures present in the weld metal. What changes are the amount of each phase in each image analyzed? For low amounts of cold wire (20% and 40%) the percentages of AF and FS(NA) have an average increase of 47% and 28%, respectively. However, for percentages of cold wire from DCW-GMAW-60%, there is a decrease in the amount of these phases, while the FS(A) increases, even tripling its composition in the case of low wire feed with 100% of cold wire. An image of the phases present in a sample of DCW-GMAW-60% for the feed speed of 12 m/min can be seen in **Figure 8c**. The work of [24] still shows that the silicon levels are drastically high, in the CW-GMAW and DCW-GMAW variants, thus also increasing the weld metal hardness levels. From average values of 155 HV to peaks of up to 190 HV.

In the case of the predominant microstructures, the phases in the forms of PF, PF(G), and FS(NA) ferrites constitute 98% of the composition of the microstructures present in the weld metal. What changes are the amount of each phase in each image analyzed? For low amounts of cold wire (20% and 40%), the percentages of AF and FS(NA) have an average increase of 47% and 28%, respectively. However, for percentages of cold wire from DCW-GMAW-60%, there is a decrease in the amount of these phases, while the FS(A) increases, even tripling its composition in the case of low wire feed with 100% of cold wire. An image of the phases present in a sample of DCW-GMAW-60% for the feed speed of 12 m/min can be seen in **Figure 8c**. The work of [24] still shows that the silicon levels are drastically high, in the CW-GMAW and DCW-GMAW variants, thus also increasing the weld metal hardness levels. From average values of 155 HV to peaks of up to 190 HV.

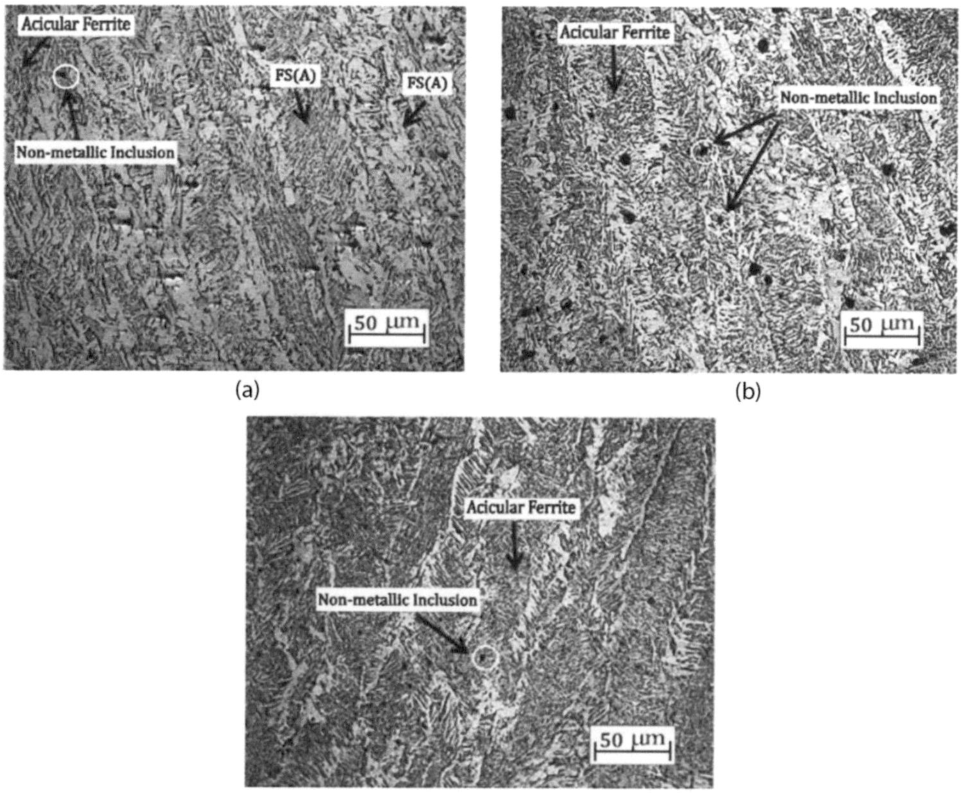

(a) (b) (c)

Figure 8.
Microstructures present in the weld metal: (a) GMAW, (b) CW-GMAW, and (c) DCW-GMAW [17].

3.3 HW-GMAW

Thinking about increasing the melting rate and the deposition rate of the GMAW process, making an adaptation in the CW-GMAW variant, the design of the HW-GMAW was arrived at. That is, the additional wire, which was previously free of energy, now has a low direct current to assist in the fusion of the filler metal, using the Joule effect as a basic principle, which, through the resistance of the metal, converts electrical energy into thermal energy. Bearing in mind that when introducing hot wire into the process, it is not intended to significantly increase the heat imposed on the part, but only to increase productivity with the design of this variant of GMAW.

The proposition of the HW-GMAW variant is relatively new, despite the similarity with more consolidated and widely used processes in the industry in general. Currently, his studies focus on the application of hard coatings on surfaces to increase wear resistance [27, 28, 33].

Firstly, in terms of welding itself, the works by [11, 34] studied the influence of generic parameters such as: welding direction, hot wire feed rates. However, remembering that the extra wire feed rates obey the ratio given according to Eq. (1), both of which are related having as a reference point a percentage of the electrode wire feed speed (m/min). Thus, [27] using 5 m/min electrode wire speed with a percentage of 140% hot wire, casing welds were performed on flat bars (9.5 mm X 56 mm X

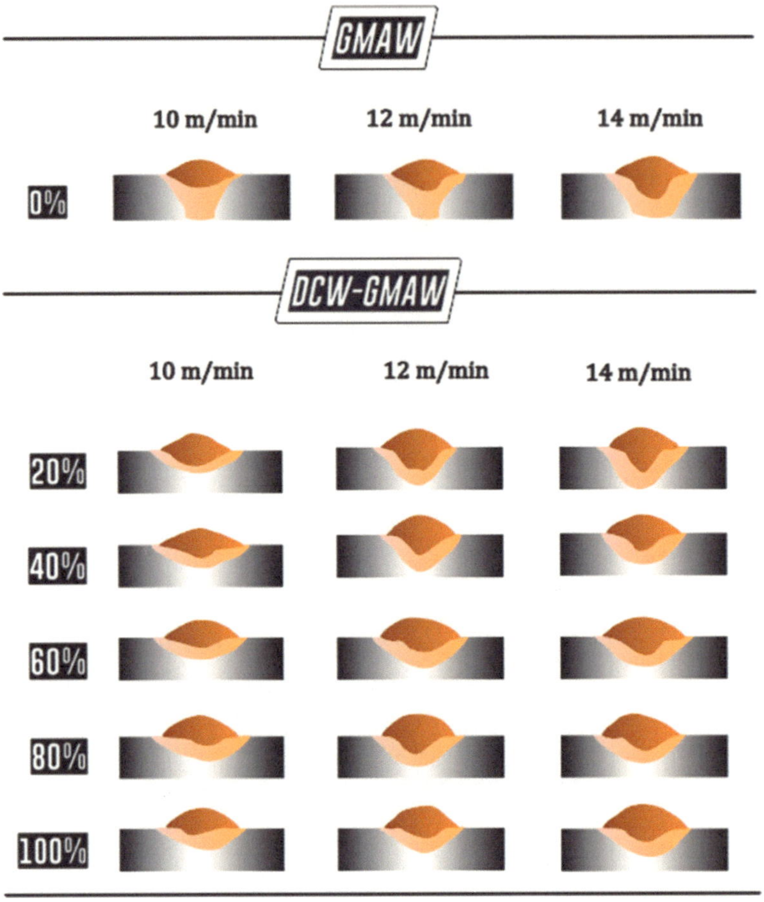

Figure 9.
Standard profile of welded beads with the GMAW and DCW-GMAW processes with percentages of 20%, 40%, 60%, 80%, and 100% of cold wire. Based on the work of [22].

225 mm) of AISI/SAE 1020 carbon steel, both wires used were of the AWS ER70S-6 class, the electrode wire having a diameter of 1.2 mm and the energized wire having a diameter of 1.0 mm. The welding parameters were a voltage of 23.6 V, current of 180 A and a contact tip-to-work distance of 15 mm, in the pulling welding technique. The variable parameter used was the direct polarity current of the hot wire at levels of 40 A, 80 A, 120 A, and 150 A. The solder used as a comparison was the CW-GMAW with 50% cold wire and parameters almost identical to those mentioned previously. The results obtained suggest that the w/h ratio has values above the previously established limit, greater than 0.3, with an average of 0.35. They may not be ideal for chamfering, but excellent for application as a coating. Penetration is slightly higher with values of up to 30% higher and the HAZ practically remains very similar. In general, the bead profiles are similar to those in **Figure 10**, in which they are based on the work of [11], in which the interference of other parameters such as the stability of the process through cyclograms, the polarity of the hot wire (on both poles: positive and negative), the welding direction (pull or push) varying wire feed rates at 20% and 100% hot wire. Emphasizing that the same material of low carbon steel and electrode wire of the same AWS class were used.

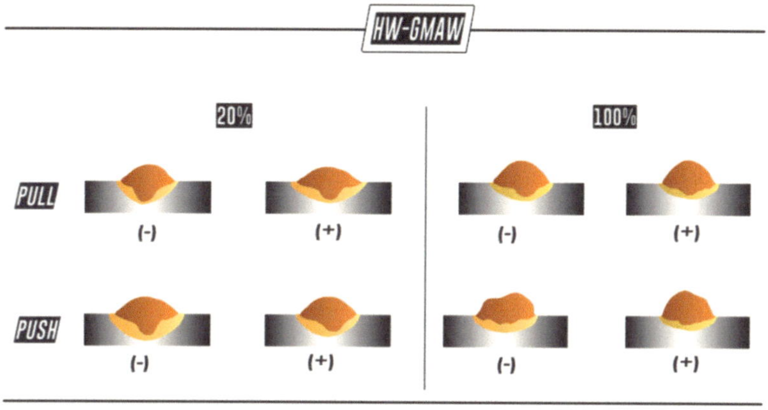

Figure 10.
Standard profile of welded beads with the HW-GMAW processes with percentages of 20% and 100% of hot wire in the torch movements: (a) pull and (b) push. Based on the work of [11].

Still in the work of [11], the authors concluded that, for high wire feed rates, the penetration can drop by up to 45% and the dilution by up to 25%, when compared with the original GMAW. It has also been shown that hot wire polarity can attract or repel the arc and, together with the HW feed rate, can change bead geometry through changes in penetration depth and bead height. On the other hand, welding directions and wire feed rates are the parameters that most affect arc stability. And finally, in most weld beads the penetration is lower than the same weld in the conventional process.

However, regarding the metallurgical issues of grain size and the formation of microstructures from HW-GMAW welding in carbon steel materials, it will be necessary to continue the research, since it has not yet been published. Noting that there are already many works that show the structures of alloys based on Ni and FeCrC, which will not be addressed in this chapter.

4. Conclusions

Modifications of the GMAW process giving rise to the CW-GMAW, DCW-GMAW, and HW-GMAW variants provide very significant results, despite a very similar trend when the percentages of extra wire become very high from 80%, regardless of whether this extra wire is energized or not. However, some observations must be addressed:

1. The CW-GMAW variant in several works presents greater versatility of implementation and use in different sectors of the industry, both for straight and narrow gap and for angled chamfers, in addition to the possibility of application for coatings, whether special or not. In general, low electrode wire feed speeds and low percentages of extra wire favor the variant in use for splices, as linear penetration remains almost unchanged. However, high wire feed speeds and high percentages of cold wire allow for a decrease in penetration, dilution, and w/h ratio, increasing the height of the reinforcement, favoring its use for coatings. This feasibility is further increased when the combination of wires used can be manipulated, generating different chemical combinations for the deposited metal.

2. In the case of DCW-GMAW, the behavior changes in relation to the chances of application for certain types of industry, being more limited for use in coating situations, since the cold wire mass drastically reduces metal penetration of weld, the dilution, favoring for low values of w/h, increasing the profile of the reinforcement. Also, there is a potential chance of decreasing residual stresses when welding with this variant, and this has already been observed in the CW-GMAW variant, but not yet proven.

3. For the application of the variant HW-GMAW, similar to the DCW-GMAW, its application is already well established for use in hard coatings, with several works already published, where this variant is used in coatings of special alloys to increase the resistance to wear. The weld bead profile of this variant is characterized by low dilution and small w/h values, being strongly influenced by the polarity of the current applied to the hot wire.

4. For all abovementioned variants:

    a. There is a significant decrease in the HAZ, which favors the implication that there is a decrease in defects in this region, which has been extensively studied and confirmed to be problematic in several situations.

    b. In the case of application in low carbon steel, it is concluded that the grain size is slightly reduced, favoring the increase of the mechanical strength of the weld metal. Where several works observed this increase in the hardness of the weld metal. In addition, it was found that there are few changes in the microstructure, where in certain cases, the formation of acicular ferrite, which is very desired in weld metals, is increased.

Acknowledgements

To Federal University of Pará, Postgraduate Program in Natural Resources Engineering in the Amazon (PRODERNA/UFPA) and to the Metallic Materials Characterization Laboratory (LCAM) for all support in carrying out the tests and coworker who ceded their scientific work.

Author details

Lino A.S. Rodrigues*, Pedro P.G. Ribeiro, Ednelson da S. Costa, Tárcio dos S. Cabral and Eduardo de M. Braga
Metallic Materials Characterization Laboratory (LCAM), Federal University of Pará, Belém, PA, Brazil

*Address all correspondence to: lino@ufpa.br

References

[1] Xiong J, Zhang G. Adaptive control of deposited height in GMAW-based layer additive manufacturing. Journal of Materials Processing Technology. 2014;**214**(4):962-968

[2] González J et al. Additive manufacturing with GMAW welding and CMT technology. Procedia Manufacturing. 2017;**13**:840-847

[3] Le VT, Mai DS, Hoang QH. A study on wire and arc additive manufacturing of low-carbon steel components: Process stability, microstructural and mechanical properties. Journal of the Brazilian Society of Mechanical Sciences and Engineering. 2020;**42**(9):1-11

[4] Cooke KO, Lhazaa A, Atieh AM. Dissimilar welding and joining of magnesium alloys: Principles and application. In: Magnesium-the Wonder Element for Engineering/Biomedical Applications. London, United Kingdom: IntechOpen; 2019

[5] Cooke KO, Shar MA, Hussain S. Micro-alloying and surface texturing of Ti-6Al-4V alloy by embedding nanoparticles using gas tungsten arc welding. Journal of Manufacturing and Materials Processing. 2020;**4**(2):29

[6] De Araujo Ribeiro, Rafael. The Cold Wire Gas Metal Arc Welding (CW-GMAW) Process: Description and Applications. Waterloo, Canada; 2020

[7] Cabral TDOSS et al. Influence of a cobalt-based wire injection in austenitic coating deposited via CW-GMAW. Journal of the Brazilian Society of Mechanical Sciences and Engineering. 2018;**40**(9):1-11

[8] Assunção C et al. Feasibility of narrow gap welding using the cold-wire gas metal arc welding (CW-GMAW) process. Welding in the World. 2017;**61**(4):659-666

[9] Costa ES et al. Residual stresses in cold-wire gas metal arc welding. Science and Technology of Welding and Joining. 2017;**22**(8):706-713

[10] Marques LFN et al. Fatigue life assessment of weld joints manufactured by GMAW and CW-GMAW processes. Science and Technology of Welding and Joining. 2017;**22**(2):87-96

[11] Ribeiro PPG et al. Metal transfer mechanisms in hot-wire gas metal arc welding. Welding Journal. Miami, Florida, USA: Elsevier; 2020

[12] Scotti A, Ponomarev V. Soldagem MIG/MAG: melhor entendimento, melhor desempenho. São Paulo: Artliber; 2008

[13] Silva FG. Estudo da viabilidade do processo de soldagem FCAW com adição de arame frio aplicado a indústria naval. Belém: Universidade Federal do Pará; 2010

[14] Lino Alberto Soares Rodrigues. Caracterização Microestrutural e de Propriedades Mecânicas de Soldas do Aço ASTM A 131 pelos Processos FCAW Convencional e com Adição de Arame Frio. 2011. 144 f. 2011. Tese de Doutorado. Dissertação (Mestrado)–Universidade Federal do Pará, Instituto de Tecnologia, Belém. 2011. Programa de Pós-Graduação em Engenharia Mecânica

[15] Ribeiro RA et al. Predicting weld bead geometry in the novel CW-GMAW process. Welding Journal. 2015;**94**(9):301-311

[16] Ribeiro RA et al. Application of cold wire gas metal arc welding for narrow

gap welding (NGW) of high strength low alloy steel. Materials. 2019;**12**(3):335

[17] Assunção PDC et al. Comparing CW-GMAW in direct current electrode positive (DCEP) and direct current electrode negative (DCEN). The International Journal of Advanced Manufacturing Technology. 2019;**104**(5): 2899-2910

[18] Vasconcelos CHM et al. High-hardness armor welded by CW-GMAW: Economic, geometric and CGHAZ analysis. Journal of the Brazilian Society of Mechanical Sciences and Engineering. 2019;**41**(6):1-11

[19] Ribeiro RA. et al. Characterization of the haz in X80 welds fabricated by the cw-gmaw process

[20] Ribeiro RA. et al. Cold-wire gas metal arc welding for automotive applications

[21] Mota CAM d et al. Revestimento de níquel depositado pela soldagem MIG e MIG com arame frio. Soldagem & Inspeção. 2017;**21**:483-496

[22] Assuncão PDC. Estudo da Viabilidade do Processo de Soldagem GMAW-DCW (double cold wire). Belém: Universidade Federal do Pará; 2013

[23] Assunção PDC et al. A preliminary study on the double cold wire gas metal arc welding process. The International Journal of Advanced Manufacturing Technology. 2020;**106**(11):5393-5405

[24] Ferreira ASA. A influência do teor de silício na microestrutura e propriedades mecânicas do aço ASTM 131-GRAU A soldado por GMAW-CW e GMAW-DCW. Belém: Universidade Federal do Pará; 2014

[25] Mendez PF et al. Welding processes for wear resistant overlays. Journal of Manufacturing Processes. 2014;**16**(1):4-25

[26] Allford D, Hebble D. Could GTAW hot-wire go mainstream? The Fabricator.com. 2019

[27] Günther K et al. Hot wire-assisted gas metal arc welding of Ni-based hardfacing. Welding Journal. 2018;**97**(208):99-107

[28] Schreiber F, Allebrodt B, Erpel T. Hardfacing material solutions for high performance coatings in wear and corrosion applications. Zavarivanje i Zavarene Konstrukcije. 2019; **64**(1):11-21

[29] Cabral T d S, Junior LP d SC. Especificação de procedimentos de soldagem MIG para o preenchimento de cavidades pela técnica de camadas sucessivas. Monografia (Graduação em Engenharia Mecânica). Belém, Brasil: Universidade Federal do Pará; 2008

[30] Doc IIW. No. IX-1533– 88: Guide to the light microscope examination of ferritic steel weld metals. Revue de la Soudure. 1990;**4**:29-41

[31] Bhadeshia H, Honeycombe R. Steels: Microstructure and Properties. Butterworth-Heinemann; 2017

[32] Takada A, Terasaki H, Komizo Y. Effect of aluminium content on acicular ferrite formation in low carbon steel weld metals. Science and Technology of Welding and Joining. 2013;**18**(2):91-97

[33] Günther K, Bergmann JP, Suchodoll D. Hot wire-assisted gas metal arc welding of hypereutectic FeCrC hardfacing alloys: Microstructure and wear properties. Surface and Coatings Technology. 2018;**334**:420-428

[34] Pedro Paulo Guimarães Ribeiro, et al. COBEM-2017-2277 influence of lorentz force on penetration of welded joints by the gmaw process with electric wire

Section 3

Applications of Welding and Joining

Chapter 12

Joining by Forming of Sheet Metals

Akash Biradar and Malayathodi Rijesh

Abstract

Sheet metal is the frequently used component geometry in industries, and the joining of sheets is inevitable. There exist numerous conventional processes for joining sheet metals, but the diverse needs of today's industries necessitate further research into alternative joining technologies. Joining by forming methods to join similar and dissimilar sheet metals has a great potential to reach current industrial requirements. The process involves plastic deformation of at least one constituent part. Among several techniques that evolved in recent times, friction-based welding, impact welding, and roll bonding are the three solid-state welding methods, often known as joining by forming. The present chapter starts with a brief overview of the various aspects of joining sheet metals by forming methods. The working principle, procedure, and the consequences of the impact-based methods such as vapor foil actuator welding, electromagnetic welding, and laser impact welding, along with the roll bonding process are discussed.

Keywords: sheet metal, joining, forming, impact welding, roll bonding

1. Introduction

Sheet metal is the foundation of most engineering today. It has a wide variety of applications, including automobiles, airplanes, machinery, equipment, home facades, and furniture. Sheet metal is one of the shapes a metal can be formed into by industrial processing. Sheet metal is defined as any metal with a thickness of 0.5–6 mm. When constructing a sheet metal product, engineers will inevitably use sheet metal parts. It's a challenging task to join several sheet metal parts in a cost-effective and secure manner.

The very commonly used conventional methods of joining sheet metals are shown in **Figure 1**; folding/tab-joint, pulling and pressing of a rivet, self-clinching, screw joint, and welding joints.

Folding/tab-joint (**Figure 1a**): Two sheets of metal are connected by folding or bending tabs in the shape of a buckle and a clamping groove. The assembly is straightforward, convenient, and completed quickly. However, full positioning is not guaranteed, and further supplementary positioning is required.

Rivet pulling and pressing (**Figure 1b**): Riveting is done in the holes that correspond to the two pieces, and the rivet gun is used to draw the rivet to expand and deform the outer rivet sleeve, therefore securing the two parts. The resulting connection will be simple, convenient, and quick. Stringers and airframe skins are virtually always joined by rivets. Although millions of rivets are used in aircraft structures (adding weight),

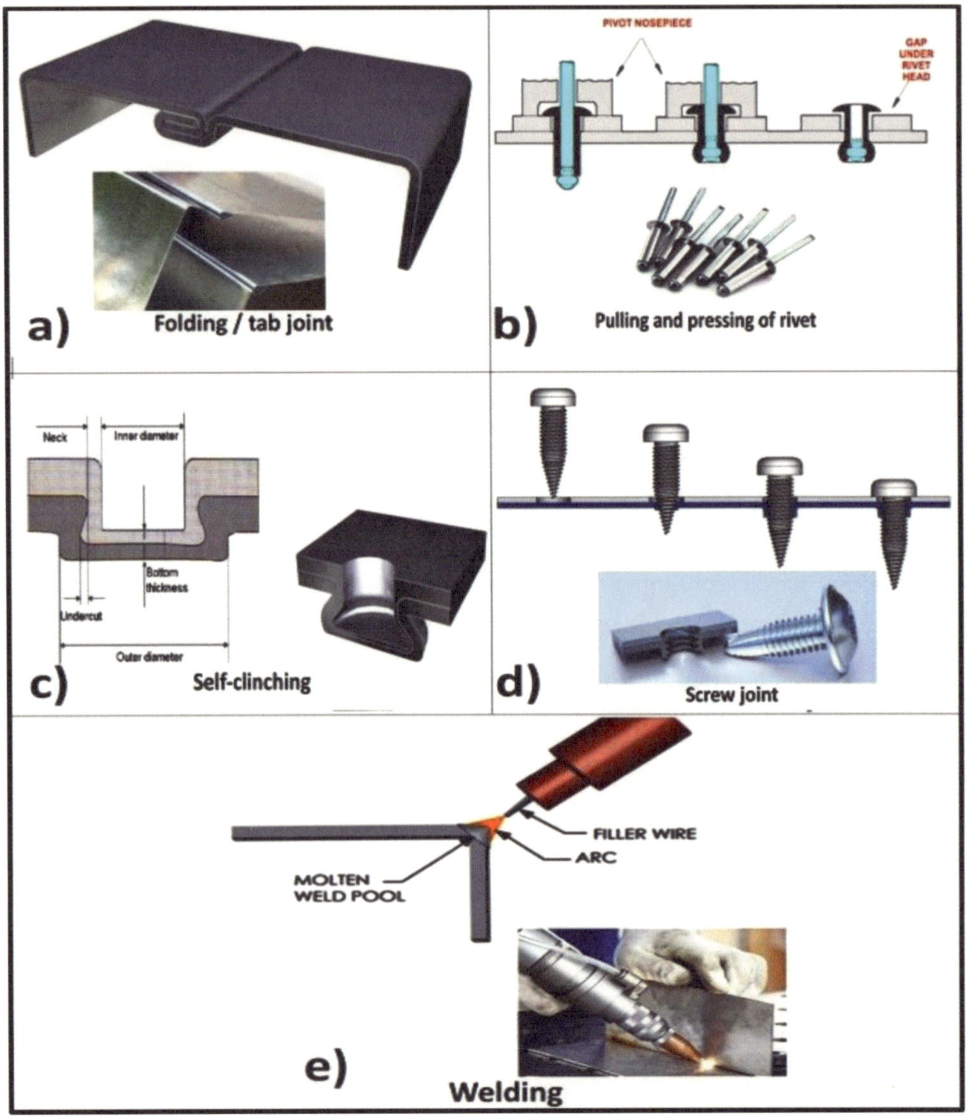

Figure 1.
Conventional methods of joining sheet metals.

the stress concentrations caused by rivet holes quadruple the skin's local stresses. One advantage of rivets is that they are more reliable, if not more efficient.

Self-clinching (**Figure 1c**): Self-clinching is also known as self-riveting, is a method of completing mutual fastening by deformation between the sheet metal. Despite its simplicity, this technology is employed frequently in regions where disassembling is not required.

Screw joint (**Figure 1d**): Self-tapping screws to directly tap the thread on a piece of sheet metal, so the fit is good, which requires disassembling.

Welding (**Figure 1d**): This is a spot or seam welding used to keep a sequence of solder joints on two sheets of metal together. At the welding head, it directly melts the local sheet metal.

During the welding of sheet metals, everything happens quickly. As a result of the rapid heating of the materials, distortion like warping gets amplified. Furthermore, the thin materials burn when heated abruptly, resulting in undesired perforations. Mechanical strength is compromised if heat-affected zones are not controlled (HAZ). Welders of thin metal materials must concentrate on reducing the warping, melt-through, and size of HAZ [1]. Selecting the most appropriate welding method, carefully managing the physical setup, creating optimum welding parameters, and precisely completing the weld are necessary steps in overcoming these problems. Precision joint preparation for a tight fit and appropriate clamping to prevent movement during the weld is part of the physical setup. Much percussion are needed to be taken care of during the process [2]. As an example, a copper backup bar is often handy to place beneath the joint. During excessive melting, this bar conducts heat away from the base materials.

Joining by forming a sheet metal involves solid-state joining methods, which are a viable alternative to traditional, fusion-based welding and is effective for combining materials with different melting temperatures or brittle intermetallic compounds [3]. Several existing and emerging solid-state (SS) welding methods can produce sound metallurgical welds across similar and dissimilar metals without melting, yet some localized and isolated pockets of melting may occur. Joining by forming sheet metals is the name given to these methods, and there has been a surge in interest in them in recent years. Solid-state joining allows combining advanced metals and dissimilar metal combinations that are difficult or impossible to join using fusion welding to avoid melting [4]. It is possible to achieve substantially greater joint efficiencies than fasteners and adhesives. Solid-state joining technologies might vary depending on usage, the materials of interest, and the topologies they use. Joining by forming methods are preferable to fusion welding from a metallurgical standpoint since no significant microstructural change occurs and a lesser risk of intermetallic formation or inter-diffusion across the weld interface.

Multi-material components are becoming increasingly important in recent industrial criteria. Conventional welding procedures are unable to meet the manufacturing needs of these components. Forming can be an alternative to joining sheet metal. Among several techniques that evolved in recent times, friction-based welding [5], impact welding [6], and diffusion bonding/roll bonding [7] are the three main solid-state welding procedures, often known as joining by forming. Since the 1950s, explosives used to perform impact welding, in which one metal sheet (called a flyer) is driven to a high velocity by a high-pressure pulse to contact another sheet (called a target) at an angle [8]. The hydrodynamic nature of the metals surfaces causes them to eject as a jet at the moment of impact. Surface oxides and other impurities are present in the jetted material, but clean metallic surfaces in high-pressure contact are left behind. As the space between the two parent sheets closes and the impact point moves closer to the edge of the sheets, a metallurgical bond forms between them. Depending on the material properties, impact angle, and velocity, the weld interface morphology can range from flat to variable degrees of waviness, and may or may not contain pores or a layer of intermetallic compounds [9, 10]. As a result, for a particular material combination, there is a "welding window" of characteristics that generate a strong weld. Explosive welding (EXW) is useful for joining centimeter-thick flyers to targets that are as thick as or thicker than that; but, at lower scales, this method performs poorly. High-velocity impact welding is characterized by a low welding temperature and fast welding speed. The process is conducted at room temperature. Furthermore, there is no external heat input during the welding process.

The following sections explain impact-based sheet metal joining methods such as electromagnetic welding, vaporizing foil actuator welding, and laser impact welding, along with the roll bonding process.

2. Impact welding

In the 1940s, Carl proposed explosives can be used to drive metal and metal collisions for metallurgical bonding, which he termed explosive welding [11]. People nowadays deploy chemical energy, electromagnetic field energy, high-energy-density light energy, high-pressure gas, and other driving sources to produce various forms of impact welding by releasing high energy transiently and driving high-speed collision of welding parts.

During the Great War, engineers noticed shrapnel unusually attached to armored tanks, not simply inserted into the tank's side (**Figure 2**). The tank material and shrapnel fused due to the force of the collision. That is, the impact resulted in the formation of a weld.

In the process of impact welding, two or more metal sheets collide at high speed. The impact begins in the vicinity at the speed of sound in the air [12]. The work-pieces experience severe plastic deformation at the contact area during impact by transforming kinetic energy into plastic deformation energy. The superficial (oxide) layers are broken apart by this plastic deformation. The required angle between the workpieces causes a line-shaped contact zone to go over the surface of the workpieces as the space between them closes. The high-pressure contact substantially removes gaps and results in a metallurgical bond between the materials. The impact welding process successfully welds dissimilar metal and does not produce a heat-affected zone. A common drawback of traditional welding processes is that the processes alter the material properties by the local temperature condition, often resulting in a softer region around the weld (heat-affected zone). Joints are by impact, the properties of the base metal are intact at the joint. The resulting interface may or may not be as robust as the base metals.

Different welding technologies, such as electromagnetic pulse welding, vaporizing foil actuator welding, and laser impact welding, are all part of the impact welding family. Although the primary working principle of these processes is a high-velocity collision between a flyer and a target, the method of accelerating the flyer differs. These methods also have a wide range of length scales, giving the impact welding

Figure 2.
Shrapnel stuck to armored tank.

group a wide range of applications. Impact welding can drive the development of numerous scientific investigations, which are necessary for optimizing current production processes by developing new welding techniques and solutions.

2.1 Electromagnetic welding

Electromagnetic welding (EMW) is a solid-state welding process that primarily joins conductive materials using high-speed electromagnetic force. The technique efficiently joins similar or dissimilar metals, as well as metals and non-metals [13]. The process utilizes very high velocity and strain rate to join several materials. The impulsive Lorenz force, produced by repelling magnetic fields due to pulse current, accelerates one or both joining materials, resulting in a high-velocity collision and joint formation. The weld contact does not melt, unlike traditional joining techniques, preserving the material characteristics.

Electromagnetic pulse welding was used by Yu et al., [14] to create aluminum-covered steel tubes. The findings suggest that the proposed EMW technique can generate strong cladding bonds to make a tubular clad component with a long axial length. P. Q. Wang et al., [15] successfully welded Al/Cu dissimilar sheet metal by EMW. Many process parameters define the mechanical performance and interfacial morphology of EMP welded joints. The investigation by C. Li et al., [16] reported a relationship between process parameters such as discharge current frequency, the Lorentz force, and the displacement in the base metals. Shaoluo Wang et al., [17] reported that to improve the weldability range by EMW, discharge energy, and the wieldable standoff distance range would be enhanced as the ascent of discharge energy could make the flyer plate have a higher collision velocity. The process has a significant effect on the microstructural changes at the bond interface as well. The interface morphology generally includes the formation of a dislocation network, mechanically induced dissolution of precipitates, and recrystallization [18]. There is little evidence one can find in the melting of base materials. In addition, it is worth expecting the significant enhancement in the strength for the post-weld heat-treatable alloys by thermomechanical processing. During service, dissimilar metal junctions will invariably experience corrosion, which results in premature failure of the welded joints. Galvanic corrosion at the joints of hybrid structures accelerates corrosion. Corrosion characteristics of EMP welded galvanized steel/aluminum sheets are reported in [19].

The recent advancement in the technique led to the commercialization of the process with many welding applications. T. Aizawa et al., [20] used EMW to provide successful metallurgical and electrical bonds between flexible printed circuit boards (FPCB). The applications are suited for tubular assembly, regular or irregular shapes, and flat shape connections. The EMW is widely used to manufacture crimped gearbox parts, crimped Al/Steel tube instrumental panel beam, hemming of aluminum pressure vessel, Aluminum lid for the pharmaceutical glass bottle etc [21].

2.1.1 Working procedure

An AC power supply charges the capacitor bank. After storing the appropriate quantity of energy in the capacitors, it is released into a coil instantly (**Figure 3**). The discharge current creates a strong transient magnetic field inside the coil, which causes eddy currents to form in the work piece. Eddy currents prevent the magnetic field from diffusing through the outer work piece, resulting in a difference in the magnitude of the magnetic field on both sides of the work piece. The magnetic field

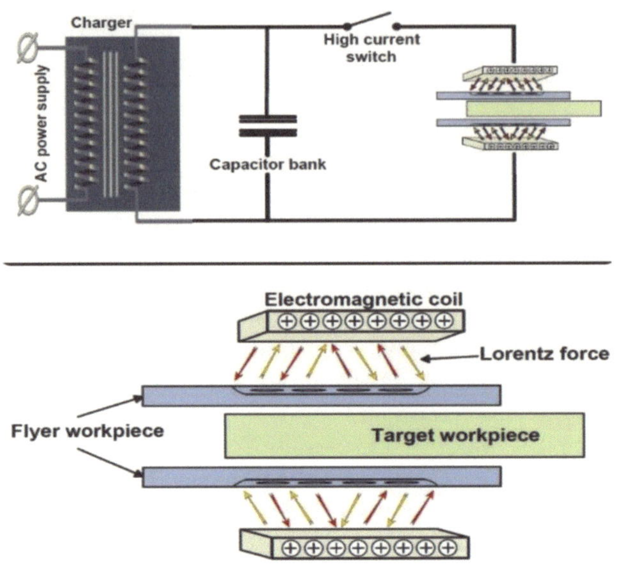

Figure 3.
Electromagnetic welding process.

causes the outer work piece to collide with the inner one. The collision of the work pieces causes bonding via a variety of mechanisms. The bonding will be intact as the distance between the atoms becomes smaller than the range of mutually attractive forces. In this instance, electrons are shared between the two materials, resulting in an intermetallic phase (potentially high hardness).

The actual bonding procedure takes less than 30 seconds. There are no shielding gases, fillers, or any auxiliary materials utilized. The electromagnetic pulse welding procedure is like a cold joining method, with very little heat generated. As a result, there is no heat-affected zone, and materials retain their qualities. The lack of heat and solid-state nature of the method allows for the joining of different materials. Aluminum to copper, aluminum to steel, and copper to brass are examples. The process EMW can weld the sheet metals with cross-sections comparable to that of an explosive weld. Since there is limited intermetallic phase generation at the interfaces, EMW produces a robust metallurgical bonded structure. EMW finds applications in tube forming, sheet metal forming, crimping, welding, and metal cutting with good results in highly conducting metals such as aluminum, copper, steel, and others.

2.2 Vaporizing foil actuator welding

Vaporizing foil actuator welding is an impact welding technique without chemical explosives. In this process, welding is achieved through a high-speed, oblique impact between the welding materials. At small-scale length scales and with similar driving pressures as explosive welding, vaporizing foil actuator welding can weld a wide range of advanced and dissimilar metal combinations. The fabrication of nano-sized particles and the structuring of high current pulses are two examples of outstanding achievements. Until the recent work by Vivek et al., [16] vaporizing foil actuators were not explored much. VFAW uses the same machinery as EMW, but instead of vaporizing a thin foil by the discharge current, the capacitor quickly vaporizes a thin layer to launch the metal flyer plate, as shown in **Figure 4**. Thermal deformation does not occur due to the low heat generated during the operation, and the properties of base metal do not deteriorate in the weld. Daehn et al., [22] developed the novel collision welding method called vaporizing foil actuator welding (VFAW). Many experimental [23, 24] and numerical simulations [25] suggest that VFAW can be a cost-effective, high-performance technology to join similar and dissimilar metal sheets. According to Hahn et al., [26] VFAW is a competitor technology to the MPW. Vivek et al., [27] successfully joined bulk metallic glass and copper sheets by VFAW. Suhani Chen et al., successfully welded Al −3003 and pure titanium by VFAW. Shuhai Chen et al., [23] reported microstructures, interfacial morphology, and mechanical property of dissimilar metals joint by the VFAW and investigated the influence of processing parameters on the anti-shear capacity of the joint produced by VFAW.

The process can be used to join steel sheets with aluminum, aluminum alloy sheets with different grade aluminum, and many dissimilar metal combinations [28, 29]. This could solve problems that automotive industries are facing in improving fuel efficiency by weight minimization.

2.2.1 Working procedure

The vaporizing foil actuator is placed against the flyer and supported by an anvil in VFAW, to direct the driving pressure toward the flyer sheet. (**Figure 4**). VFAW works by sending a strong electrical pulse into a foil. The foil is sandwiched between an anvil and a flyer. The flyer is forced into another base material called the target. An electrical pulse travels from the capacitor bank to the foil (actuator), inducing enough energy in the foil to cause it to vaporize, transforming it from a solid to a plasma gas in an instant that strains the system. The flyer placed over the foil moves a short distance before impacting the target. The flyer and target may be overlapping portions (as in a lap joint) or base metals overlapping in a different arrangement (as in a flange weld). A modest standoff distance exists between the base metals, which is crucial.

For the required pressure distribution of the flyer, a suitably shaped foil is required. A minor constriction in the foil can create a spot pressure for developing an impact spot weld. The foil is cut into variable weld shape, a longitudinal seam, stitch geometry, or even many weld places in a limited region. The quality of the weld depends on how foil explodes when a large amount of energy is transferred into it. A few kilojoules (up to 10) from a capacitor bank are deposited in 10 microseconds and directed to a small location.

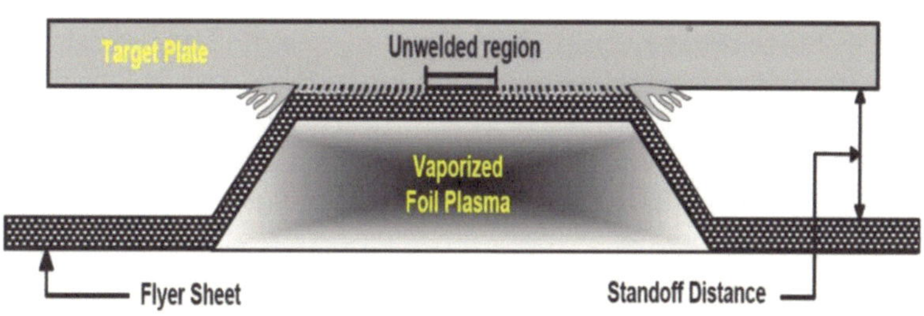

Figure 4.
Schematic diagram of vaporizing foil actuator welding.

2.3 Laser impact welding

Laser impact welding (LIW) is designed and developed at Ohio State University to join similar and dissimilar material combinations using the energy provided by high-velocity impact.

Joining dissimilar metals for small-scale parts like those used in medical devices and microelectronics could be one of the leading applications. Due to impact welding, the flyer plate collides with the base plate by a high-pressure shock wave created by an intense pulse laser. The basic principle of the technique is that an intense pulse laser beam focused by the lens will generate a specific spot diameter. The absorbent layer then evaporates instantly at a high temperature when exposed to laser irradiation. The vapor absorbs the laser energy, forming high-temperature, high-pressure plasma between the confinement layer and the flyer plate. Between the flyer plate and the confinement layer, the plasma continues to collect laser light and expands faster. As a result of the confinement layer's activity, high surface pressure is created, which propagates through the flyer plate as a shock wave [30].

One notable advantage of LIW over other impact welding processes is that the impact can be confined to a precise spot (sub-micron precision) in a precise time segment (precision of <10-5 seconds). Furthermore, the quantity of energy required for LIW is low, on the order of a few joules. As a result, LIW is an excellent method for generating welds in micro/nano-interface applications. Preliminary studies have been carried out on similar and dissimilar combinations of aluminum, titanium, copper, nickel, and iron. The goal is to understand the underlying bonding mechanisms and to discover if metallurgical reactions that typically lead to embrittlement in dissimilar metal systems (i.e. intermetallic formation) may be avoided. Good bonding can be accomplished in all cases with a typical "wavy" bond contact.

Many investigations on laser impact welding have recently been published. A novel laser high-speed impact spot welding method was proposed by Liu et al., [31]. They used laser impact spot welding to join Ti and Cu metal foils and used scanning electron microscopy (SEM) and energy-dispersive x-ray spectroscopy (EDS) to examine the microstructure of the bonding interface. The LIW of aluminum alloy 1100 and low carbon steel 1010 was investigated by Zhang et al., [32]. The welding joint had a relatively gentle curved bonding interface. Wang et al., [33, 34] optimized the flyer plate system in the laser impact welding device (confinement layer, absorption layer, and connecting layer). Wang et al., [35] proposed a parallel laser impact spot welding method. LISW was used to successfully weld dissimilar metal foil plates Cu/Al and Ti/Al. The impact of various welding parameters on welding quality was then carefully examined [35].

LIW process finds applications when thin metal sheets and foils of micron size (at least 25 μm) are to be joined. One distinctive advantage of this approach is that it appears applicable to arbitrarily small foil thicknesses and length scales, and does not rely on the intrinsic electrical conductivity of the flyer. This makes the method well suited for the manufacture and assembly of micro-devices such as micro-electro-mechanical systems [36, 37]. Wang et al., [34] reported laser impact welding of aluminum foil to titanium which can be used in medical devices, for example, the battery of the heart peacemaker.

2.3.1 Working procedure

The schematic diagram of LIW is shown in **Figure 5**, and the process involves the following stages:

a. Excitation stage: The ablation layer is vaporized into plasma as the laser irradiates it through the confinement layer. The reaction force of the plasma drives the flyer to emit due to the constraint of the confinement layer.

b. Flight phase: The flyer passes a pre-determined flight distance or standoff distance before colliding with the target at a specific speed and angle.

c. Welding stage: From the onset of the metallurgical bonding until the end position, the flyer and the target collide at a specific angle to complete welding

The benefits and drawbacks of Impact welding are summarized as follows:

- Compared to traditional thermal welding procedures, the technology offers significant advantages because it uses pressure rather than heat to achieve the bond.

- Quick and cost-effective joining of sheet metals typically dissimilar material joints.

- Manufacturing sheet metal products that were previously difficult to process using traditional joining methods.

- Impact welding such as magnetic pulse welding is referred to as a "cold" joining technique. Because the temperature increase is localized (on the order of 50 m), the work pieces only reach 30–50°C at the outside surfaces. So pieces can be unloaded and dealt with using simple equipment immediately after welding.

- One can achieve consistent joint quality; high repeatability.

- It is possible to achieve a high production rate.

- Contact-free: no forming tool marks, coatings, or sensitive materials can be processed.

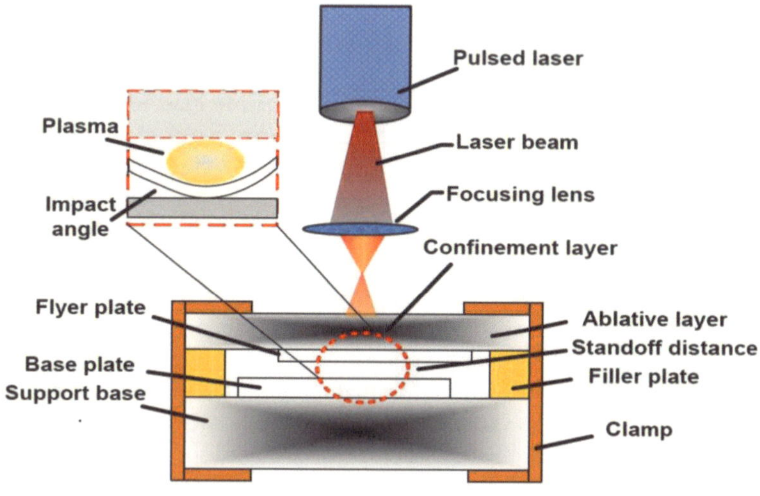

Figure 5.
Schematic diagram explaining LASER impact welding process.

- When compared to traditional welding processes, the technology has a substantially reduced negative environmental impact and is far more environmentally friendly:
- There is no heat, radiation, gas, or smoke thus the operator is protected by a shielding gas. Machines can execute the joining process in adverse situations, reducing further expenses in operator safety.
- Because the technology is environmentally friendly, it is feasible to improve the working conditions of the welder or operator.
- The magnetic pulse welding process is energy-efficient.

The workpieces to join, magnetic pulse welding has various limitations

- In EMPW one of the sheet materials must be a good electrical conductor; otherwise, a conductive 'driving' material should be utilized to enhance impact velocity.
- In the impact welding process, one workpiece must endure the impact of the other, a mandrel or support is required to prevent distortion.
- Due to the size of the welding equipment, it's done in a workshop; nevertheless, this is not necessarily a drawback when considering the prospective applications, which are largely factory-made parts.

3. Roll bonding

Cold roll bonding (CRB) is a solid phase method of bonding similar or dissimilar metal sheets by rolling at room temperature, which has been widely employed in the production of large multilayer composite sheets and foils. Cold pressure welding by rolling (CRB) is a sort of pressure welding or solid-state welding technology in which bonding is created by joint plastic deformation of the metals to be welded, suggesting that the degree of deformation is one of the major criteria [7]. CRB can be applied to a wide range of materials, cold bonding also works effectively with materials that cannot be fused by standard fusion. Rolling generates the high interfacial pressure required for bonding between two metal parts. Tensile shear test, slide shear test, multistep shear test, peeling test, and T-peel test is some of the procedures used to assess the bond strength of layered materials. Many investigations on the parameters influencing bonding have been conducted to understand the mechanisms of the complicated nature of bonding and the process conditions that have been established empirically [38]. The roll bonding parameters to be considered are, the type of metal under consideration, the amount of deformation, the bonding temperature, the amount of pressure applied, the bonding time, the metal purity, the lattice structure, the surface preparation conditions, the geometry of the deformation zone (shape factor), the stacking sequence, the number of layers, the layer thickness, and the type of post-heat treatment, which have been reported in the literature. The removal of contaminated layers from the surface using chemical and mechanical treatments is critical in the CRB process (surface preparation).

This usually entails washing and prepping surfaces to eliminate any impurities (bonding barriers) from the surfaces of the two metals to be bonded.

The complicated nature of the bonding mechanism involved in the roll bonding process has been the subject of numerous recent research studies [39]. Due to significant work-hardening by scratch-brushing, the metal surface gets hardened to a depth [40]. The cover layer fracture when exposed to sufficient surface expansions. Fracturing due to high interface pressure opens up new crack surfaces. Extrusion begins when the normal pressure is high enough and the surface expansion (or crack width) is significant. A connection is established when the asperities of the decovered virgin material of the two opposing surfaces make contact. This is the first bonding process mentioned [41]. The materials like aluminum, copper is easy to cold roll bond. Gold, silver, and platinum are other FCC metals that can be cold-bonded easily [7]. The bonding characteristics of hexagonal metals, such as magnesium, cadmium, and zirconium, are inferior to those of FCC materials.

Some of the applications of roll bonded sheet metal parts are, Al-Cu roll bonded sheets used in cooking utensils, heat exchangers, roof and wall plates. Al-Fe bonded sheets are used for electric heater reflectors, automobile silencers. Al roll bonded to stainless steel is used in automobile trims, Al-Steel-Al sheets used in the automobile exhaust system. Ti-Stainless steel-Ni found applications in the bipolar electrode in the fuel cell, Cu to Stainless steel bonded sheets have been widely used in communicator plate, and so on.

3.1 Working procedure

The schematic explaining the process of roll bonding is shown in **Figure 6**. The strips to be roll bonded are initially cleaned thoroughly by acetone to remove the surface contaminate layer. The cleaned strips are scratch brushed using a wire brush. The wire brushing on any one side produces a rough surface which will help in bonding the sample in two ways. One strain hardens the surface forming a hard brittle cover layer, and the other is it brings the fresh virgin metal surface to open up. Now, the two scratched samples are stacked and clamped as shown in the **Figure 6**. The stacked samples are passed between the pair of rollers to produce bonding. The bonding between

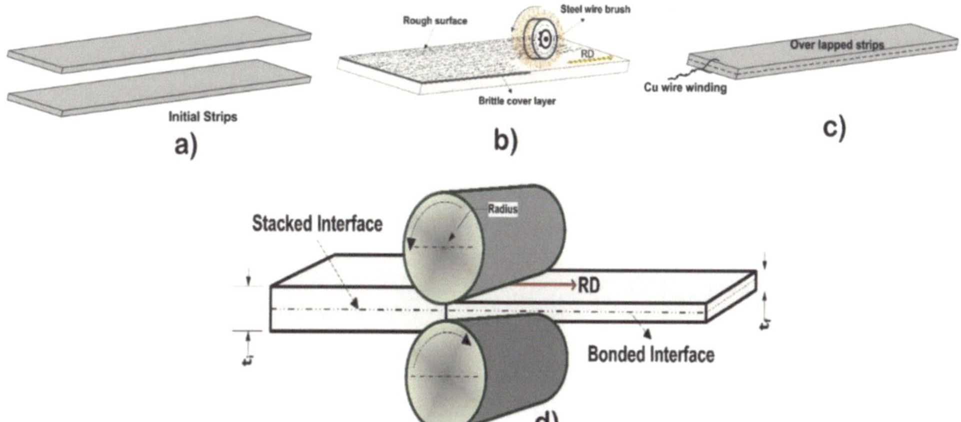

Figure 6.
Schematic diagram of roll bonding process.

the samples is possible only if the reduction given will be more than the threshold reduction [42]. It is feasible to heat the sheets near or above their recrystallization temperature to get improved bond strength.

4. Conclusion

The chapter elucidates the importance of sheet metal joining by forming (plastic deformation). The technique developed so far have immense potential and significance to join similar and dissimilar metal sheets without melting, which allows achieving substantially greater joint efficiency than conventional sheet joining methods. The impact-based solid-state techniques, EMW, VFAW, and LIW share a similar mechanism to join sheet metal, but they differ in the welding energy sources they utilize as indicated by their names. In the roll bonding, the overlapped sheets are joined by joint plastic deformation; the squeezing action of rollers creates sufficient pressure at the interface to create a bond.

The joining by forming methods provide consistent joint quality, high repeatability, and higher production rates; which are the prerequisite factors that industries look for. The processes are environmentally friendly, as there is no emission of heat, radiation, or smoke during the process. Hence, it is feasible to use and develop these forming methods to join sheet metals.

Conflict of interest

The authors declare no conflict of interest.

Author details

Akash Biradar and Malayathodi Rijesh*
Department of Manufacturing Engineering, School of Mechanical Engineering, Vellore Institute of Technology, Vellore, Tamil Nadu, India

*Address all correspondence to: rijesh.m@vit.ac.in

References

[1] Davim JP. Welding Technology. Berlin, Germany: Springer International Publishing; 2021

[2] Division USNA and SATU. Selected Welding Techniques. Washington, DC: National Aeronautics and Space Administration; 1963

[3] Hovanski Y, Upadyay P, Kleinbaum S, Carlson B, Boettcher E, Ruokolainen R. Enabling dissimilar material joining using friction stir scribe technology. Journal of Metals. 2017;**69**(6):1060-1064

[4] Cai W, Daehn G, Vivek A, Li J, Khan H, Mishra RS, et al. A state-of-the-art review on solid-state metal joining. Journal of Manufacturing Science and Engineering, Transactions of the ASME. 2019;**141**:MANU-18-1431

[5] Kumar Rajak D, Pagar DD, Menezes PL, Eyvazian A. Friction-based welding processes: Friction welding and friction stir welding. Journal of Adhesion Science and Technology. 2020;**34**:2613-2637. DOI: 10.1080/01694243.2020.1780716

[6] Wang H, Wang Y. High-velocity impact welding process: A review. Metals. 2019;**9**:1-18

[7] Li L, Nagai K, Yin F. Progress in cold roll bonding of metals. Science and Technology of Advanced Materials. 2008;**9**:023001

[8] Weman K. Pressure welding methods. Weld Process Handb. 2012:119-132

[9] Wu X, Shang J. An investigation of magnetic pulse welding of Al/Cu and interface characterization. Journal of Manufacturing Science and Engineering, Transactions of the ASME. 2014;**136**:MANU-13-1052

[10] Cowan GR, Bergmann OR, Holtzman AH. Mechanism of bond zone wave formation in explosion-clad metals. Metallurgical and Materials Transactions B: Process Metallurgy and Materials Processing Science. 1971;**2**:3145-3155

[11] Crossland B. Explosive Welding of Metals and Its Application. Oxford, UK: Clarendon Press; 1982

[12] Watanabe M, Kumai S. High-speed deformation and collision behavior of pure aluminum plates in magnetic pulse welding. Materials Transactions. 2009;**50**:2035-2042

[13] Chen S, Jiang X. Microstructure evolution during magnetic pulse welding of dissimilar aluminium and magnesium alloys. Journal of Manufacturing Processes. The Society of Manufacturing Engineers. 2015;**19**:14-21. DOI: 10.1016/j.jmapro.2015.04.001

[14] Yu H, Fan Z, Li C. Magnetic pulse cladding of aluminum alloy on mild steel tube. Journal of Materials Processing Technology. 2014;**214**:141-150. DOI: 10.1016/j.jmatprotec.2013.08.013

[15] Wang PQ, Chen DL, Ran Y, Yan YQ, She XW, Peng H, et al. Electromagnetic pulse welding of Al/Cu dissimilar materials: Microstructure and tensile properties. Materials Science and Engineering A. 2020;**792**:139-842. DOI: 10.1016/j.msea.2020.139842

[16] Li C, Zhou Y, Wang X, Shi X, Liao Z, Du J, et al. Influence of discharge current frequency on electromagnetic pulse welding. Journal of Manufacturing Processes. 2020;**57**:509-518. DOI: 10.1016/j.jmapro.2020.06.038

[17] Wang S, Xu L, Sun T, Li G, Cui J. Effects of process parameters on

mechanical performance and interfacial morphology of electromagnetic pulse welded joints between aluminum and galvanized steel. Journal of Materials Research and Technology. 2021;**10**:552-564. DOI: 10.1016/j.jmrt.2020.12.047

[18] Li Z, Beslin E, den Bakker AJ, Scamans G, Danaie M, Williams CA, et al. Bonding and microstructure evolution in electromagnetic pulse welding of hardenable Al alloys. Journal of Materials Processing Technology. 2021;**290**:116965

[19] Wang S, Luo K, Sun T, Li G, Cui J. Corrosion behavior and failure mechanism of electromagnetic pulse welded joints between galvanized steel and aluminum alloy sheets. Journal of Manufacturing Processes. 2021;**64**:937-947. DOI: 10.1016/j.jmapro.2021.02.039

[20] Aizawa T, Okagawa K, Kashani M. Application of magnetic pulse welding technique for flexible printed circuit boards (FPCB) lap joints. Journal of Materials Processing Technology. 2013;**213**:1095-1102. DOI: 10.1016/j.jmatprotec.2012.12.004

[21] Sapanathan T, Raoelison RN, Buiron N, Rachik M. Magnetic pulse welding: An innovative joining technology for similar and dissimilar metal pairs. Joining Technologies; 2016. pp. 243-273. DOI: 10.5772/63525

[22] Vivek A, Hansen SR, Liu BC, Daehn GS. Vaporizing foil actuator: A tool for collision welding. Journal of Materials Processing Technology. 2013;**213**:2304-2311. DOI: 10.1016/j.jmatprotec.2013.07.006

[23] Chen S, Daehn GS, Vivek A, Liu B, Hansen SR, Huang J, et al. Interfacial microstructures and mechanical property of vaporizing foil actuator welding of aluminum alloy to steel. Materials Science and Engineering A. 2016;**659**:12-21. DOI: 10.1016/j.msea.2016.02.040

[24] Meng Z, Gong M, Guo W, Liu W, Huang S, Hua L. Numerical simulation of the joining interface of dissimilar metals in vaporizing foil actuator welding: Forming mechanism and factors. Journal of Manufacturing Processes. 2020;**60**(Nov):654-665. DOI: 10.1016/j.jmapro.2020.11.009

[25] Wang K, Shang SL, Wang Y, Vivek A, Daehn G, Liu ZK, et al. Unveiling non-equilibrium metallurgical phases in dissimilar Al-Cu joints processed by vaporizing foil actuator welding. Materials and Design. 2020;**186**:108-306. DOI: 10.1016/j.matdes.2019.108306

[26] Hahn M, Weddeling C, Taber G, Vivek A, Daehn GS, Tekkaya AE. Vaporizing foil actuator welding as a competing technology to magnetic pulse welding. Journal of Materials Processing Technology. 2016;**230**:8-20

[27] Vivek A, Presley M, Flores KM, Hutchinson NH, Daehn GS. Solid state impact welding of BMG and copper by vaporizing foil actuator welding. Materials Science and Engineering A. 2015;**634**:14-19. DOI: 10.1016/j.msea.2015.03.012

[28] Hansen SR, Vivek A, Daehn GS. Impact welding of aluminum alloys 6061 and 5052 by vaporizing foil actuators: Heat-affected zone size and peel strength. Journal of Manufacturing Science and Engineering, Transactions of the ASME. 2015;**137**:1-6

[29] Chen S, Huo X, Guo C, Wei X, Huang J, Yang J, et al. Interfacial characteristics of Ti/Al joint by vaporizing foil actuator welding. Journal of Materials Processing Technology. 2019;**263**:73-81. DOI: 10.1016/j.jmatprotec.2018.08.004

[30] Liu H, Shen Z, Wang X, Wang H, Tao M. Numerical simulation and experimentation of a novel micro scale laser high speed punching. International Journal of Machine Tools and Manufacture. 2010;**50**:491-494. DOI: 10.1016/j.ijmachtools.2010.02.003

[31] Liu H, Gao S, Yan Z, Li L, Li C, Sun X, et al. Investigation on a novel laser impact spot welding. Metals. 2016;**6**:1-15

[32] Zhang Y, Babu SS, Prothe C, Blakely M, Kwasegroch J, Laha M, et al. Application of high velocity impact welding at varied different length scales. Journal of Materials Processing Technology. 2011;**211**:944-952. DOI: 10.1016/j.jmatprotec.2010.01.001

[33] Wang H, Taber G, Liu D, Hansen S, Chowdhury E, Terry S. Laser impact welding: Design of apparatus and parametric optimization. Journal of Manufacturing Processes. 2015;**19**:118-124. DOI: 10.1016/j.jmapro.2015.05.007

[34] Wang H, Vivek A, Wang Y, Taber G, Daehn GS. Laser impact welding application in joining aluminum to titanium. Journal of Laser Applications. 2016;**28**:032002-032007. DOI: 10.2351/1.4946887

[35] Wang X, Gu Y, Qiu T, Ma Y, Zhang D, Liu H. An experimental and numerical study of laser impact spot welding. Materials and Design. 2015;**65**:1143-1152. DOI: 10.1016/j.matdes.2014.08.044

[36] Moagar-Poladian GM, Illyefalvi-Vitez Z, Balogh B, Ulieru D. Coraci A: Laser Applications in the Field of MEMS. Washington, DC: The International Society for Optical Engineering; 2008. pp. 7007-7010

[37] Wang H, Liu D, Lippold JC, Daehn GS. Laser impact welding for joining similar and dissimilar metal combinations with various target configurations. Journal of Materials Processing Technology. 2020;**278**:116498. DOI: 10.1016/j.jmatprotec.2019.116498

[38] Jamaati R, Toroghinejad MR. Investigation of the parameters of the cold roll bonding (CRB) process. Materials Science and Engineering A. 2010;**527**:2320-2326

[39] Eizadjou M, Danesh Manesh H, Janghorban K. Mechanism of warm and cold roll bonding of aluminum alloy strips. Materials and Design. 2009;**30**:4156-4161. DOI: 10.1016/j.matdes.2009.04.036

[40] Jamaati R, Toroghinejad MR, Jamaati R, Toroghinejad MR. The role of surface preparation parameters on cold roll bonding of aluminum strips. Journal of Materials Engineering and Performance. 2011;**20**:191-197

[41] Zhang W, Bay N. Cold welding – Theoretical modeling of the weld formation. Welding Journal. 1997;**76**:417

[42] Eizadjou M, Danesh Manesh H, Janghorban K. Investigation of roll bonding between aluminum alloy strips. Materials and Design. 2008;**29**:909-913

Chapter 13

Welding Defect Detection with Deep Learning Architectures

Paolo Tripicchio and Salvatore D'Avella

Abstract

Welding automation is a fundamental process in manufacturing industries. Production lines integrate welding quality controls to reduce wastes and optimize the production chain. Early detection is fundamental as defects at any stage could determine the rejection of the entire product. In the last years, following the industry 4.0 paradigm, industrial automation lines have seen the introduction of modern technologies. Although the majority of the inspection systems still rely on traditional sensing and data processing, especially in the computer vision domain, some initiatives have been taken toward the employment of machine learning architectures. This chapter introduces deep neural networks in the context of welding defect detection, starting by analyzing common problems in the industrial applications of such technologies and discussing possible solutions in the specific case of quality checks in fuel injectors welding during the production stage.

Keywords: deep learning, visual inspection, industry 4.0, welding defects, imbalanced data, transfer learning

1. Introduction

The Fourth Industrial Revolution, or Industry 4.0, aims at automating traditional manufacturing and industrial practices exploiting the most recent technologies depicted in **Figure 1**. Integrating artificial intelligence (AI) and robotics with traditional practices, the world of manufacturing processes is undergoing a transformation from activities that rely on human experience and skills into flexible environments, including objective decisional systems fully integrated within the industrial process. Advanced robotics is meant to develop autonomous and intelligent systems that could reduce the intervention of human workers [1] in many of the crucial and repetitive tasks that represent the core business of companies. Augmented and virtual reality can give operators more information about their tasks [2] and help them to alleviate mental stress during some jobs. Additive manufacturing [3] can speed up the production process. Internet of things (IoT) [4] allows new forms of communication between machines, giving rise to smart devices that can help humans achieve their objectives. Radiofrequency identification (RFID) technologies are used for efficient logistics and inventory warehouse management [5] reducing costs while increasing quality and competitiveness.

IntechOpen

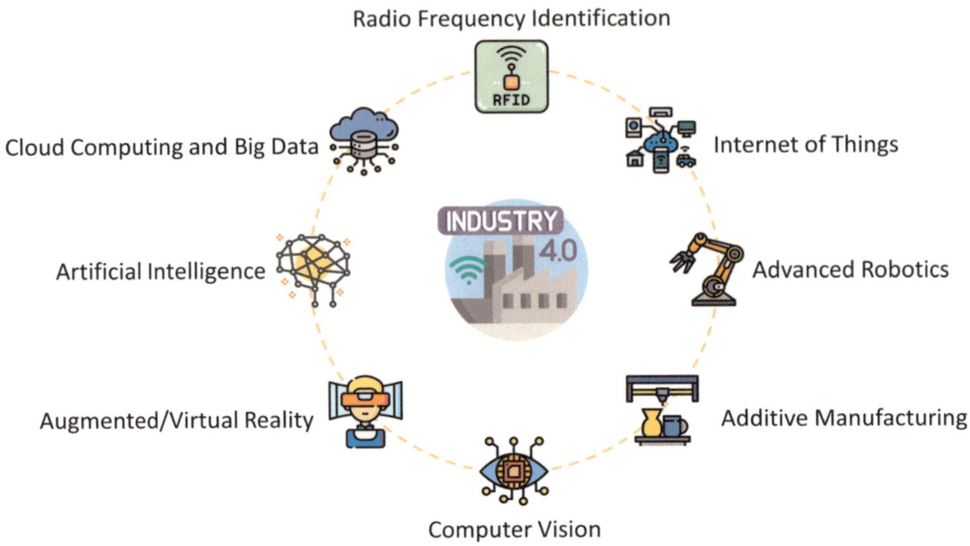

Figure 1.
Industry 4.0 pillar technologies.

Among all the aforementioned technologies, AI is perhaps the one that received more interest during the years. Indeed, nowadays, the industrial interest in AI applications in various sectors is undeniable. However, for industries, artificial intelligence is both a source of enthusiasm and skepticism. One reason is that deep learning (DL) is a technology based on data, and problems solved using AI are as good or as bad as the data they are trained on. In addition, companies perceive AI as a black box and would prefer understandable and explainable processes [6]. Both these aspects should be taken into consideration when developing industrial AI solutions.

Current automation-assisted production is mostly open-loop and relies on specific checkpoints to perform product quality analysis. Early systems based on vision date back to the nineties. Such an approach suits best when critical issues can be formally expressed by taking advantage of geometrical measurements or well-known features on the inspected objects. Unfortunately, these techniques cannot perform many quality-control activities because they need a predefined sequence of actions where quality checks should be designed carefully to meet the precise production requirements. Moreover, human nature shows formidable efficiency in learning simple checks even if it would be difficult to formalize such operations with a sequence of rules. Indeed, experience plays a relevant role in human evaluation for products quality assessment. Similarly, vision inspection processes performed by automated machines will require the development of novel algorithms that should be trained and improved with time and experience.

The introduction of automation systems in the production lines that exploit AI techniques has reduced the need for human intervention in the manufacturing process of many products. This innovation had a major impact on many industrial applications, and visual inspection is by far the activity that has profited most. Thanks to deep neural networks (DNNs), difficult computer vision tasks, such as object classification or detection and image segmentation, have been addressed recently using an adequate number of training data. DNNs are scalable, experience-based, and have similar performance to human workers. Since the development of AlexNet [7], solutions based on deep learning have been encouraged, and convolutional neural networks (CNN) also have been extensively utilized for automating optical quality inspections. However, since such networks need a huge amount of labeled data for training their parameters, it is difficult to have an adequately large set of faulty samples with well-optimized industrial processes for creating a *balanced dataset* for efficiently training the network to defect classification. Therefore, in most cases, the objective of the training moves from defect classification to anomaly detection.

Welding is a fundamental activity in many industrial manufacturing processes, such as automotive, shipbuilding, aerospace, and electronics. It is a crucial operation for the overall quality of the production line because a defect not detected in the early stages can determine the rejection of the entire product. This chapter introduces deep neural networks in the context of welding defect detection, starting by analyzing common problems in the industrial applications of such technologies and presenting in detail a solution for quality checks in fuel injectors welding during the production stage.

2. Background

Inspection analysis can be classified into one of the following categories [8]—*structural quality*, which searches for the presence of unnecessary components or lack of required parts; *surface quality*, where objects surfaces are inspected for wear, scratches, cracks, and other defects; *dimensional quality*, where the dimensions of the objects are checked to fall within given tolerances; *operational quality*, which evaluates the correctness of the quality inspection processes.

As of today, different methods have been proposed for inspecting the welding process online [9]. Their design is suited to diverse defects types and differ in the data processed during the evaluation. Among the sensing technologies employed in literature, optical detectors [10], acoustic measurements [11], and vision analysis [12] are surely the most utilized. While, for classification applications, artificial neural networks [13–15] and fuzzy inference systems [16, 17] are usually preferred thanks to the wide range of problems and diversity of defects they could cope with as in the case of classification of steel strip defects [18, 19].

However, the focus of these works is on defects classification and not on their detection. Therefore, they could not cope with feature understanding problems such as discriminating between good samples and defective ones. A different approach is proposed by Ak et al. [20] where X-ray images are used to detect defects in metal castings.

Recent literature is plenty of research addressing the problem of welding localization employing off-the-shelf DL architectures or introducing slight modifications on the tail of popular networks. These approaches are mostly based on the R-CNN [21], Faster R-CNN [22], and YOLO [23] architectures. The reason behind their adoption is that these architectures usually require little fine-tuning procedures for efficiently localizing welding areas and spots. Such efficiency is strictly related to the presence of plain metal surfaces in the surrounding area of the welding by enabling simple and accurate segmentation of the feature under inspection. This is the case of resistance spot welding (RSW) processes typically employed to connect metal sheets at a low cost and in a short time.

Concerning detection approaches, early methods based on traditional computer vision techniques [24] require hand-crafted features and complex threshold settings to adapt to environmental conditions. However, approaches based on deep learning allow increasing the robustness of the detection coping with environmental noise and the sensitivity of the welding processes.

The majority of approaches are built upon the above-mentioned architectures for welding spots localization. Fast R-CNN [25] is a region proposal network that computes the region of interest (ROI) on the feature map, thus improving upon the R-CNN architecture. Faster R-CNN integrates convolutional layers for object classification, feature extraction, bounding box regression, and region proposals into a network, further improving the detection performance but still not reaching real-time capabilities. Unlike the R-CNN family, which has a two-stage detection architecture, YOLO implements a regression network with a grid of bounding boxes and associated class probabilities, thus enabling real-time detection with recent hardware. In the race for timing performance, YOLOv2 [26] borrowed the anchor mechanism from SSD [27] and Faster R-CNN, which also enhanced the network *accuracy*. Focusing on small object detection (like welding spots), YOLOv3 [28] builds upon a backbone network combined with a feature pyramids network (FPN) [29] improving multi-scale prediction. The efficiency of the detection depends on the selected backbone network. Common choice are VGG [30], ResNet [31], DenseNet [32], and MobileNet [33]. These architectures differ in the computational complexity, the number of parameters, and inference speed.

Considering the reduced dimension of small spot welds, low-resolution feature maps in the backbone, and convolution strides dimension could cause an information leak. To face this issue, the work proposed by Dai et al. [34] introduces a modified MobileNEtV3 [35] architecture obtaining a good tradeoff between *accuracy* and *timing*.

Focusing on the classification and detection of defects over the welding area or joint, off-the-shelf solutions are no more efficient by themselves, and some issues need to be faced to enable the use of DNNs. Clustering and segmentation become difficult because the feature to be recognized are not easily separable. This chapter introduces some of the most common issues in the employment of DL for industrial quality inspection discussing the practical case of detection of welding defects in diesel injectors heads.

3. Inspection pipeline

Quality inspection systems based on vision techniques in most cases follow the workflow depicted in **Figure 2**. The process starts by collecting the sample images using a set of cameras or sensors exploiting an adequate source of illumination. Such samples are then processed to improve images quality. Therefore, once the features are extrapolated, the evaluation of the quality and the classification of the defect are performed. Measurement and classification could either be implemented with traditional computer vision algorithms, with modern DNN architectures, or with a fusion of both of them, as in the case presented in the following. Usually, the inspection system also provides an actuation step that triggers actions, depending on the analysis result, to the production lines that directly communicate with the control unit (commonly based on programmable logic controllers (PLCs)).

The work discussed in the study by Sassi et al. [36] originated from industrial demands with the specific target of detecting welding defects on diesel injectors in the production line. Such a project focused on realizing the most effective combination of

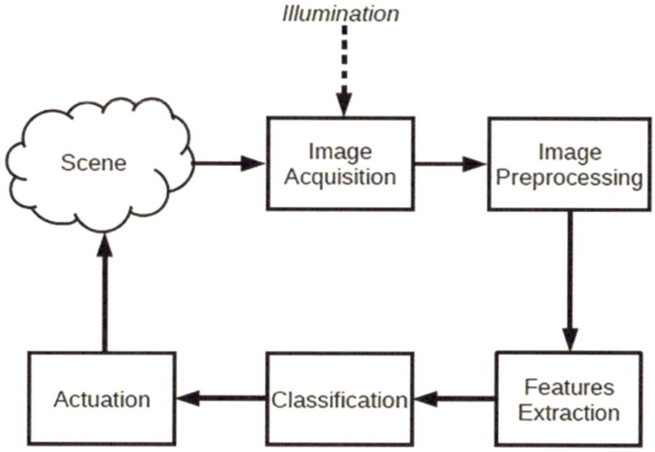

Figure 2.
Typical visual inspection workflow.

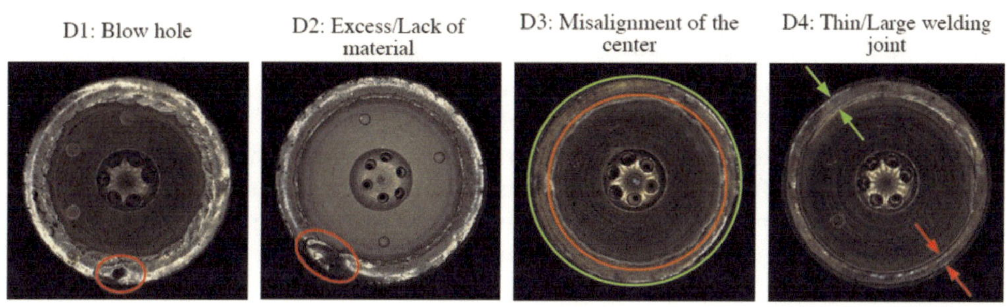

Figure 3.
Examples of defect classes. IN D3, green and red circles show the detected inner and outer edges of the welding joint. In D4, the red arrows highlight thin welding, while the green ones are standard ones.

traditional computer vision methods and deep neural network architecture for identifying the defects in the welding. In particular, the aim was to substitute the existing vision inspection system extending the classes of detectable defects in the analysis phase.

Welding joint defects may appear in different typologies: some are related to anomalies on the surface of the joint, while others are related to its geometrical properties, such as its thickness and position. Four categories have been defined for the analysis of the welding joint, as depicted graphically in **Figure 3** showing an example from each category:

- D1 (*Blowhole*): This defect corresponds to the joint area in which the material is blown up, thus, generating a loss of welding tightness and cavity;
- D2 (*Excess/Lack of material*): Blowholes are usually accompanied by a lack of material, while when they are unexploded with an excess of it;
- D3 (*Misalignment of the welding center*): When the injector head is not correctly aligned with the laser source, the resulting welding joint position is not centered;

- D4 (*Large/Thin welding joint*): When the amount of melt welding material is excessive or limited, the joint could be larger or thinner.

Defects D3 and D4 are quantitative measurable and are examined employing an algorithm based on traditional computer vision techniques (similar to the existing commercial solution). On the contrary, the others (D1 and D2) are more qualitative and are recognized through a method based on deep learning.

Furthermore, the analysis of the defects must be performed within a time slot that depends on the actual production line (1.8 seconds cycle time in the depicted scenario) to avoid interferences with the manufacturing process. This amount of time is required for the actuation system and the welding stage to process a new injector as input to the system.

4. Common challenges and possible countermeasures

During dataset preparation, the ideal case is the one in which several samples (in the order of thousands or more) are available for each class to be detected, the classes have balanced data, and they are well separated from each other. In such an ideal case, it is possible to give the network a representative set of samples of the whole input space for the training and avoid confusing the network with an uneven distribution of the inputs or the similarities between the classes.

Unfortunately, industrial production lines having well-optimized processes are usually present with few defective products and much more good samples. Therefore, it is often unfeasible to get sets of defective samples large enough to train CNNs for classification purposes. In the majority of the cases, the objective of the training moves from defect classification to anomaly detection. The worst-case scenario is the one presenting an *imbalanced dataset* with scarce availability of defect samples and classes that are not easily separable. Deep metric learning uses DNNs to directly learn a similarity metric, rather than creating it as a byproduct of solving a classification task [37]. They are well suited for tasks where the number of object classes is perhaps endless, and classification is not applicable. The approach is to compute a certain distance metric between input samples and reference prototypes. Moreover, the training will not even require defective samples if the class features are well defined and distinct from each other. Unfortunately, textured objects present surface appearance and properties that are stochastic.

Different sampling strategies could be implemented to deal with *imbalanced datasets*. When the minority class represents the defective pieces, it could be convenient to use as many elements from the majority class as the available defective ones. This approach is known as ***undersampling*** in literature. It can certainly be applied when the amount of defective samples is adequate for the training task. The alternative for not reducing the majority class is to give the network the available defective samples multiple times, trying to get the same amount of the good ones. Such an approach is called ***oversampling*** in literature. It is important to notice that this method could be risky as it can be easy to overfit the network due to the scarce representation of the input space that usually cannot completely cover the possible scenarios. Nevertheless, there are cases in which the beforementioned solutions are valuable tools to enhance the performance of the classifier as the work proposed by Yap et al. [38].

An alternative approach that is often used to increase the robustness of the classification is ***data augmentation***. Traditional techniques involve operations on the input images, such as scaling, cropping, rotation, mirroring, and color shift [39]. The samples are augmented based on the available data with the risk of a strong correlation between the original samples and the augmented ones that could probably lead to overfitting scenarios on a small dataset. However, if the augmentation is correctly managed, it can boost the performance of the classifier. Indeed, data augmentation has been employed with success in many defect inspection methods [40, 41].

Other ways for enlarging the dataset have been experimented like passing the input data through an encoder-decoder network that applies different transformations featured with random noise [42]. Another approach worth mentioning is the generation of ***virtual samples***. For example, the work presented by Leng et al. [43] successfully exploits virtual samples for face reconstruction.

Virtual data generation could be obtained by producing synthetic images with the intent to cover the whole input feature space. Generative adversarial network (GAN) [44] or the most recent conditional GAN (cGAN) [45] could be alternatively used for this purpose. However, this is computationally expensive and requires taking into account all possible configurations and boundary conditions for generating samples as close as possible to real ones. Domain randomization techniques [46] could be applied to synthetically generated data for improving the generalization capabilities and the robustness of the network.

Similar to humans, when learning new concepts or rules, if not clearly defined, the training can lead to fuzzy assumptions, possibly resulting in wrong outcomes. Additionally, when dealing with data obtained by a sensing apparatus, it is important to check the correctness of the acquired data samples to avoid possible causes of classification errors. A cleaning process should remove outliers (wrong data association of a sample with a class) and spurious samples that could confuse the learning process. Industrial processes often rely on qualitative evaluation, and unfortunately, different quality experts in the same industrial process classify the same product as belonging to different classes. If the same confusion is transferred to the DL architecture, the learning process will probably worsen the decision process. For this reason, a preprocessing stage on the data is essential. In most cases, the help of professionals of the sector for interpreting, filtering, and preprocessing the data is welcome.

A last and quite important aspect is the adoption of correct performance metrics and *loss functions* enabling successful training with *imbalanced datasets*. In this context, Mower [47] proposes a balanced accuracy statistic that mediates the *recall* and *specificity* metrics. A more general approach is to directly scale the confusion matrix terms based on the relative support of each class as proposed by Tripicchio et al. [48]. Other studies modify the *loss function* to account for class imbalance. In particular, *binary cross-entropy loss* is a common choice for classification tasks. In a study by Xi and Tu [49] a balanced *cross-entropy* is introduced where, differently from the *binary cross-entropy loss*, the contribution of the dominant class is multiplied by the fraction of the less dominant class. However, the method does not differentiate between easy/hard examples. A different approach is proposed by Lin et al. [50] where the authors focus the training on hard negatives, down-weighting the *loss* assigned to well-classified examples. The resulting *loss* is called *focal loss*.

5. Transfer learning for defect detection

It has been seen that the first layers of CNNs learn kernels acting as color blob detectors or Gabor filters. Such a property seems to be very general and the features learned do not appear to be strictly dependent on the particular training set that has been adopted. As humans can learn from experience and transfer the notion learned in diverse application domains, similarly, a DL architecture can transfer the features learned on a particular dataset to another CNN, which will be trained on a different one [51]. Such a technique is called ***transfer learning*** and is a worthy tool for solving the problem of scarcity of defective samples. This paradigm involves pretraining the network on a dataset (usually larger) for learning feature extraction layers and afterward fine-tuning the classification pipeline with the relevant dataset for the specific task. Knowledge transfer breaks the fundamental assumption that the data presented to the network during the training phase must be in the same feature space as the ones presented in the inference phase. Feature extraction layers obtained applying the *transfer learning* paradigm would be able to extract generic convolution features that could be exploited in different tasks.

Following the *transfer learning* approach, the work by Sassi et al. [36] yielded a 97% *accuracy* during testing in the laboratory and proved successful during operation in a real production line, reaching an *accuracy* of 99% after subsequent training.

The work combines a traditional computer vision pipeline together with a DL architecture. This pipeline was necessary to maintain the compatibility with classical production lines and provide a correct input to the welding defect detection phase. The algorithm receives the raw image as input, converts it from Bayer format to grayscale, and improves the edge detection by equalizing the levels and applying a Gaussian blur. In a successive step, since different kinds of injectors can be analyzed by the same system, the type of injector is identified, and the position of its center is obtained. The algorithm proceeds to detect the outer shell of the injector head by estimating an external radius that approximates the detected blob. Then, using the extracted information, the algorithm performs an area search for welding points and estimates a welding circle on the joint. Subsequently, the algorithm collects statistics about the number of welding points found and their positions. In traditional industrial systems, a set of thresholds decided by the manufacturing company is used to evaluate the welding quality from the measured quantities.

A schematic overview of the algorithm is shown in **Figure 4**. The algorithm's output gives quantitative information about the welding and produces a processed image to be given as input to the second analysis stage. The extracted information allows evaluating the continuity of the welding in a certain area on the injector's head, verifying the centering of the inner part of the injector with respect to the outer one,

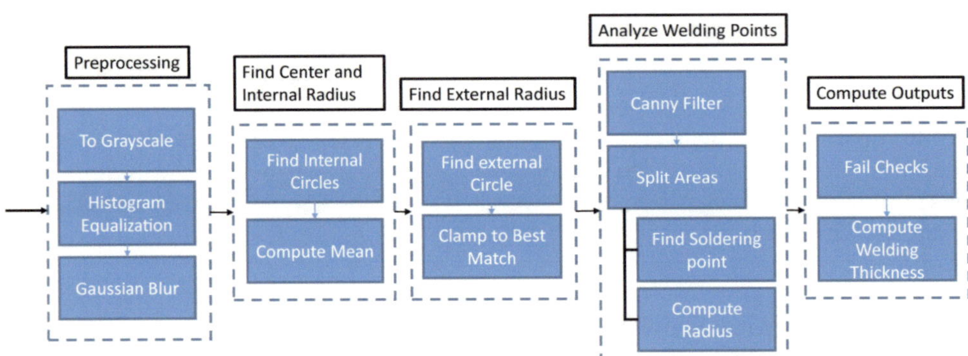

Figure 4.
Schematics of the components of the geometrical analysis pipeline.

Figure 5.
Schematic representation of the layers and blocks in the DenseNet-121 deep learning architecture.

and eventually the welding thickness. This information is also beneficial to clean the image from unnecessary data for the subsequent analysis and to center the injector images to obtain more controlled conditions on the input of the successive stage.

The DL architecture chosen in that work is the DenseNet-121. **Figure 5** depicts the structure of the network. DenseNet efficiently simplifies the connectivity pattern between layers guaranteeing maximum information flow by reusing the features through the network. Concerning the training phase, every layer has direct access to the gradients from the original input image and the *loss function*. Different from the first feedforward neural networks that connect the output of each layer to the subsequent layer after applying a composite of operations, DenseNet concatenates the output feature maps of the layers to obtain the equation $x_l = H_l([x_0, x_1, \dots, x_{l-1}])$. The network is formed by dense blocks, which have a constant size of the feature maps within a block but a varying number of filters, and transition layers that connect the blocks combining batch normalization, $1x1$ convolution, and $2x2$ pooling.

In the approach presented by Sassi et al. [36], the *transfer learning* technique has been employed and evaluated by comparing the results achieved when the features are transferred from a pretrained model on the *Material in Context* (MINC) [52] dataset. Such dataset contains 2,996,674 patches obtained from 436,749 images labeled according to 23 material classes. A binary classification problem has been set up by selecting positive samples as scrap injectors and negative samples as good injectors. The results of the classification problem are shown as a confusion matrix having four possible values—*true negative* (tn), *true positive* (tp), *false negative* (fn), and *false positive* (fp). The metrics that better estimate the quality of the defect analysis are the *recall* $\left(\frac{tp}{tp+fn}\right)$, which describes the ability to detect faulty pieces, and the *accuracy* $\left(\frac{tp+tn}{tp+tn+fp+fn}\right)$ that describes the overall quality of the analysis. The *precision* $\left(\frac{tp}{tp+fp}\right)$ is important to not discard too many injectors, but it is not crucial as the *recall* since not detecting a defect could be dangerous if it proceeds through the assembly line.

Unfortunately, the MINC dataset is highly unbalanced. Therefore, three classes, that is, *plastic*, *metal*, and *others*, have been selected as a subset for the pretraining stage to alleviate training problems. The idea of using such a dataset for *transfer learning* was to exploit the metallic features that could resemble the ones in the welding images, and the dataset class reduction does not affect the learned features on metallic materials.

6. Managing production variability

Sometimes, during production lines maintenance or innovations, the replacement of a machine, the change of a supplier, or the change in a manufacturing process, could lead to a significant variation on the usual production procedure in terms of the visual quality of the products. Such situations could vanish the capacity of a machine computation to return the expected results.

In this context, continuing on the problem of detecting welding defects on injectors heads, the work presented by Tripicchio et al. [48] proposes possible solutions to this issue without requiring an architectural change in the learning architecture. The new case had to handle some modifications concerning the parameters associated with the welding process, producing input samples with specific artifacts that the previously designed and trained network did never encounter. In particular, such new inputs were correlated to a variation in the substance used for the soldering that generated gold-violet spots on the injector head in random positions. Such noise introduces a novel complexity in the detection of the defects because the spots can hide or visually resemble the presence of bumps and holes in the welding layer. The followed approach was to make fewer changes as possible in the architecture of the network, operating a smart preprocessing and applying filtering techniques.

The results show the ability to train a network with almost 7 million parameters on just 306 training images belonging to the new alteration, achieving a *recall* of 100.00% and an *accuracy* of 97.22%.

Such a result has been achieved leveraging on two important aspects. The first is the design of a custom preprocessing and filtering stage, while the second is the adoption of a novel data balancing strategy.

A preprocessing stage is needed on the input images with the aim of erasing or smoothing the chromatic nuances that could confuse the feature learning process. In particular, three filtering approaches have been proposed and tested (**Figure 6**). The first filter (*constant filter*) detects regions on the image in the gold and violet ranges of the HSV space filling such regions with a constant RGB color resembling the chromatic value of the injector contour. The second kind of filter (*median filter*), once selected the regions, fills them with the median RGB value of each channel. In the third filtering approach (*patch filter*), a 4×4 patch is virtually generated to resemble a part not affected by defects, and it is used to fill the detected gold and violet regions. In particular, every pixel of the regions is substituted with the value of the corresponding pixel of the synthetic image, adding the median value of the original image.

Different analyses have been done to assess the performance improvement given by such filters. As a result, a *patch filter* was selected as the method for the subsequent tuning of the network.

Concerning data imbalance, an exploration of different unbalanced splits has been performed. To prevent overfitting and lead the learning process toward generalization, the authors propose to compute the performance metrics at each evaluation step considering the input imbalance. In particular, metrics like *specificity* or *recall* are not affected by the imbalance of the data different from other metrics, like *accuracy*, which should be revised.

Figure 6.
Different filters applied on a sector of the same injector contour image. (a) No filter. (b) Median fill filter. (c) Patch filter.

Defective injectors were chosen as positive samples and *false negatives* and *true positives* values were weighted depending on the imbalance since the defective class was the smaller one. The imbalance is compensated by multiplying these values by the proportion of the input dataset. Consequently, the confusion matrix presents a more balanced indication of the performance of the training.

Cross-validation has been applied to improve generalization concerning the stochastic gradient descent optimization. The network has been trained multiple times by combining different variations of the proportions between defective and good samples and changing the numbers of epochs. During the training phase, each epoch is compared with all previous epochs for obtaining the one with the highest performance in terms of *recall* and *accuracy*.

The *F-score* was chosen as the best multi-performance metric to evaluate the testing achieved on the different variations of the training. Concerning the imbalance, the obtained performances give that an *unbalanced dataset* could provide better results if the imbalance is considered while training.

7. Conclusions

This chapter highlights the importance of the employment of deep learning architectures in the context of future industrial applications with a focus on welding and welding defects detection. The industrial sector and especially the manufacturing industry pose several challenges to the design of efficient and robust quality inspection processes. The most common issues are discussed in detail, and possible countermeasures are suggested to overcome such issues. In particular, the problem of data imbalance, scarcity of examples, environmental noises, change in the nominal conditions of the process, or the presence of artifacts are discussed. Application examples from previous works of the authors are proposed to clarify how the suggested countermeasures can be put into practice. Although many industries are still scared of adopting deep learning approaches due to a lack of knowledge of their internal processes or reasoning, extensive use of artificial intelligence applications is envisaged for the near future.

Abbreviations

AI	Artificial Intelligence
CNN	Convolutional Neural Network
DL	Deep Learning
DNN	Deep Neural Network
GAN	Generative Adversarial Network
ICT	Information and Communication Technologies
IoT	Internet of Things
PLC	Programmable Logic Controller
ROI	Region of Interest
RSW	Resistance Spot Welding

Author details

Paolo Tripicchio* and Salvatore D'Avella
Department of Excellence in Robotics and AI, Institute of Mechanical Intelligence, Scuola Superiore Sant'Anna, Pisa, Italy

*Address all correspondence to: paolo.tripicchio@santannapisa.it

References

[1] D'Avella S, Tripicchio P. Supervised stowing as enabling technology for the integration of impaired operators in the industry. Procedia Manufacturing. 2020; **51**:171-178

[2] Fraga-Lamas P, FernáNdez-CaraméS TM, Blanco-Novoa ÓS, Vilar-Montesinos MA. A review on industrial augmented reality systems for the industry 4.0 shipyard. IEEE Access. 2018;**6**:13358-13375

[3] Tofail SAM, Koumoulos EP, Bandyopadhyay A, Bose S, O'Donoghue L, Charitidis C. Additive manufacturing: Scientific and technological challenges, market uptake and opportunities. Materials Today. 2018;**21**(1):22-37

[4] Manavalan E, Jayakrishna K. A review of internet of things (IoT) embedded sustainable supply chain for industry 4.0 requirements. Computers & Industrial Engineering. 2019;**127**:925-953

[5] Tripicchio P, Unetti M, D'Avella S, Buffi A, Motroni A, Bernardini F, et al. A synthetic aperture uhf rfid localization method by phase unwrapping and hyperbolic intersection. IEEE Transactions on Automation Science and Engineering. 2021. DOI: 10.1109/TASE.2021.3057433

[6] Tripicchio P, D'Avella S. Is deep learning ready to satisfy industry needs? Procedia Manufacturing. 2020;**51**: 1192-1199

[7] Krizhevsky A, Sutskever I, Hinton GE. Imagenet classification with deep convolutional neural networks. In: Proceedings of the 25th International Conference on Neural Information Processing Systems – Volume 1, NIPS'12. USA: Curran Associates Inc.; 2012. pp. 1097-1105

[8] Malamas EN, Petrakis EGM, Zervakis M, Petit L, Legat J-D. A survey on industrial vision systems, applications and tools. Image and Vision Computing. 2003;**21**(2):171-188

[9] Shao J, Yan Y. Review of techniques for on-line monitoring and inspection of laser welding. Journal of Physics: Conference Series. 2005;**15**(1): 101-107

[10] Zheng K, Li J, Lei Tu C, Song Wang X. Two opposite sides synchronous tracking x-ray based robotic system for welding inspection. In: 2016 23rd International Conference on Mechatronics and Machine Vision in Practice (M2VIP). Piscataway, NJ: IEEE; 2016. pp. 1-5

[11] Zhao J, Sheng H, Zhou X. Study on the application of acoustic emission testing technique in monitoring 16 mn steel welding defects. In: 2016 International Conference on Advanced Mechatronic Systems (ICAMechS). Piscataway, NJ: IEEE; 2016. pp. 318-321

[12] Fang Z, Xu D, Tan M. A vision-based self-tuning fuzzy controller for fillet weld seam tracking. IEEE/ASME Transactions on Mechatronics. 2011; **16**(3):540-550

[13] Podržaj P, Čebular A. The application of lvq neural network for weld strength evaluation of RF-welded plastic materials. IEEE/ASME Transactions on Mechatronics. 2016; **21**(2):1063-1071

[14] Senthil Kumar G, Natarajan U, Ananthan SS. Vision inspection system for the identification and classification of defects in MIG welding joints. International Journal of Advanced

Manufacturing Technology. 2012;**61** (9–12):923-933

[15] Zahran O, Kasban H, El-Kordy M, Abd El-Samie FE. Automatic weld defect identification from radiographic images. NDT & E International. 2013;**57**:26-35

[16] Yan Z, Zhang G, Wun L. Simulation and controlling for weld shape process in p-gmaw based on fuzzy logic. In: 2011 IEEE International Conference on Mechatronics and Automation. Piscataway, NJ: IEEE; 2011. pp. 2078-2082

[17] Zammar IA, Mantegh I, Huq MS, Yousefpour A, Ahmadi M. Intelligent thermal control of resistance welding of fiberglass laminates for automated manufacturing. IEEE/ASME Transactions on Mechatronics. 2015; **20**(3):1069-1078

[18] Masci J, Meier U, Ciresan D, Schmidhuber J, Fricout G. Steel defect classification with Max-Pooling Convolutional Neural Networks. In: 2012 Int. Jt. Conf. Neural Networks. Piscataway, NJ: IEEE; 2012. pp. 1-6

[19] Zhou S, Chen Y, Zhang D, Xie J, Zhou Y. Classification of surface defects on steel sheet using convolutional neural networks. Materials and Technology. 2017;**51**:123-131

[20] Ronay Ak, Max Ferguson, Yung-Tsun T. Lee, and Kincho H. Law. Automatic localization of casting defects with convolutional neural networks. In: 2017 IEEE International Conference on Big Data (BigData 2017). Boston, MA: IEEE; 2017

[21] Girshick R, Donahue J, Darrell T, Malik J. Rich feature hierarchies for accurate object detection and semantic segmentation. In: Proceedings of the IEEE Conference on Computer Vision and Pattern Recognition. Piscataway, NJ: IEEE; 2014. pp. 580-587

[22] Ren S, He K, Girshick R, Sun J. Faster R-CNN: Towards real-time object detection with region proposal networks. Advances in Neural Information Processing Systems. 2015;**28**:91-99

[23] Redmon J, Divvala S, Girshick R, Farhadi A. You only look once: Unified, real-time object detection. In: Proceedings of the IEEE Conference on Computer Vision and Pattern Recognition. Piscataway, NJ: IEEE; 2016. pp. 779-788

[24] Ruisz J, Biber J, Loipetsberger M. Quality evaluation in resistance spot welding by analysing the weld fingerprint on metal bands by computer vision. The International Journal of Advanced Manufacturing Technology. 2007;**33**(9):952-960

[25] Girshick R. Fast R-CNN. In: Proceedings of the IEEE International Conference on Computer Vision. Piscataway, NJ: IEEE; 2015. pp. 1440-1448

[26] Redmon J, Farhadi A. Yolo9000: better, faster, stronger. In: Proceedings of the IEEE Conference on Computer Vision and Pattern Recognition. Piscataway, NJ: IEEE; 2017. pp. 7263-7271

[27] Liu W, Anguelov D, Erhan D, Szegedy C, Reed S, Cheng-Yang F, et al. Ssd: Single shot multibox detector. In: European Conference on Computer Vision. New York, NY: Springer; 2016. pp. 21-37

[28] Redmon J, Farhadi A. Yolov3: An incremental improvement. In: Computer Vision and Pattern Recognition (CVPR). Berlin/Heidelberg, Germany: Springer; 2018. pp. 1804-2767. arXiv preprint arXiv:1804.02767

[29] Lin T-Y, Dollár P, Girshick R, He K, Hariharan B, Belongie S, editors. Feature pyramid networks for object detection. In: Proceedings of the IEEE Conference on Computer Vision and Pattern Recognition. Piscataway, NJ: IEEE; 2017. pp. 2117-2125

[30] Simonyan K, Zisserman A. Very deep convolutional networks for large-scale image recognition. In: The 3rd International Conference on Learning Representations. San Diego, CA: ICLR; 2015. arXiv preprint arXiv:1409.1556

[31] He K, Zhang X, Ren S, Sun J. Deep residual learning for image recognition. In: Proceedings of the IEEE Conference on Computer Vision and Pattern Recognition. Piscataway, NJ: IEEE; 2016. pp. 770-778

[32] Huang G, Liu Z, Van Der Maaten L, Weinberger KQ. Densely connected convolutional networks. In: Proceedings of the IEEE Conference on Computer Vision and Pattern Recognition. Piscataway, NJ: IEEE; 2017. pp. 4700-4708

[33] Howard AG, Zhu M, Chen B, Kalenichenko D, Wang W, Weyand T, Andreetto M, Adam H. Mobilenets: Efficient convolutional neural networks for mobile vision applications. 2017. arXiv preprint arXiv:1704.04861

[34] Dai W, Li D, Tang D, Jiang Q, Wang D, Wang H, et al. Deep learning assisted vision inspection of resistance spot welds. Journal of Manufacturing Processes. 2021;**62**:262-274

[35] Howard A, Sandler M, Chu G, Chen LC, Chen B, Tan M, et al., editors. Searching for mobilenetv3. In: Proceedings of the IEEE/CVF International Conference on Computer Vision. Piscataway, NJ: IEEE; 2019. pp. 1314-1324

[36] Sassi P, Tripicchio P, Avizzano CA. A smart monitoring system for automatic welding defect detection. IEEE Transactions on Industrial Electronics. 2019;**66**(12):9641-9650

[37] Ren R, Hung T, Tan KC. A generic deep-learning-based approach for automated surface inspection. IEEE Transactions on Cybernetics. 2018; **48**(3):929-940

[38] Yap BW, Rani KA, Rahman HAA, Fong S, Khairudin Z, Abdullah NN. An application of oversampling, undersampling, bagging and boosting in handling imbalanced datasets. In: Herawan T, Deris MM, Abawajy J, editors. Proceedings of the First International Conference on Advanced Data and Information Engineering (DaEng-2013). Singapore: Springer Singapore; 2014. pp. 13-22

[39] Cubuk ED, Zoph B, Mané D, Vasudevan V, Le QV. Autoaugment: Learning augmentation policies from data. In: Proceedings of the IEEE/CVF Conference on Computer Vision and Pattern Recognition. Piscataway, NJ: IEEE; 2019. pp. 113-123. ArXiv, abs/1805.09501

[40] Wang T, Chen Y, Qiao M, Snoussi H. A fast and robust convolutional neural network-based defect detection model in product quality control. The International Journal of Advanced Manufacturing Technology. 2018;**94**(9):3465-3471

[41] Chen J, Liu Z, Wang H, Núñez A, Han Z. Automatic defect detection of fasteners on the catenary support device using deep convolutional neural network. IEEE Transactions on Instrumentation and Measurement. 2018;**67**(2):257-269

[42] Jorge J, Vieco J, Paredes R, Sánchez J-A, Benedí J-M. Empirical evaluation of

variational autoencoders for data augmentation. In: VISIGRAPP. Setúbal, Portugal: SCITEPRESS; 2018

[43] Leng B, Yu K, Jingyan QIN. Data augmentation for unbalanced face recognition training sets. Neurocomputing. 2017;**235**(C):10-14

[44] Goodfellow IJ, Pouget-Abadie J, Mirza M, Xu B, Warde-Farley D, Ozair S, et al. Generative adversarial nets. In: NIPS. Red Hook, NY: Curran Associates, Inc.; 2014

[45] Mirza M, Osindero S. Conditional generative adversarial nets. 2014. cite arxiv:1411.1784

[46] Tobin J, Fong R, Ray A, Schneider J, Zaremba W, Abbeel P, editors. Domain randomization for transferring deep neural networks from simulation to the real world. In: 2017 IEEE/RSJ International Conference on Intelligent Robots and Systems (IROS). Piscataway, NJ: IEEE; 2017. pp. 23-30

[47] Mower JP. Prep-Mt: Predictive RNA editor for plant mitochondrial genes. BMC Bioinformatics. 2005; **6**(1):96

[48] Tripicchio P, Camacho-Gonzalez G, D'Avella S. Welding defect detection: Coping with artifacts in the production line. The International Journal of Advanced Manufacturing Technology. 2020;**111**(5):1659-1669

[49] Xie S, Zhuowen T. Holistically-nested edge detection. In: Proceedings of the IEEE International Conference on Computer Vision. Piscataway, NJ: IEEE. 2015; pp. 1395-1403

[50] Lin TY, Goyal P, Girshick R, He K, Dollár P. Focal loss for dense object detection. In: Proceedings of the IEEE International Conference on Computer Vision. Piscataway, NJ: IEEE. 2017. pp. 2980-2988

[51] Yosinski J, Clune J, Bengio Y, Lipson H. How transferable are features in deep neural networks? In: NeurIPS. Red Hook, NY: Curran Associates, Inc.; 2014

[52] Bell S, Upchurch P, Snavely N, Bala K. Material Recognition in the Wild with the Materials in Context Database. Piscataway, NJ: IEEE; 2014

Chapter 14

Center Stir Zone Investigations of Dissimilar AA6082, AA2014 and AA7075 Welds

K.T. Thilagham and S. Muthukumaran

Abstract

The study compares the mechanical and metallurgical properties of AA6082, AA2014, and AA7075 dissimilar friction stir welded aluminum 6 mm plates. The alloys AA2014 and AA7075 are aerospace grade, whereas AA6082 is structural grade. The AA6082/AA7075, AA6082/AA2014, and AA2014/AA7075 joints were formed with optimized parameters of 2° tilt angle, 900 rpm rotational speed, and 80 mm/min feed rate with a constant axial force of 20 kN. Then, to investigate the stir zone properties of the joints, the tensile strength, microstructural, and hardness variations across the weld were revealed. Despite the fact that the strength of each joint was varied, the fine grain in the stir zone across the weld and advancing side weld/HAZ failure in tensile failure were studied for all welds. Further EBSD analysis revealed fine grains for the formation of its center stir zone due to dynamic recovery recrystallization during welding.

Keywords: friction stir welding, dynamic recovery and recrystallization, metallurgical and mechanical properties, stir zone

1. Introduction

Generally, the AA7075-T651 and AA2014-T6 are aerospace alloys, which are relatively high in strength compared to AA6082-T6 structural grade alloy. These heat-treatable grades AA2xxx, 6xxx, and 7xxx alloys have more difficulties with fusion welding. Mainly Kah et al. [1] stated that the low melting eutectic at the grain boundary of the heat-affected zone for precipitation-hardenable AA7075 or AA2014 alloys continued to crack during welding. Under extreme conditions, the cracks might be formed along with the fusion zone interface also. So, it is identified that the lower strength in HAZ leads to hot cracking or liquation crack due to the redistribution of intermetallic in fusion welding.

Thomas [2] showed that a sudden decrease in hardness in the HAZ when compared with the fusion zone and the unaffected base metal zone during welding of Al6061-T6 alloy. This HAZ degradation is due to the transformation of the intermetallic precipitates to the non-strengthening or coarsened precipitates at the temperature range between 290 and 400°C, while the whole weld was experienced the same temperature range during welding. This increases when heat input increases.

Beiranvand et al. [3] experimented that the percentage of the added alloying elements in the alloy was responsible for the solidification crack sensitivity. The AA6082 alloy is crack sensitive because the alloys contain approximately 1.0% magnesium silicide (Mg_2Si), which gives greater chances of solidification crack sensitivity. The Mg_2Si content invariably produces cracking during welding.

Ogbonna et al. [4] highlighted that the key cause of porosity in aluminum is hydrogen, which has a very high solubility in molten aluminum but very low solubility in solid. However, in MIG welding, minimizing the risk of porosity was accomplished by changing the cooling rate to allow the gas to escape, raising the welding current, and/or lowering the travel speed to increase the heat input. Thus, the fusion welding shows as liquation cracking, porosity, solidification cracks, poor dead profile, etc. As a result, friction stir welding is generally used to combine the alloys mentioned above.

This friction stir welding utilizes a tool shoulder-pin primarily to join the metals, which requires heat and pressure, similar to other solid-state welding processes [5]. It is therefore advantageous to combine similar/dissimilar materials without the addition of an auxiliary arrangement using a non-consumable stirring tool. This could be useful for joining many structural metals and their alloys such as aluminum, copper, lead, and steel plates [6–10]. The process is ecologically beneficial and nature-friendly in light of the fact that no toxic gases are produced during this process.

Mishra and Ma [5] investigated the FSW of AA7075 with process parameters of 350 rpm, 152 mm/min for 6.32 mm thick plate for cylindrical tool pin. Kumar et al. [11] experimented on the combination of dissimilar friction stir welding of AA7075-T651 and AA6061-T651 alloys to resolve fusion welding difficulties for similar alloy applications. Trials were carries out on the parameters such as the rotational speed of 800–1000 rpm, welding speed of 90–110 mm/min with various tool profiled tapered cylindrical threaded, simply square and taper square threaded tools. On considering this, here the comparative study on the dissimilar FSW on 6 mm AA2014-T6, AA6082-T6, and AA7075-T651 were studied using constant optimal parameters such as tilt angle (2°), tool rotation speed (900 rpm), and transverse feed of (80 mm/min) with the conical tapered pin tool.

The microstructural grain size and grain orientation relationships that control the strength properties and performance of the crystalline material could be characterized and analyzed by EBSD nowadays. These diffraction patterns helped to understand the plastic deformation across the weld interface [12]. The geometry-based rotations in the friction stir weld for the local shear during pin rotation changed the textures. So, it caused the low and/or high boundaries between the grains as (LAB)/(HAB) in the weld center [13, 14]. The dynamic variations across grain size can be clearly explained using electron back scattered diffraction (EBSD) analysis [15, 16].

The main objective of this study is a comparative study of various dissimilar AA6082/AA7075, AA6082/AA2014, and AA2014/AA7075 welds, which has been detailed here.

2. Materials and methods

Here, the dissimilar welding of the joints was carried out using friction stir welding machine 3 T/HYD machine. **Figure 1** shows the schematic representations of the dissimilar friction stir welding process. The tool was plunged into the weld centerline for the production of the frictional heat and plastic flow. Using the above process parameters, the aluminum plates were joined together by extreme thermo-mechanical conditions to alter the microstructure of the welding. The AA7075 and AA2014 alloys

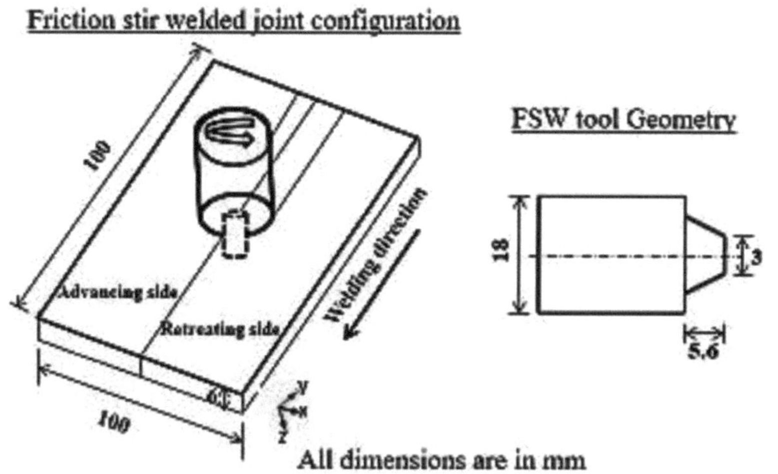

Figure 1.
Friction stir welding process.

Aluminum alloy	Zn (%)	Fe (%)	Cu (%)	Mn (%)	Mg (%)	Cr (%)	Si (%)	Ti (%)	Al (%)
AA6082	0.002	0.169	0.035	0.841	0.54	0.005	1.16	0.017	Bal.
AA7075	5.476	0.224	1.916	0.06	2.253	0.193	0.095	0.021	Bal.
AA2014	0.076	0.193	4.171	0.597	0.639	0.003	0.681	0.051	Bal.

Table 1.
Chemical composition of the alloys.

were placed in the retreating side, AA6082 was placed in the advancing side to obtain better property in the weldments.

The materials AA7075-T6, AA6082-T6, and AA2024-T6 chemical compositions are in **Table 1**. **Table 2** illustrates the mechanical properties of the alloys. **Table 3** is illustrating process parameters used for dissimilar FSW of AA7075, AA6082, and AA2024. **Figure 2** is showing the welded samples of the dissimilar alloys. Then, the as welded samples were prepared to carry out testing. Further the tensile tests were carried out as per ASTM E 384-99. Then, for the microstructure studies the specimens were prepared using consecutive 220, 400, 600, and 800 grade emery paper, and followed by cloth polishing. After that the specimens were etched with freshly prepared Keller's reagent of solution containing 190 ml water, 5 ml HCL, 3 ml HF, and 2 ml HNO_3. And the microstructures were taken using METZER microscope.

Properties	AA7075	AA6082	AA2014
Yield strength	105 MPa	301 MPa	414 MPa
Tensile strength	317 MPa	330 MPa	483 MPa
Elongation at break	17%	14%	13%
Hardness Vickers (HV)	175	100	155

Table 2.
Mechanical properties of alloys.

Factors	Values
Rotational speed (rpm)	900
Travel speed (mm/min)	80
Tilt angle (°)	2

Table 3.
FSW process parameters used.

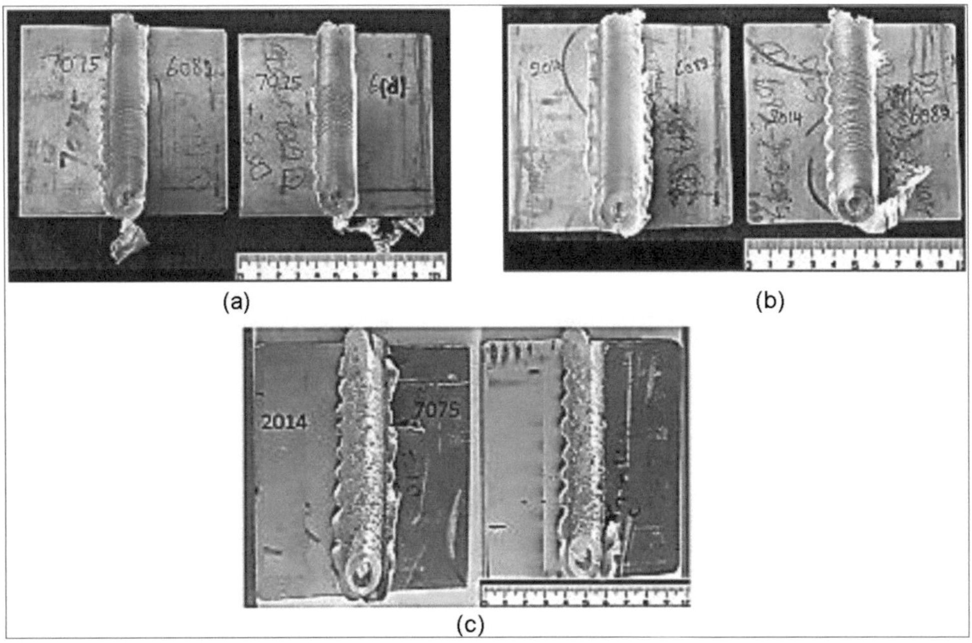

Figure 2.
Dissimilar FSW samples. (a) AA6082&2014, (b) AA6082&7075, and (c) AA2014&7075.

The microhardness readings were carried out using a MATSUZAWA MMTX3 microhardness tester with 300 gm of weight at 15 seconds of dwell time. The grain orientation and its size were studied using EBSD analysis. So as to carry out EBSD analysis, the samples were emery polished and electro polished to 100 μm thick to carry out EBSD scans using SEM-FEI Quanta with TSL-OIM software.

3. Results and discussions

3.1 Tensile test

During friction stir welding, the maximum temperature with plastic deformation causes an alteration in the distribution of the precipitate present in the base materials. So, the strength and ductility of the joints are also altered by the heat and temperature distribution in the welding process [17, 18]. Moreover, the FSW joints produce better joint efficiency compared to fusion welding processes, but there is a considerable difference between the base metal and the weld metal strength values.

Table 4 and **Figure 3** is showing the tensile test properties of the dissimilar joints welded using the FSW parameters of 80 mm/min, 900 rpm with tool tilt angle of 2°. It showed the AA2014/AA7075 joint with high tensile strength of 312 MPa than AA6082/AA7075 with the low tensile strength of 203 MPa. The AA6082/AA2014 weld showed an intermediated strength of 218 MPa. However, the AA6082/AA2014 dissimilar joint had a weld efficiency of 67%, which was slightly higher than the AA6082/AA7075 dissimilar joint, which had a weld efficiency of 62%. The higher tensile efficiency was seen for AA2014/AA7075 joint. Further, the AA6082/AA2014 joint had showed a 13% elongation percentage, whereas the AA6082/7075 joint had only a 7.5% elongation percentage.

Figure 4 is showing the tensile failure samples. Due to the increased heat along the advancing side, it was also noted that all of the weld failures were located near the advancing side of the dissimilar joint. In addition to tensile failure, the SEM analysis were revealed approximately 98% of fine and shallow dimples confirmed the ductile mode of fracture of the weld (see **Figure 5**).

3.2 Macrostructure

Mostly friction stir welded macrostructure reveal geometrical effect of the tool in the central stir region as the reason for the onion ring formation in the weld [19]. Similarly the forces around pin forming center zone [5, 20] and defects such as tunnel, groove, kissing bonds, lack of diffusion, and flush in heat input or stirring during the process [21–23].

Here, **Figure 6** is showing macrostructure of the dissimilar joints of AA7075, AA6082, and AA2024. It shows a clear stir zone with defect-free joints. The dissimilar joints clearly showed the two materials mixed up in the joint's central stir zone. It is also observed the clear stir zone (SZ), thermo-mechanically affected zone (TMAZ), heat affected zone (HAZ), and base metal region (BM) on both sides of the joint material. Here, the basin shape center stir zone could be seen for all the dissimilar AA6082/7075, AA6082/2014, and AA2014/AA7075 joints.

3.3 Microstructure

Figures 7 and **8** are showing the microstructure variations of advancing and retreating sides of the AA6082/7075 weld. Also **Figures 9** and **10** are showing the microstructure of advancing and retreating sides of the AA6082/2014 weld. **Figures 11** and **12** show the microstructure of the advancing and retreating sides of the AA2014/7075 weld. Here stir zone, TMAZ, HAZ, and base material zones were revealed on the advancing and retreating sides of the microstructures.

Samples	Yield strength (MPa)	Tensile strength (MPa)	Elongation (%)	Weld strength efficiency (%)	Fracture location
6082/2014	164	219	13.00	67	HAZ
6082/7075	159	203	9.50	62	WELD
2014/7075	186	312	7.5	75	HAZ

Table 4.
Tensile test properties of the dissimilar friction stir welded samples.

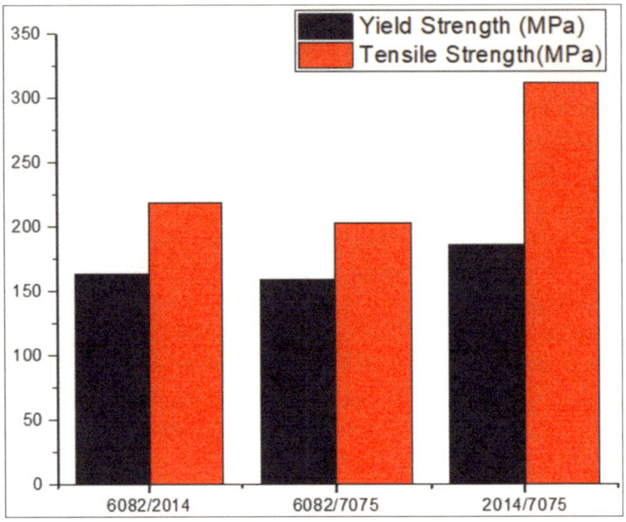

Figure 3.
Tensile test of dissimilar welds.

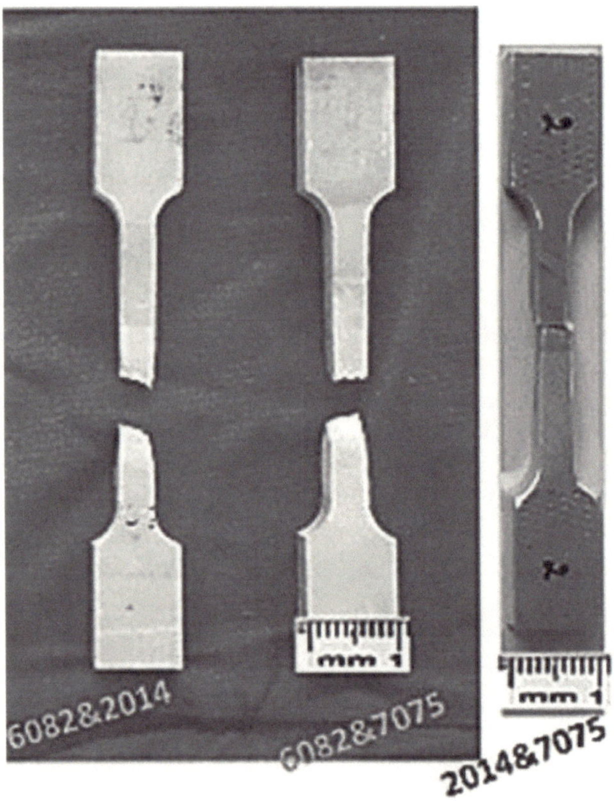

Figure 4.
Tensile failured samples.

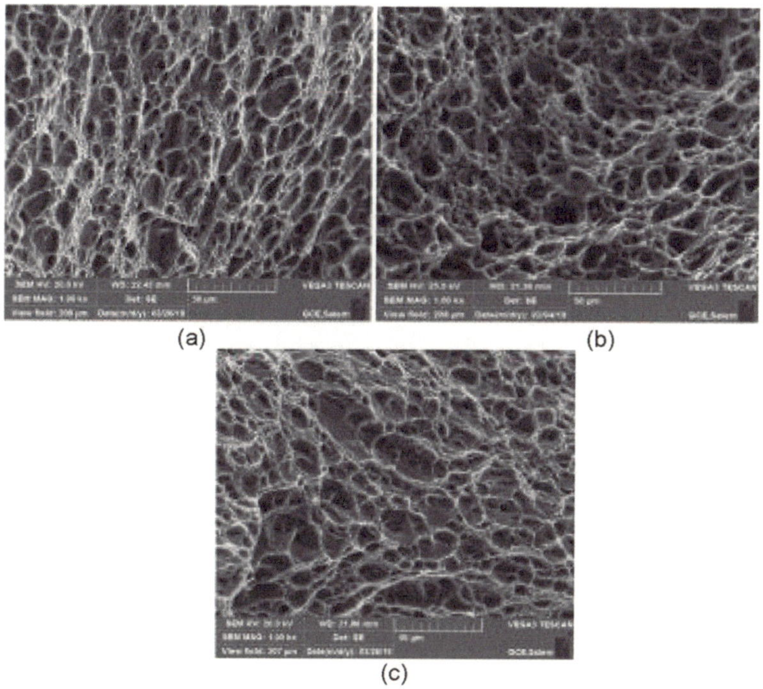

Figure 5.
SEM fractographs of dissimilar welds. (a) AA6082&2014, (b) AA6082&7075, and (c) AA2014&7075.

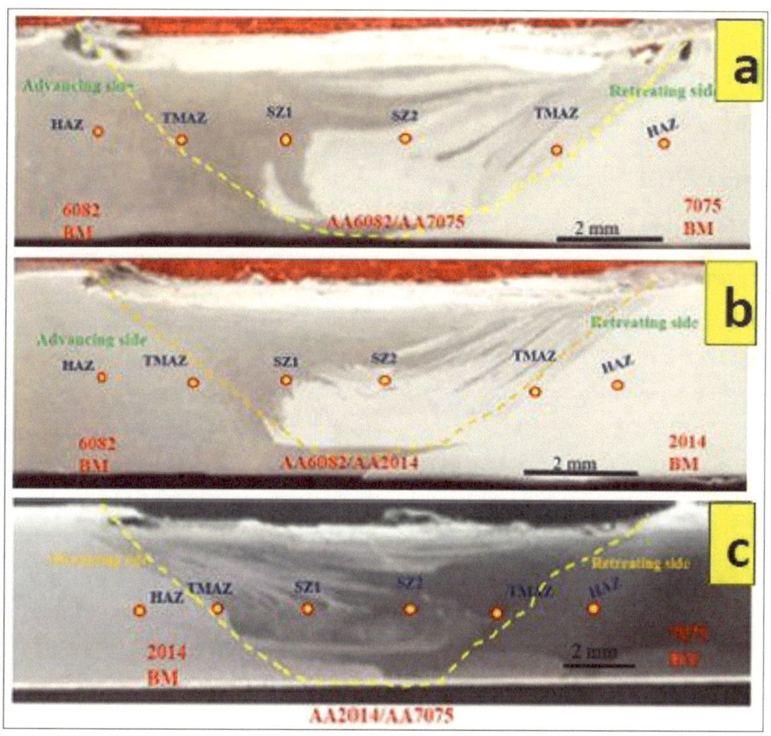

Figure 6.
Macrostructure of the dissimilar joints (a) AA6082/7075, (b) AA6082/2014, (c) AA2014/7075.

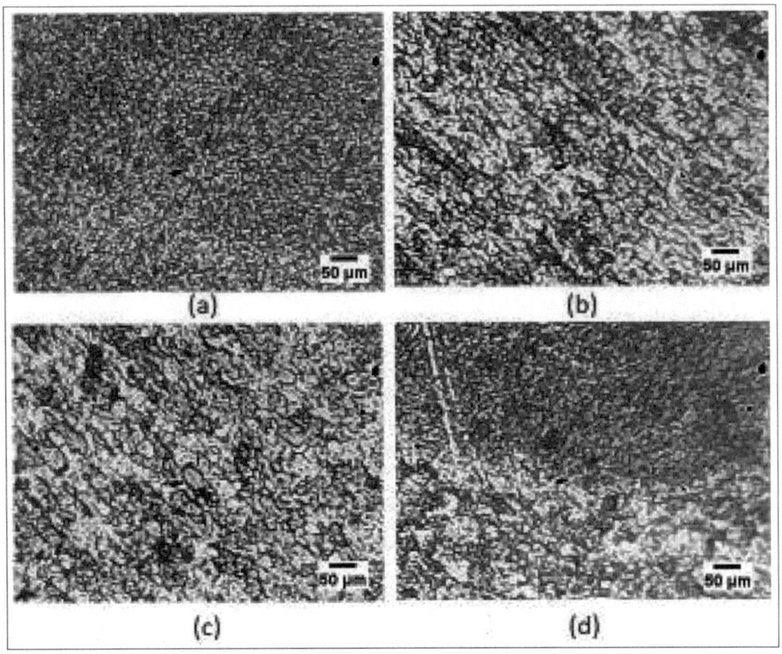

Figure 7.
Microstructure of advancing side of the AA6082/7075 weld. (a) AA6082BM, (b) HAZ, (c) TMAZ, and (d) stir zone.

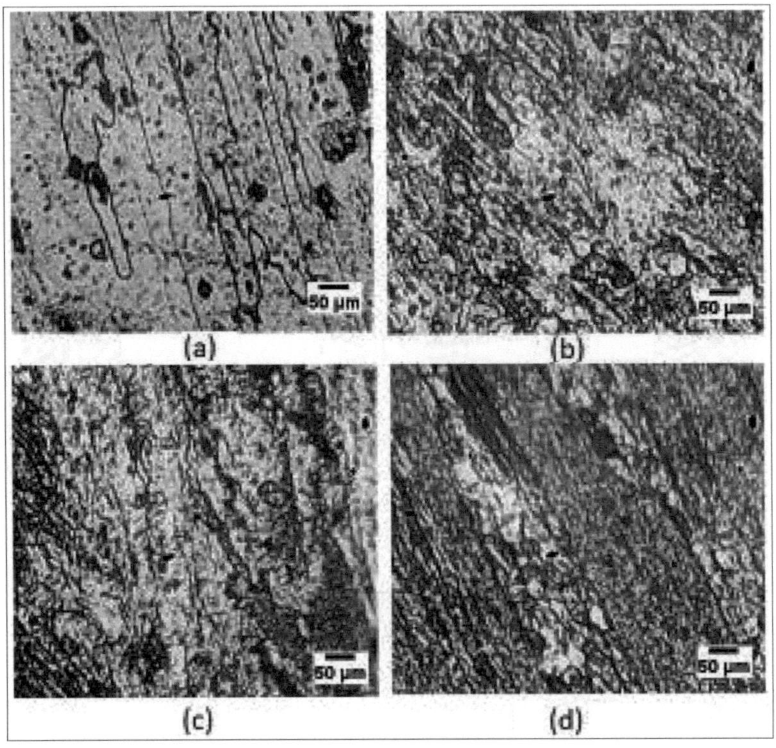

Figure 8.
Microstructure of retreating side of the AA6082/7075 weld. (a) AA7075BM, (b) HAZ, (c) TMAZ, and (d) stir zone.

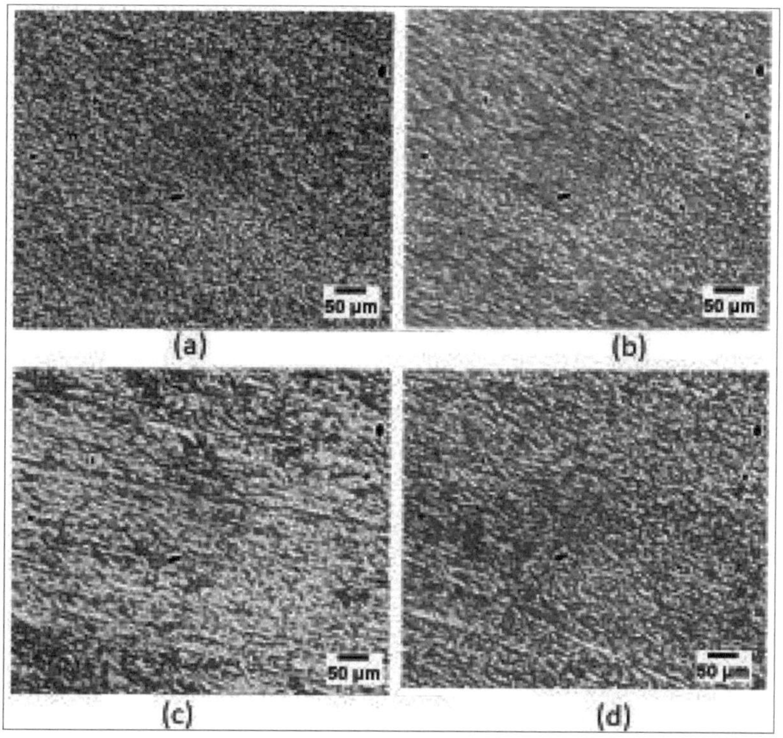

Figure 9.
Microstructure of advancing side of the AA6082/2014 weld. (a) AA6082BM, (b) HAZ, (c) TMAZ, and (d) stir zone.

Jata et al. [24] clarified that the evolution of the microstructure of the Al-Li alloy welding zone consisted of 9 µm of grains due to dynamically recrystallized grains. Su et al. [25] established that the different zones of the FSW AA7050-T651 alloy as dynamically recrystallized zone (DXZ), thermo-mechanically affected zone (TMAZ), and heat-affected zone (HAZ).

In the microstructures below, the base metals AA6082 showed α-Al-Si-Mg solid solutions with their intermetallics. Similarly, AA7075 and AA2014 showed Al-Cu and Al-Zn α-solid solutions with their intermetallics. It was due to dynamic recovery and recrystallization in stir zone caused by high-speed rotation of the pin from top to bottom of the stir zone, revealed fine grain structure. The TMAZ zone has experienced heat and metal flow to reveal aligned intermetallics with deformed grains along the pin rotation are visible. Then, the HAZ zone with similar grains as that of base materials, which was influenced by the heat changes of welding, leading to some microstructure changes of coarse intermetallics.

3.4 Metal flow in stir zone

Mukhopadhyay [26] highlighted the XRD and EDAX analysis of the precipitate phase-type/shape, habitat plane, and solute content for AA7075, AA2014, and AA6082. It was showed that the AA7075 (Al-Zn-Mg-Cu) with (Mg_2Zn)/Plate in {111} plane, AA2014 (Al-Cu-Mg-Si) with S (Al_2CuMg)/lath and GPZ/rod in {210} plane, and AA6082 (Al-Mg-Si) with (Mg_2Si)/rod in <100> direction. Carron [27] highlighted the high peak aged β^{II} (Mg_5Si_3-needles) up to 40 × 40 × 350 Å precipitates formation, before

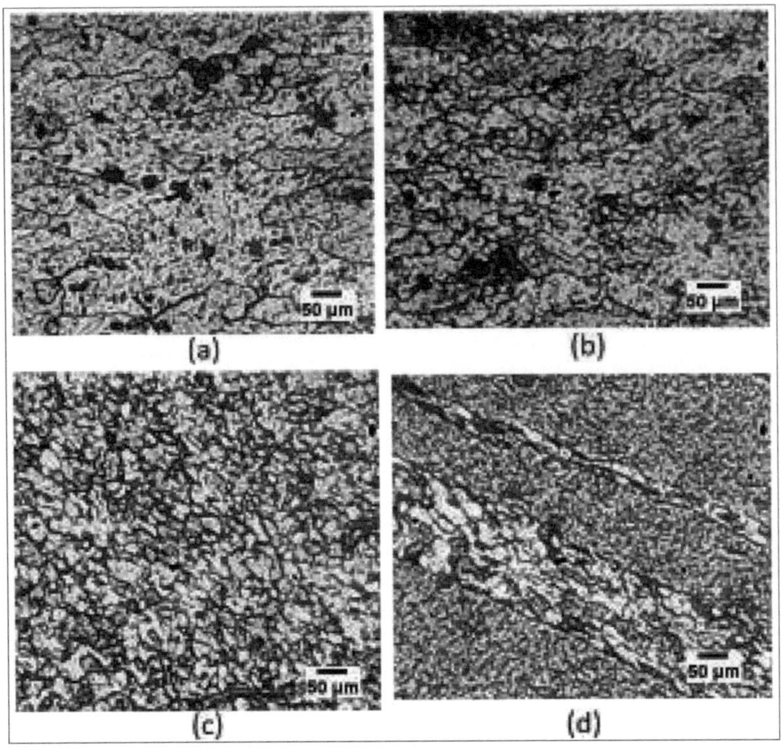

Figure 10.
Microstructure of retreating side of the AA6082/2014 weld. (a) AA2014BM, (b) HAZ, (c) TMAZ, and (d) stir zone.

β' precipitate. After that, 10–20 μm of over aged β (Mg_2Si-rods) precipitates will form. The common precipitates sequences of the base metals are as follows: for the AA6082 supersaturated solution solid solution (SS) → Guinier-Preston (GP) zone (spherical) → β'' (Mg_5Si_3-needle) → β' (Mg_2Si-rod) → β, for AA7075 supersaturated solution solid solution → GP zones → η′ ($MgZn_2$) → η ($MgZn_2$), for the AA2014 supersaturated solution solid solution (SSSS) → GP (rod) → θ^1 (Al_2CuMg-lath) → θ (Al_2CuMg) [28].

Mironov et al. [29] investigated the formation of high angle grain boundaries (HABs) on the Ti-6AL-4V alloy using friction stir processing (FSP) in microstructural studies. These LABs/HABs were clearly defined using electron back scattered diffraction (EBSD) studies. It was also stated that the α-titanium which was at room temperature in the HCP structure has to slip along the HCP crystal's basal plane to produce more HABs than LABs with dynamic recovery and recrystallization. So, it was therefore understood assertion that the material shear or strain low to moderate to high during deformation was mostly dependent on how the plane slides along a particular slip plane to create a substructure of low angle boundaries (LAB)/high angle boundaries (HAB).

Similarly in friction stir welding also the high speed rotation of the pin in the center stir zone shears or strains the material low to moderate to high deformation was mostly dependent on how the plane slides along a particular slip plane to create a substructure of low angle boundaries (LAB)/high angle boundaries (HAB). Kalemba-Rec et al. [13] explained the misorientation angle laid in between 2 and 15° was classified as low angle grain boundary (LAGBs) and whereas the misorientation angle greater than 15° was classified as a high angle grain boundary (HAGBs). So the grain

DOI: http://dx.doi.org/10.5772/intechopen.102652

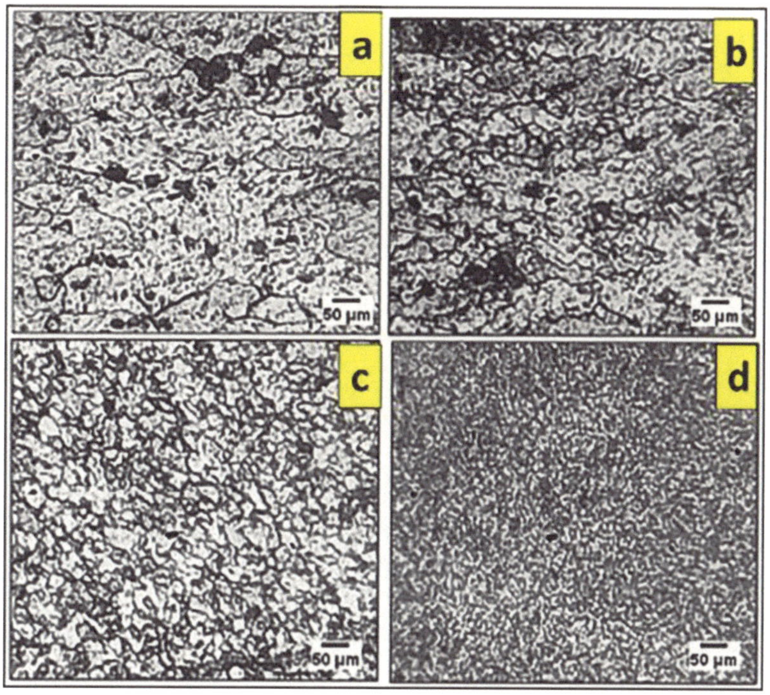

Figure 11.
Microstructure of advancing side of the AA2014/7075 weld. (a) AA2014BM, (b) HAZ, (c) TMAZ, and (d) stir zone.

size and its orientation variations could be clearly identified by electron back scattered diffraction (EBSD) scan images. Here, EBSD scan on zones such as SZ, TMAZ, HAZ, and base metal zones across the weldment gives the grain size variation across the AA6082/7075 and AA6082/2014 welds.

Figure 13(a) and **(c)** were showing the EBSD maps of the as-received materials of the base material AA6082-T6 with 53.3% of LABs and 46.7% of HAGs and the AA7075-T651 with 21.8% of LABs and 78.2% of HAGs. After welding base materials were not affected by any microstructural changes, thereby these base materials were not affected by the mode of metal flow. Moreover, in **Figure 13(b)** and **(d)** showing 42.5% of LABs and 57.5% of HABs of misorientation variations were seen on the advancing AA6082WN and 43.2% of LABs and 56.8% of HABs on the AA7075WN. In addition to misorientation variations, **Figure 14(a)** and **(b)** by EBSD analysis were showed 10–50 μm of grains in AA7075BM and 4–3 μm of grains in AA6082BM. Also **Figure 14(c)** and **(d)** were revealed the grains of 4–10 μm of grains on the AA6082 stir zone and 4–10 μm of grains on the AA7075WN were seen. Thus, the high speed rotating pin modified the grains to 4–10 μm of fine recrystallized grains in the stir zone were revealed in AA6082/7075 weld.

Similarly, the EBSD image quality figure with grain boundary orientation map of the AA6082-T6 and AA2014-T6 base metal used in the present study was shown in **Figure 15(a)** and **(c)**. It revealed that the total fraction of LAGBs was 48.3% and HAGBs was 51.7% on the advancing side of the AA6082, whereas the total fraction of LAGBs was 29.8% and HAGBS was 70.2% on the retreating side of the AA2014 side. In **Figure 15(b)** and **(d)**, the center stir zone revealed that the total fraction of LAGBs was 42.9% and HAGBs was 57.1% on the advancing side of the AA6082, whereas the

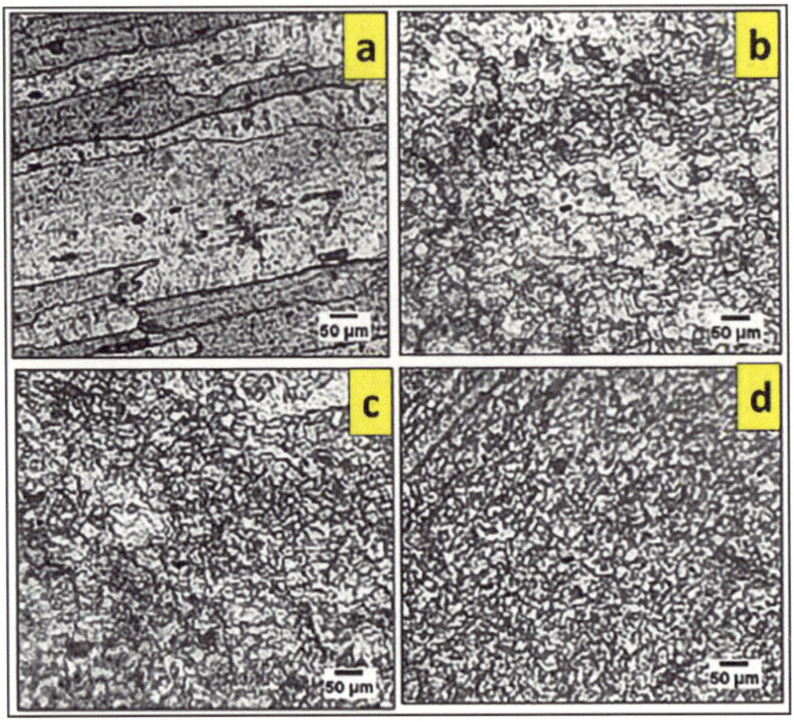

Figure 12.
Microstructure of retreating side of the AA2014/7075 weld. (a) AA7075BM, (b) HAZ, (c) TMAZ, and (d) stir zone.

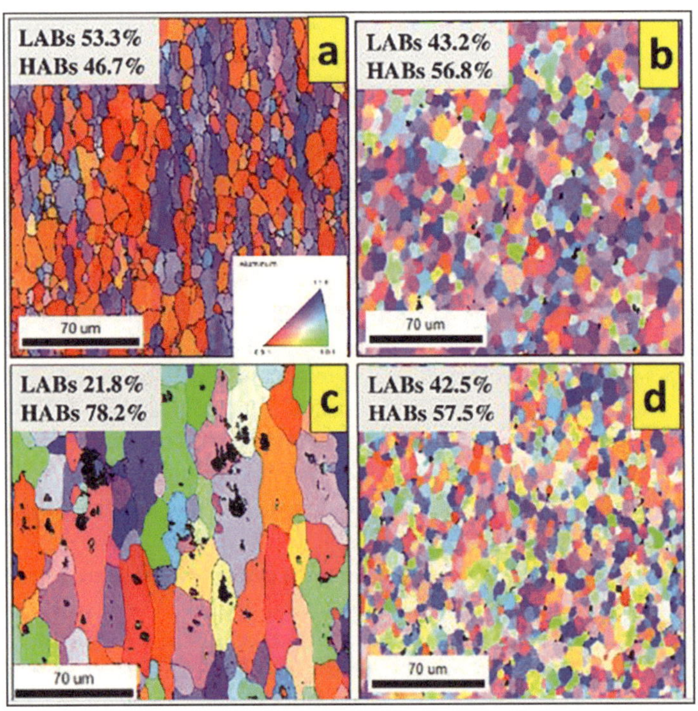

Figure 13.
EBSD misorientation images on: (a) AA6082BM, (b) AA6082WN, (c) AA7075BM, and (d) AA7075WN.

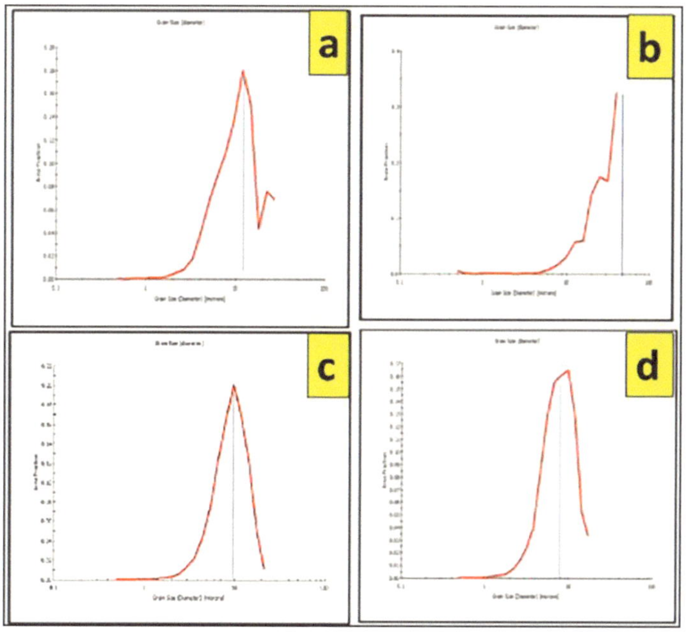

Figure 14.
Grain size images on: (a) AA6082BM, (b) AA7075BM, (c) AA6082WN, and (d) AA7075WN.

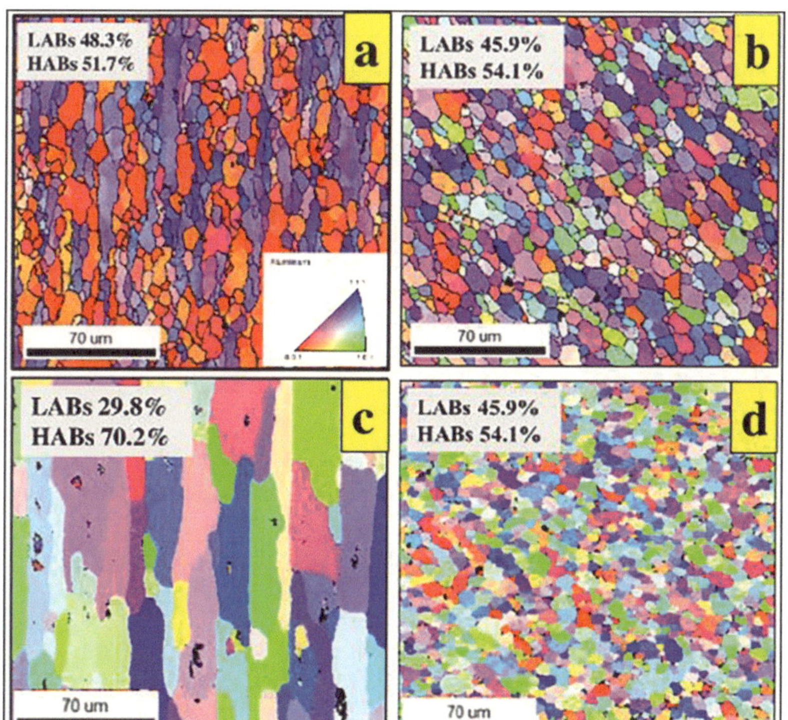

Figure 15.
EBSD inverse pole figure (IPF) maps: (a) AA6082BM, (b) AA2014BM, (c) AA6082WN, and (d) AA2014WN.

total fraction of LAGBs was 45.9% and HAGBS was 54.1% on the retreating side of the AA2014 side. The **Figure 16(c)** and **(d)** also confirms the 4–10 μm of fine dynamically recrystallized grains in the stir zone in AA6082/2014 weld.

Here, in this research, the AA7075 base material would deform along the {111} habit plane, AA2014 slips along {210} planes and AA6082 slips along the <100> directions, respectively by dynamically recovery and recrystallization. Thus, more grain substructures within the grain would evolve as a result of this dynamic recrystallization, which is surrounded by the high-energy grain boundaries of the weld matrix. As a result, substantial low angle grain boundary substructures are relatively common in the stir zone.

A vast hardness variations could be seen across the FSW weld. Since it results in advancing and retreating side BM, HAZ, TMAZ, and WN depend not only on the material and its rolled condition but also on the plastic deformation, frictional temperature and pressure applied during stir welding. Thomas et al. [15] and Threadgill et al. [16] studied the microhardness profiles measured in the welded material frequently exhibit a W-shaped characteristic. Moreover, Giraud et al. [30] experimented on asymmetric "W" shaped hardness profile of all joints, but it became cyclic on the welding path on the similar alloys AA7020-T651 to AA6060-T6.

In **Figure 17(a)-(c)** illustrated the hardness variations across the AA6082/2014, AA6082/7075, and AA2014/AA7075 weld, respectively. The hardness across weld was taken at t/2 (t-thickness) middle layer of the weld. The base metal hardness was 103.9 HV0.3 on AA6082BM and 121.5 HV0.3 on AA7075BM. Whereas, in **Figure 17(b)** 88.6 HV0.3 of hardness was observed in the center SZ, which was lower hardness than the base metal regions. This decreased hardness was due to the dynamic recrystallization with frictional heat in that zone. Consequently, 74.3 HV0.3 of hardness was

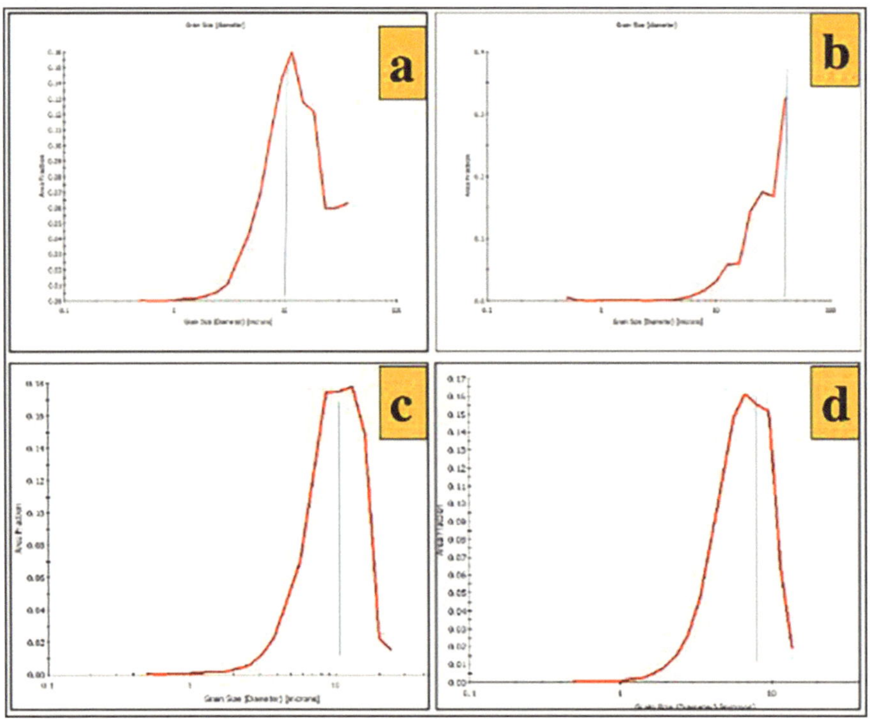

Figure 16.
Grain size on: (a) AA6082BM, (b) AA2014BM, (c) AA6082WN, and (d) AA2014WN.

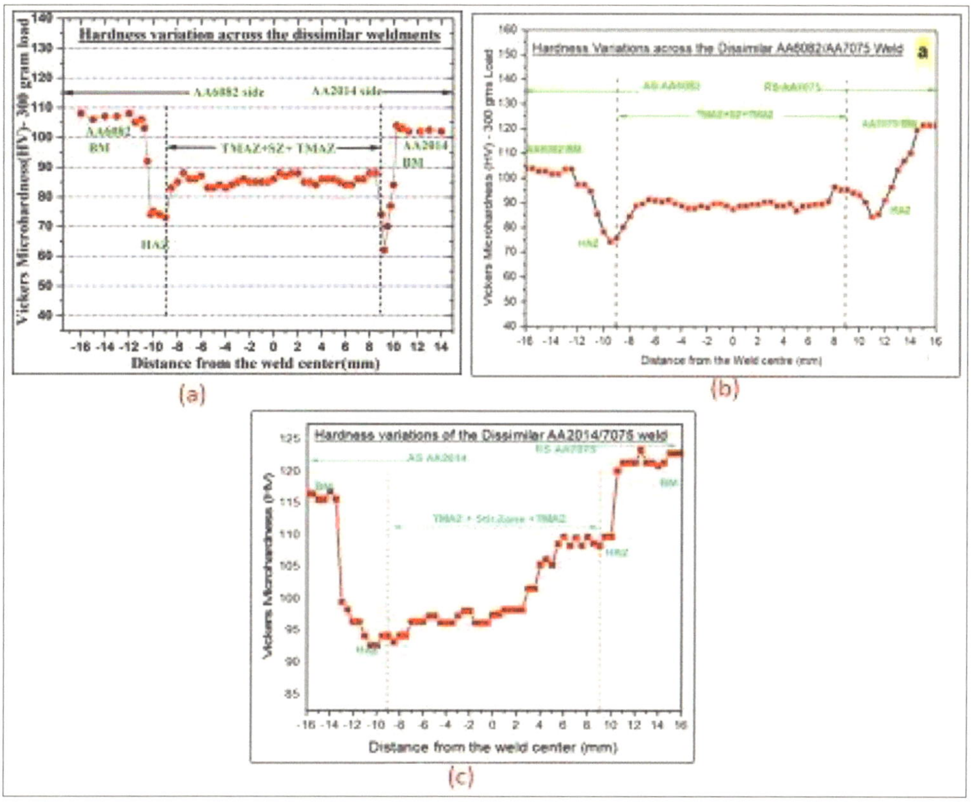

Figure 17.
Hardness variations across the dissimilar joints. (a) AA6082&2014, (b) AA6082&7075, and (c) AA2014&7075.

noted on the advancing side AA6082HAZ, due to heat, which coarsened the grains with its precipitates in that zone. Thus the low hardness led to the tensile failure on AA6082HAZ side.

Similarly, in **Figure 17(a)** the initial observation showed that the hardness of the AA6082-T6 base metal was comparatively higher hardness than that of AA2014-T6 base metal. The hardness of the SZ matrix of the dissimilar weldment was lesser than that of BMs. Irrespective of the advancing side and retreating side of the dissimilar weldment of HAZ showed relatively lesser hardness than that its weld and its respective base metal. It was mainly attributed to the heating effect during FSW and thereby the disappearance of GP zones and the formation of overaged precipitates.

In **Figure 17(c)** the AA7075 base metal region showed high hardness than AA2014 BM side. Also the center stir zone showed the increased hardness from AA2014 advancing side to AA7075 retreating side. Also the hardness drop near AA2014 HAZ of 92.5HV0.3 had showed weld failure also.

The significant hardness drops in the HAZ were conceded that the maximum peak temperature evolved in the HAZ of the dissimilar weldment was touched above the β^1 and θ^1 and thereby the partial reversion occurs [31]. The literature confirmed that the main strengthening precipitates for the AA6082 AA were Mg_5Si_6 which was more stable at temperatures lesser than 200°C [32].

Generally, the precipitates were more stable and exist in the unaffected base material matrix and were absent in the SZ matrix and HAZ of the weldment. It was mainly attributed to the evolution of higher peak temperature in the SZ matrix during the heating cycle of FSW and it caused the dissolved strengthening in the weld matrix.

4. Conclusions

The following were the conclusions of the above study:

1. The AA6082/AA7075, AA6082/AA2014, and AA2014/AA7075 joints were welded using process parameters of 900 rpm, 80 mm/min, and 2° tilt angle showed the higher tensile strength of 312 MPa for the AA2014/AA7075 and the lower tensile strength of the dissimilar joint was 203 MPa for AA6082/AA2014 joint.

2. The joints microstructures were revealed α-solid solution of aluminum with its intermetallic in base metal regions. The stir zone is showing micrograph of fine re-crystallized grains. In the EBDS analysis also showed the stir zone with more fractions of LAGs than HAGBs. Fine grains is due to the dynamic recovery recrystallization which evolved by the high speed rotation of the tool pin in stir zone.

3. The lower hardness was observed in the stir zone than base metal zones. Among four different zones, the fine grains in the stir zone were due to the dynamic recovery and recrystallization. On the other hand, the HAZ had the lowest hardness because there was more heat conduction along the advancing side, which could lead to coarsening of precipitates.

Acknowledgements

Author express heartfelt thanks to National Institute of Technology, Tiruchirappalli, India for hardness measurement, OIM-Lab, IITB for their timely help for EBSD analysis and Government College of Engineering, Salem, India to carry out extensive FSW welding for this research work.

Author details

K.T. Thilagham[1*] and S. Muthukumaran[2]

1 Government College of Engineering, Salem, Tamil Nadu, India

2 National Institute of Technology, Tiruchirappalli, Tamil Nadu, India

*Address all correspondence to: thilagham.met@gmail.com

References

[1] Kah P, Rajan R, Martikainen J, Suoranta R. Investigation of weld defects in friction-stir welding and fusion welding of aluminium alloys. International Journal of Mechanical and Materials Engineering. 2015;**10**(1):1-10

[2] Thomas AW. Parameter development for the MIG welding of high strength aerospace aluminium alloys [thesis]. Cranfield, England: Cranfield University; 2000

[3] Beiranvand ZM, Ghaini FM, Moosavy HN, Sheikhi M, Torkamany MJ. Solidification cracking susceptibility in pulsed laser welding of Al–Mg alloys. Materialia. 2019;7:100417

[4] Ogbonna OS, Akinlabi SA, Madushele N, Mashinini PM, Abioye A. Application of MIG and TIG welding in automobile industry. Journal of Physics: Conference Series. 2019;**1378**:42-65

[5] Mishra RS, Ma YS. Friction stir welding and processing. Materials Science and Engineering R. 2005;**50**:1-78

[6] Çam G, İpekoğlu G. Recent developments in joining of aluminium alloys. The International Journal of Advanced Manufacturing Technology. 2017;**91**(5):1851-1866

[7] Çam G, Güçlüer S, Çakan A, Serindağ HT. Mechanical properties of friction stir butt-welded Al-5086 H32 plate. Journal of Achievements in Materials and Manufacturing Engineering. 2008a;**30**(2):151-154

[8] Çam G, İpekoğlu G, Tarık Serindağ H. Effects of use of higher strength interlayer and external cooling on properties of friction stir welded AA6061-T6 joints. Science and Technology of Welding and Joining. 2014;**19**(8):715-720

[9] Çam G, Ipekoglu G, Küçükömeroglu T, Aktarer SM. Applicability of friction stir welding to steels. Journal of Achievements in Materials and Manufacturing Engineering. 2017;**80**(2):65-85

[10] Çam G, Serindağ HT, Çakan A, Mistikoglu S, Yavuz H. The effect of weld parameters on friction stir welding of brass plates. Materialwissenschaft und Werkstofftechnik. 2008b;**39**(6):394-399

[11] Kumar SR, Rao VS, Pranesh RV. Effect of welding parameters on macro and microstructure of friction stir welded dissimilar butt joints between AA7075-T651 and AA6061-T651 alloys. Procedia Materials Science. 2014;**5**:1726-1735

[12] Kalemba I, Muszka K, Wróbel M, Dymek S, Hamilton C. EBSD analysis of friction stir welded 7136-T76 aluminium alloy. Solid State Phenomena. 2013;**203**:258-261

[13] Kalemba-Rec I, Wróbel M, Kopyściański M. Investigations of friction stir welds between 5083 and 7075 aluminium alloys using EBSD and X-ray techniques. Acta Physica Polonica A. 2016;**130**(4):996-999

[14] López MG, Brewer LN, Massone JM, Boeri RE. EBSD analysis of the primary austenite grains in lamellar graphite cast iron. Metallography, Microstructure, and Analysis. 2019;**8**(3):386-392

[15] Thomas WM, Threadgill PL, Nicholas ED. Feasibility of friction stir welding steel. Science and Technology of Welding and Joining. 1999;**4**(6):365-372

[16] Threadgill PL, Leonard AJ, Shercliff HR, Withers PJ. Friction stir welding of aluminium alloys. International Materials Reviews. 2009;**54**(2):49-93

[17] Pradeep Kumar P, Ahmmad Basha SK, Sai Kumar S. Optimization of friction stir welding process parameters of aluminium alloy AA7075-T6 by using Taguchi method. International Journal of Innovative Technology and Exploring Engineering. 2019;**8**:12-20

[18] Ramanjaneyulu K, Reddy G, Gokhale H. Optimization of process parameters of aluminium alloy AA 2014-T6 friction stir welds by response surface methodology. Defence Technology. 2015;**51**

[19] Krishnan KN. On the formation of onion rings in friction stir welds. Materials Science and Engineering A. 2002;**327**(2):246-251

[20] Li Y, Murr LE, McClure JC. Flow visualization and residual microstructures associated with the friction-stir welding of 2024 aluminium to 6061 aluminium. Materials Science and Engineering A. 1999;**271**(1-2):213-223

[21] Bayazid SM, Farhangi H, Ghahramani A. Effect of pin profile on defects of friction stir welded 7075 aluminium alloy. Procedia Materials Science. 2015;**11**:12-16

[22] Kim YG, Fujii H, Tsumura T, Komazaki T, Nakata K. Three defect types in friction stir welding of aluminium die casting alloy. Materials Science and Engineering A. 2006;**415**(1-2):250-254

[23] Hou X, Yang X, Cui L, Zhou G. Influence of joint geometry on defects and mechanical properties of friction stir welded AA6061-T4 T-joints. Materials and Design. 2014;**53**:106-117

[24] Jata KV, Sankaran KK, Ruschau JJ. Friction-stir welding effects on microstructure and fatigue of aluminium alloy 7050-T7451. Metallurgical and Materials Transactions A: Physical Metallurgy and Materials Science. 2000;**31**(9):2181-2192

[25] Su J-Q, Nelson TW, Mishra R, Mahoney M. Microstructural investigation of friction stir welded 7050-T651 aluminium. Acta Materialia. 2003;**51**(3):713-729

[26] Mukhopadhyay AK. Microstructure and properties of high strength aluminium alloys for structural applications. Trans IIM. 2009;**62**:113-122

[27] Carron D, Bastid P, Yin Y, Faulkner RG. Modelling of precipitation during friction stir welding of an Al-Mg-Si alloy. Technische Mechanik - European Journal of Engineering Mechanics. 2010;**30**(1-3):29-44

[28] ASM Databook. Published in Mer. Prog. 1979;**116**(1)

[29] Mironov S, Sato YS, Kokawa H, Hirano S, Pilchak AL, Semiatin SL. Microstructural characterization of friction-stir processed Ti-6Al-4V. Metals. 2020;**10**(7):976

[30] Giraud L, Robe H, Claudin C, Desrayaud C, Bocher P, Feulvarch E. Investigation into the dissimilar friction stir welding of AA7020-T651 and AA6060-T6. Journal of Materials Processing Technology. 2016;**235**:220-230

[31] Ringer SP, Hono K. Microstructural evolution and age hardening in aluminium alloys: Atom probe field-ion microscopy and transmission electron microscopy studies. Materials Characterization. 2000;**44**(1-2):101-131

[32] El-Danaf EA, El-Rayes MM. Microstructure and mechanical properties of friction stir welded 6082 AA in as welded and post weld heat treated conditions. Materials & Design. 2013;**46**:561-572

Chapter 15

Challenges and Advances in Welding and Joining Magnesium Alloy to Steel

Shamsu Tukur Auwal, Murtala Sule Dambatta, Singh Ramesh and Tan Caiwang

Abstract

Hybrid structures built using Mg/steel are expected to have an increasing impact on the future developments of the manufacturing sector, especially where lightweight structures are required in order to reduce fuel consumption, greenhouse gases and improve efficiency of energy-converting systems. To this end, there is a pressing need for a joining technology to produce effective and low-cost dissimilar Mg/steel joints. Joining of these materials has always been a challenging task for researchers, due to the wide discrepancies in physical properties and lack of metallurgical compatibilities that make the welding process difficult. Based on the existing literature, a successful joint between magnesium alloys and steel can be achieved by inserting an interlayer at the interface or mutual diffusion of alloying elements from the base metal (BM). Thus, intermetallic phases (IMCs) or solid solutions between Mg and the interlayer and also the interlayer and Fe formed at the interface. However, the interfacial bonding achieved and the joints performance depend significantly on the intermediate phase. This paper reviewed the research and progress in the area of joining of Mg alloys to various grades of steel by variety of welding processes, with focus on the techniques used to control the morphology and existence state of intermediate phase and improving the mechanical properties.

Keywords: review, Mg/steel dissimilar welding, intermetallic compounds (IMCs), microstructure, and mechanical properties

1. Introduction

With the growing concerns on global warming and energy prices, the demand for environmentally friendly vehicles with better fuel efficiency is increasing [1–3]. These issues can be addressed by reducing car weight, lowering travel resistance, advancing drive-train efficiency, developing new sources of power, and so on. Vehicle weight reduction using advanced lightweight structural materials, such as Mg, Al, and Ti, is considered as one of the most promising strategies to address these issues [4–10]. Generally, for every 10% weight reduction, the specific fuel consumption could reduce by 3–7%, while maintaining the same functionality [11, 12]. Thus, the use of lightweight alloys as structural materials is considered as the factor for development of aerospace and automotive manufacturing sectors in the future. To achieve lightweight, safety, and low cost, the multi-material structure using steels, Al, Mg alloys is considered to be efficient [9]. Therefore, effective dissimilar joining process of such light metals and steels is essential.

As the lightest structural material, Mg alloys receive great attention due to their high specific strength, sound damping capabilities, hot formability, good castability, recyclability among others [13–18]. Potential applications of Mg alloys in an automobile include seat components, bracket carrier, roof, bonnets, cylinder head, wheels, etc. [8, 19]. Steel is currently the automaker's material of choice, due to its inherent properties, including high strength and toughness, good ductility, and low cost [8, 20–22]. Recently, it has been demonstrated through the next-generation vehicle project that stainless steels are promising candidates for vehicle construction, and they can be used replaced carbon steels, especially in crash-relevant components such as door pillars.

Therefore, for practical applications in automotive industries, Mg alloys will have to be joined with existing steel parts. Recently, many automotive components have been produced using a combination of Mg alloys and steels, but the major issues arise from the joining techniques and corrosion of the joined parts [23–26]. Thus, attaining reliable Mg/steel hybrid joints is paramount for facilitating lightweight industrial fabrication and expanding the industrial applications of Mg alloys in automotive industries [21, 27–29].

Joining Mg alloys directly to steel is extremely difficult because of the huge differences in their physical and metallurgical properties, the lattice mismatch between Fe and Mg is very large and there is almost zero solubility between Mg and Fe [27, 29–35]. Hence, an appropriate technique that overcomes the aforementioned problems is very much desired.

Generally, a successful joint between Mg and steel can be achieved by inserting an intermediate material at the interface or diffusion of alloying elements from the BM. At present, several authors have focused on joining magnesium alloys to different grades of steel, using various welding technology, such as friction stir welding (FSW), ultrasonic spot welding (USW), diffusion and eutectic bonding, resistance spot welding, laser welding brazing, laser-TIG hybrid welding and gas metal arc weld-brazing. In these studies, various interlayer elements and alloys such as Zn, Ni, Cu, Cu-Zn, Sn, Al, and Ag have been explored. In contrast to direct joining of magnesium to steel, which is mainly a mechanical bonding, with insertion of the interlayer elements, formation of intermetallic phases or solid solutions between Mg and the interlayer and also the interlayer and Fe indicated that metallurgical bonding is achieved. However, the joint performance and the interfacial bonding achieved depend significantly on the IMC phase formed [31, 36–38]. To control the morphology and existence state of the intermediate phase, the selection of suitable interlayer material and joining techniques are essential. Generally, choosing the suitable interlayer for joining Mg alloys to steel largely depends on the interlayer composition that gives excellent wetting and bonding without generating thick layers of hard and brittle IMCs at the joint interface [31, 35, 39–41]. Moreover, when choosing the joining process that will be used, minimization of the thickness of any brittle intermetallic compounds along the interfaces of the magnesium alloy-interlayer-steel joint and minimization of intermixing between the Mg and Fe in the molten-state are the main factors that must be considered [8, 21, 42, 43].

Currently, a great deal of research has been conducted on the interface characteristics and mechanical performance of Mg alloys to steel joints, particularly under static loading. Under optimized processing conditions, excellent static strength has been achieved, even surpassing that of Mg alloy base metal with insertion of Ag, Cu, and Ni intermediate elements [38, 44]. However, few experiments have been carried out on the corrosion behavior of the jointed parts and the joints performance under dynamic loading [45–48].

With the continuously increasing usage of Mg alloys in industries and the large number of potential applications of Mg/steel hybrid structures, two of the specific areas of concerns for broader utilization of magnesium alloys are reliable joining techniques and corrosion behavior of the joined parts. To better understand and address these challenges, there is a need to comprehensively review the research conducted so far and provide the most efficient strategies to address the challenges. This paper presents a review on Mg alloys/steel joining techniques, with focus on the techniques used to control the morphology and existence state of intermetallic compound (IMC) and improving mechanical properties. The general motives behind this review are to obtain a better understanding on the weldability issues associated with joining magnesium alloys to steel. It would also establish global, state-of-the-art welding techniques of Mg alloys to steel.

2. Weldability of Mg alloys to steel

Some of the inherent properties of Mg include high thermal conductivities and coefficients of thermal expansion, large solidification temperature ranges, strong tendency to oxidize, low viscosity and surface tensions, high solidification shrinkage, low melting and boiling temperatures, a tendency to form low melting point constituents and high solubility for hydrogen in the liquid state [8, 30, 49]. It is obvious that the properties of Mg differ significantly from those of Fe. For instance, the melting points of magnesium and iron are 649°C and 1536°C, respectively. This wide discrepancy in melting points makes it very difficult to melt the base materials at the same time as might be required in fusion-welding process [21].

In addition, the crystal structure of Iron at room temperature is body-centered cubic (BCC), whereas that of magnesium is close-packed hexagonal (HCP). Crystallographic analysis has shown that the lattice mismatch of Fe and Mg is very large [34, 50, 51]. Although the welding process itself is a non-equilibrium process, phase diagram has always been an effective tool to predict the reactions formed during welding process and serves as a reference to examine the feasibility of achieving a metallurgical bonding between the metals. According to the Mg/Fe binary phase diagram, the maximum solid solubility of iron in magnesium is 0.00043 wt.% while that of magnesium in iron is nil, and the Mg concentration at the eutectic point is estimated to be less than 0.008 at.% [34, 52–54]. Therefore, magnesium and steel are immiscible (neither the formation IMC nor atomic diffusion occurs between them after solidification), thereby presenting difficulty in joining them together.

3. Joining of Mg to steel: state of the art

In recent years, numerous techniques have been applied to bond Mg alloys to various grades of steels. These techniques can be broadly classified into solid-state joining (friction stir welding, ultrasonic spot welding, diffusion-welding processes) and fusion welding (resistance spot welding, laser welding brazing, laser hybrid welding, and arc welding). Review of the literature reveals that metallurgical bonding along the Mg/steel interface can be achieved with addition of suitable interlayers (which possess a substantial solid solubility in both Mg and Fe) or inter-diffusion of the alloying element from the BM. Therefore, the joint quality is significantly influenced by the interlayer characteristics (forms, thickness, and compositions). Among the joining techniques, a variety of thin interlayers such as Zn, Cu, Al, Ni, Sn, Cu-Zn, and Ag has been reported to improve the interfacial reaction between magnesium alloys and steel. In this section, the potential of several methods for joining magnesium and steel will be discussed.

3.1 Solid-state joining processes

Solid-state joining technology has been applied to bond Mg alloys to steel and get high-quality joints than fusion-welding processes because of the added advantage of minimal oxidation because of the solid-state nature of the process. Solid-state joining processes, such as friction stir welding (FSW), ultrasonic spot welding (USW), and diffusion and eutectic bonding, have been used to join Mg alloys to steel. Generally, for solid-state bonded Mg alloys-interlayer-steel joints, the joint performance is influenced by the intimate contact between the dissimilar materials, and the micro-structure, particularly the formation of IMCs [55, 56].

3.1.1 Friction stir welding (FSW)

FSW is a solid-state welding technique invented by Thomas et al. [57]. The combined action of pressure and stirring during the FSW led to mutual diffusion of the alloying elements from the BM, which promoted metallurgical bonding at the interface and improved the joint performance. Some of the unique advantages of FSW include low distortion and residual stresses due to low heat input and absence of melting, filler metal is not required and the heat efficiency is very high relative to traditional fusion-welding processes [58, 59]. Joining Mg alloys to steel by FSW has been extensively studied.

Watanabe et al. [60, 61] investigated the weldability of AZ31 Mg alloy to uncoated SS400 steel joints by FSW. Under optimum joining condition (0.1 mm tool pin offset toward steel plate, 1250 rpm pin rotation speed, and welding speed of 100 mm/min), a joint with maximum strength of about 70% of the Mg BM was obtained. The low joint strength was associated with insufficient plasticization and presence of steel fragments in the Mg matrix. Considerable number of authors observed that the presence of zinc coating on the steel surface played an important role in the bond formation between the Mg alloys and steel [21, 27, 56, 62, 63]. For instance, Schneider et al. [63] and Jana et al. [27] joined AZ31B alloy to Zn-coated steel by FSW. It was found that the Zn coating enhanced the bondability of the Mg alloy/steel. The authors noted that the strength of the FSW welded AZ31B/galvanized steel was significantly superior to that of AZ31B/uncoated steel. The presence of Zn coating promoted the formation of liquid low melting Mg-Zn eutectic products at the interface. The liquid products, broken oxides, and other contaminants were forced out of the joining interface by the high pressure produced by the tool, exposing the fresh interfaces. As a result, mutual diffusion between magnesium alloy and steel was achieved.

Friction stir spot welding (FSSW) a variant of FSW, has also been used to join Mg alloys to steel. Liyanage et al. [29] joined AM60 to DP600 dual-phase steel by FSSW, and the welds showed no evidence of intermetallic formation. They reported that a Zn layer on DP600 steel resulted in melted eutectic material (αMg + MgZn) and cracking in the joints. The feasibility of joining 3 mm thick AZ31 Mg alloy to 1 mm 302 stainless steel by FSW was also reported [64]. The results showed that the combined action of pressure and stirring led to mutual diffusion of the alloying elements from the BM at the interface. Under optimum condition, the joints shear strength reached a maximum of 96.3 MPa. However, void and microtype defects at the interface were observed.

Joo et al. [65] reported that an acceptable AZ31B/SS400 steel joints with higher strength, sufficient material flow and tool wear reduction were obtained using hybrid gas tungsten arc welding (GTA) and friction stir welding. The tensile strength of the hybrid friction stir welds increased to 237 MPa about 91% of the Mg BM, compared to 226 MPa for conventional friction stir welds.

The prior research proved that refill friction stir spot welding (RFSSW) has many advantages for joining of Mg alloys/steel dissimilar welds [48, 66, 67]. Zhang et al. [67] joined 3 mm AZ31B to 1 mm galvanized steel joints by friction stir keyholeless spot welding (FSKSW) and reported that the stacking sequence of the workpieces played a significant role in determining the mechanical properties. Under optimum joining condition, the maximum joint strength of 8.7 kN was achieved. Furthermore, 1.53 mm ZEK100 Mg alloy was welded to 1 mm Zn-coated DP600 steel with 10 μm Zn coating by RFSSW [66]. The authors observed that the sleeve did not plunge into the bottom steel. A thin interfacial layer with thickness of <100 nm identified as $FeAl2$ by TEM was observed at the interface, which accommodates bonding between the immiscible Mg and Fe and appeared to have originated from Zn-based galvanized coating on steel. Under optimum process parameters (tool speed of 1800 rpm, welding time of 3.0 s, and penetration into the upper ZEK100 of 1.5 mm), the joints shear strength of 4.7 kN was achieved.

However, numerous studies focused on characterizing the Mg alloys to steel joints under dynamic loading condition [20, 48, 68]. For example, the fatigue behavior of FSW-produced by AZ31 and two types of Zn-coated steels (HSLA steel or mild steel (MS)) lap joints has been investigated [20]. It was found that the performance of friction stir welded joints under fatigue load is limited due to 'hook'-like features formed along the magnesium/steel interface, which act as stress raisers. Uematsu et al. [68] investigated the static and dynamic behavior of dissimilar lap joining of 2 mm AZ31 Mg alloy to 2 mm cold-rolled low carbon steel by FSSW. The results showed that fatigue crack grew through the interface, regardless of load levels. The authors noted that the effective nugget size could be the controlling factor of the fatigue strengths of dissimilar welds, and it is essential to increase the effective nugget size to improve the fatigue performance of dissimilar welds. Shen et al. [48] compared the static and dynamic behavior of 1.5 mm ZEK100 Mg alloy/0.9 mm galvanized DP600 steel dissimilar spot welds to Mg/Mg similar welds produced by RFSSW. It was observed that the static and dynamic behaviors of the magnesium/steel welds were superior to that of Mg/Mg similar spot welds. The higher joint performance observed for Mg to Zn-coated steel was associated with an increase in effective bonded area, compared to Mg/Mg spot welds due to the presence of displayed Zn-coated layer. Furthermore, the analysis of the dynamic behavior of the Mg alloys to galvanized steel joints revealed that the Zn brazing quality directly influences the fatigue life. Therefore, optimization of the welding parameters to facilitate this brazing to magnesium is necessary [48].

Comparison of the Mg alloys/steel joints maximum tensile shear strength produced by FSW and FSSW are shown in **Table 1**. In FSW welding, prior studies placed Mg alloys as the top sheet while the tool plunge depth was carefully controlled to avoid contacting the steel tool to the bottom ferrous substrate. However, this arrangement resulted in a low joint strength [64]. Interestingly, with recent RFSSW, the sleeve did not lunge into the bottom steel, which significantly increased the effective bond area and improved the joints performance [66–71]. During FSW and FSSW, the processing heads impose size and shape limitation, the process is best used with long and straight welds, and keeping the interlayer at the interface between the steel and Mg alloy is very difficult due to the stirring action of the pin and material flow with high plasticity along the interface. Thus, limited the wide range of application of these techniques in industries [48, 72]. The comparison of the joint properties reveals that good static weld strength can be obtained between magnesium alloys to galvanized steel by a diffusion and braze-bonding mechanism [48, 66]. Most studies focused on using Zn interlayer, and there is a need to experiment with other interlayers. Furthermore, no study focused on the corrosion behavior of the joined parts.

Techniques	Materials	Joint design	Transition material	Maximum tensile shear strength (MPA)	Failure mode	References
FSW	2 mm AZ31B-O/2 mm SS400 mild steel	Butt (Offset 0.1 mm toward steel)	No interlayer	178.5 MPa	Interfacial	[60, 61]
FSW	1.6 mm AZ31/0.8 mm Zn-coated steel	Lap (Mg on top)	Zn coating	3.7kN	Interfacial	[21, 62]
	1.6 mm AZ31/0.8 mm Brushed finished steel		Uncoated	2.9kN	Stir zone	
FSW	2.33 mm AZ31B/1.5 mm HDG steel	Lap (Mg on top)	Hot dip Zn coating	6.3 ± 1.0 kN	Interfacial	[27]
	2.33 mm AZ31/0.8 mm HSLA electrically galvanized steel		Electro-galvanized Zn coating	5.1 ± 1.5 kN		
FSW	2 mm AZ31B/2 mm DX54D	Lap (Mg on top)	Uncoated	98 MPa	Stir zone	[63]
			Zn coating (15 μm thick)	158 MPa		
FSSW	1.2 mm AM60/1.8 mm DP600	Lap (Mg on top)	Zn coating	2.4 ± 0.5kN	AM60 BM	[29]
FSSW	2 mm AZ31 Mg alloy/2 mm cold-rolled low carbon steel	Lap (Mg on top)	No interlayer	32 MPa	Interfacial	[68]
FSW	3 mm AZ31/1 mm SUS302	Lap (Mg on top)	No interlayer	96.3 MPa	Stir zone	[64]
GTA-FSW	3 mm AZ31B/3 mm SS400	Butt	No interlayer	237 MPa	Interfacial	[65]
FSW				226 MPa		
FSKSW	3 mm AZ31B/1 mm Q235	Lap (Steel on top)	Zn coating	8.7kN	Not reported	[67]
RFSSW	1.55 mm ZEK100/1 mm DP600	Lap	Zn coating	4.7kN	Interfacial	[66]
RFSSW	1.5 mm ZEK100/0.90 mm DP600	Lap	Pure Zn coating (10 μm thick)	3.6kN	Interfacial	[48]

Table 1.
Comparison of the Mg alloys/steel joints maximum tensile shear strength produced by FSW and FSSW.

3.1.2 Ultrasonic spot welding (USW)

USW is also a solid-state welding technique that generates coalescence through a concurrent application of localized high-frequency vibratory energy and slight clamping force [73]. At present, the research conducted on joining magnesium alloys to steel using USW is limited.

Santella [74, 75] joined 1.6 mm thick AZ31B-H24 to 0.8 mm HDG mild steel by USW. The authors noted that the presence of Zn improved the bonding mechanism, but the Mg-Zn phases were completely squeezed from the spot weld and only thin Al5Fe5 phase was formed at the interface. Under optimum welding parameters, a joint with maximum lap shear strength of 4.2 kN was achieved. In another related study, Patel et al. [73] also noted that Mg and Zn combined to form Mg-Zn IMCs, while Fe and Zn combined to form a solid solution to create the weld joint. The results of these studies could suggest that no melting on the steel side occurs during USW. In comparison, Patel et al. [76] observed that the shear strength of the magnesium alloy/bare steel with the addition of Sn interlayer was higher than that of magnesium alloy/bare steel and magnesium alloy/galvanized steel joints, due to the solid solutions of Sn formed with magnesium and iron, as well as Sn and Mg2Sn eutectic.

On the other hand, it was demonstrated that corrosion could impair the mechanical performance of Mg alloy/steel USW joints [46]. However, the details of corrosion mechanisms required further study.

Table 2 compares maximum tensile shear strength of the Mg alloys/steel joints produced by USW. The USW techniques involved the use of an interlayer to achieve interfacial reactions. The highest joint strength (about 88% of Mg/Mg ultrasonic spot weld) was obtained using Sn interlayer due to the solid solutions of Sn formed with magnesium and iron, as well as Sn and Mg2Sn eutectic structure. Therefore, using Sn interlayer resulted in better joining mechanism and mechanical performance and is thus more suitable [76]. Thus, choosing of suitable interlayer is essential for successful joining. The possibility of using different transition materials should be explored. Furthermore, studies on dynamic and corrosion behavior of the USW joint parts should be given attention.

Techniques	Materials	Joint design	Transition material	Maximum tensile shear strength	Failure mode	References
USW	1.6 mm AZ31B-H24/0.8 mm HDG mild steel	Lap	Zn coating (9 μm thick)	4.2kN	Through the AZ31 BM	[74, 75, 77]
USW	2 mm AZ31B-H24/0.8 mm HSLA	Lap	Zn coating (10 μm thick)	47 MPa	Interfacial (cohesive failure)	[76]
			Uncoated	Sample failed during specimen mounting	—	
			Uncoated with Sn interlayer (50 μm thick)	71 MPa	Partial nugget pull-out	

Table 2.
Comparison of the Mg alloys/steel joints maximum tensile shear strength produced by USW.

3.1.3 Diffusion and eutectic bonding

Diffusion methods have been used to successfully join both similar and dissimilar materials combinations, including Mg alloys/steel [37, 78]. During the diffusion process, formation of uniform and thin IMC along the Mg alloy/steel interface is the key for successful bonding. The number of research suggested that interfacial bonding could be achieved by either addition of interlayer or some alloying elements in the Mg BM. The benefits of using Cu [78–80], Ni [37, 80], and Ag [44, 81] interlayers during diffusion bonding of Mg alloy/steel have been demonstrated.

Tachibana et al. [82] studied the influence of the Zn insert on the bondability of Mg/steel lap joints. It was found that the AZ31 and cold-rolled steel plate (SPCC) could not bond due to the oxide films formed on the AZ31 that prevented the bonding, while AZ31/Zn-coated steel (GI) was bonded successfully. The presence of zinc coating enhanced the bondability of Mg alloy/steel and removed the oxide films on AZ31. The weldability of 5 mm AZ31 Mg alloy to 1 mm thick 316 L steel joints using a diffusion brazing process with addition of Cu and Ni interlayers was also reported [37, 79, 80, 83]. Solid-state diffusion of the interlayer into the magnesium alloy, eutectic formation, and the formation of ternary IMCs was observed at the joint interface. The interfacial reaction was intense in the liquid state, inducing the excessive formation of brittle and thick IMC layers, which was detrimental to the joint strength. The maximum joint strength of 57 MPa (69% of that of AZ31 BM) with Cu interlayer was obtained compared to 32 MPa for Ni added joints. The high interfacial bond obtained for Cu-added joint was associated with confined intermetallics at the bonded interface compared to dispersed intermetallics for Ni added joint [80].

Some authors focused on bonding magnesium alloys to steel using reactive transient liquid phase bonding (rTLP) to improve the joints strength between dissimilar metals. During the rTPL process, eutectic melting and subsequent isothermal solidification occurred between the interlayer and the magnesium, while the formation of a thin continuous layer between the melt and the steels resulted in interfacial bonding of the steel with the magnesium substrate [44, 81]. The bondability of AZ31 to low carbon steel with the addition of Ag interlayer using rTPL process showed that isothermal solidification of the eutectic melt was formed at the Mg alloy side through the diffusion of Ag into the magnesium BM. Thin and uniform Fe_2Al_5 layer was observed at the steel side, which significantly improved the joint strength to 201 MPa [44, 81]. In contrast, coarse, non-uniform IMC intermittently formed at the interface without Ag interlayer, which deteriorates the joint performance.

Table 3 gives a comparison of the Mg alloys/steel joints maximum tensile shear strength produced by diffusion and eutectic bonding. The literature reveals that the interlayer material has significant influence on the bondability of the Mg alloys to steel by diffusion and eutectic bonding process. The use of Cu, Ni, and Ag as intermediate elements has been studied. The major challenge associated with this technique is that the formation of ternary IMCs which could not be controlled and have detrimental effects on the joint performance [80]. A comparison of the joints strength shows that excellent static strength of Mg/steel joints has been achieved using rTPL techniques, even surpassing that of AZ31B Mg alloy BM with addition of Ag interlayer [44]. The rTPL unique qualities of short process time, coupled with forming a thin and continuous intermediate layer through the formation of a transient liquid interlayer, were responsible for the high joint performance obtained. The possibility of improving the joint performance using more interlayers should be explored.

Techniques	Materials	Joint design	Transition material	Maximum tensile shear strength (MPa)	Failure mode	References
Diffusion bonding	AZ31/SPCC	Lap	Uncoated	—	No reported	[82]
	AZ31/GI		Pure Zn coating (6 μm thick)	40		
	AZ61/GI		Pure Zn coating (6 μm thick)	70		
Diffusion brazing	5 mm AZ31/1 mm 316 L	Lap	Pure Cu (20 μm)	57	Mg BM	[80]
			Pure Ni (20 μm)	32	Steel/Ni interface	
rTPL bonding	AZ31B/Low carbon steel	Lap (Mg on top)	—	40	Interfacial	[44, 81]
			Pure Ag (1 μm)	201	Mg BM	

Table 3.
Comparison of the Mg alloys/steel joints maximum tensile shear strength produced by diffusion and eutectic bonding.

3.2 Fusion-welding processes

Fusion welding involves joining the surface of the materials through melting and solidification to produce the bonding. Fusion-welding technologies based on resistance spot welding [4, 32, 84–86] laser braze [41, 43, 87–92], laser hybrid [33, 35, 36, 38–40, 47, 93–98] and arc welding [72, 99–102] have been investigated for joining magnesium alloys to steels. In general, the insertion of an interlayer or mutual diffusion of alloying elements from the BM has been adopted to improve interfacial bonding. The addition of suitable interlayer improves the spreadability and the nucleation of magnesium on steel.

3.2.1 Resistance spot welding (RSW)

RSW is the most widely used welding technology in the auto industry [103] due its low cost, high speed, ease of operation, and automation [17, 32, 86]. Despite the inherent advantages of RSW, very limited work has been published in joining Mg alloy to steel. This include AZ31B Mg alloy/Zn-coated DP600 steel [32], AZ31 Mg alloy/HDG steel with pre-coated nanoscale Fe_2Al_5 layer [34], AZ31B-H24 Mg alloy/HDG HSLA steel [4], AZ31B Mg alloy/HDG HSLA steel [84] and AZ31B Mg alloy/electro-galvanized DP600 steel [86].

Most of the works focused on joining magnesium alloys to galvanized steel. Liu and his co-workers developed a novel technique (asymmetric electrode) to lap weld 1.5 mm AZ31B to 1.2 mm HDG DP600 steel successfully by RSW [32]. It was found that the Zn interlayer was squeezed out of the bond region producing deal condition for intimate contact of fresh Mg and steel surfaces. The joining mechanism consisted of braze welding, solid-state bonding, and soldering. Under optimum parameters, a joint with tensile shear strength of 5.0 kN was obtained. Xu et al. [4] compared the microstructure and mechanical performance of AZ31B to HDG HSLA steel joints by RSW and weld-bonding (WB), which combines the RSW and adhesive bonding. It was found that the peak shear load and energy absorption of the weld-bonded magnesium to steel joints were higher than that of resistance spot welded magnesium to steel joints. In another similar study, AZ31B was joined to HDG HSLA steel by RSW under dynamic loading [84]. It was found that the microstructure of the Mg/steel spot welds was different from that of Mg/Mg spot welds, but owing to similar crack propagation and failure mode, both welds had an equivalent fatigue resistance. Interestingly, Feng et al. [86] joined AZ31 to electro-galvanized DP600 steel by RSW process with and without HDG Q235 interlayer. Contrary to the previous studies [4, 32, 84] that used hot-dip Zn coating, the thin and compact features of electro-galvanized zinc layer prevented the zinc-coated layer on the steel to be squeezed out of the nugget. However, with insertion of HDG Q235 interlayer, the zinc-coated layer was squeezed out of the nugget and a peripheral soldered region was formed during the welding process, which significantly improved the joint tensile shear load from 4.14 kN to 5.49 kN. The feasibility of joining AZ31B Mg alloy to DP600 steel via pre-coated nano-scaled Fe_2Al_5 interlayer was also investigated [34]. The analysis of the interface characteristics revealed that metallurgical bonding was achieved due to the formation of the semi-coherent interfaces of Mg/Fe_2Al_5/Fe with well matching lattice.

However, RSW of Mg alloy to stainless steel is more challenging because of the absence of any Zn coating. However, Mg alloy to stainless steel joints has been reported [85, 104]. Min et al. [104] investigated the 0.4-mm thick AZ31B sheets, and 0.4-mm thick 443 ferritic stainless steel welded joints using the RSW with 443 ferritic stainless steel cover plates. It was found that Fe-Al IMC layer was formed at the interface, and the molten Mg could wet the surface of the interface layer. However, cracking was observed in the nugget. The crack was associated with the possible thermal behavior of various elements in the nugget during spot-welding process. Recently, Manladan [85] compared the microstructure and mechanical performance of 1.5 mm thick AZ31 Mg alloy/0.7 mm thick 316 L austenitic stainless steel joints by RSW and resistant element welding (REW). In comparison with two-zone FZ, consisting of peripheral FZ on the ASS side and main FZ observed for REW, the RSW joints were produced through welding-brazing mode, in which the Mg alloy melted and spread on the solid steel, forming the nugget only on the Mg side. The RSW produced the weak joint with a peak load of 2.23 kN and energy absorption of 1.14 J, whereas REW produced the strong joint with a peak load of 3.71 kN and energy absorption of 10.2 J.

Generally, in the RSW process of Mg alloys, large electric currents are always needed due to its high electrical and thermal conductivity, which commonly results in expulsion and electrode stick. In addition, RSW created only a localized joint, which may not be particularly strong [105].

A comparison of the Mg alloys/steel joints maximum tensile shear strength produced by RSW is shown in **Table 4**. It can be seen that limited research has been conducted on magnesium alloys/steel. The presence of zinc layer was crucial for successful welding of magnesium to steel. The predominant failure mode observed is interfacial (IF). This is the kind of failure mode commonly observed when conducting tensile shear tests on spot welds, in which the crack propagates through the nugget. A comparison of the joints performance shows that good static strength has been achieved, almost 95% Mg/Mg spot joint strength by RSW [32]. However, the research on the fatigue performance of Mg/steel dissimilar welds is still at its infancy due to the special geometry of the spot welds; it is hard to predict the crack initiation and propagation rate as reported by Liu et al. [84]. Among the RSW techniques, weld bonding produced the joint with highest mechanical resistance with joint failure on the Mg

Techniques	Materials	Joint design	Transition material	Maximum tensile shear strength (kN)	Failure location	References
RSW	1.5 mm AZ31/ 1.2 mm Zn-coated DP600	Lap (Mg on top)	HDG Zn-coated (9 μm to 12 μm)	5.0	Interfacial failure (IF) mode	[32]
RSW	1.5 mm AZ31/ 1.2 mm HDG DP600	Lap (Mg on top)	Bare steel	0.0	Not reported	[34]
			Fe_2Al_5	4.8	Not reported	
RSW	1.5 mm AZ31B/0.77 mm HSLA steel	Lap (Mg on top)	HDG Zn-coated (11 μm)	2.0	Interfacial failure (IF) mode	[84]
RSW & WB	2 mm AZ31B-H24/0.7 mm HSLA steel	Lap (Mg on top)	HDG Zn-coated (11 μm)	RSW 4.80 WB 10.0	IF (RSW) Mg BM (WB)	[4]
RSW	2 mm AZ31B/ 1.2 mm Electro-galvanized DP600	Lap (Mg on top)	No transition material	4.14	IF	[86]
			HDG Q235 (0.6 mm)	5.49	IF	
RSW & REW	1.5 mm AZ31/0.7 mm 316 SS	Lap (Mg on top)	No transition material	RSW 2.23	IF	[85]
				REW 3.71	PO	

Table 4.
Comparison of the Mg alloys/steel joints maximum tensile shear strength produced by RSW.

BM. Weld-bonding joining techniques has the advantages of low manufacturing costs, higher static and fatigue performance, and improved corrosion resistance [106, 107].

3.2.2 Laser beam welding

Laser beam welding presents a viable option for welding Mg alloys to steel due to its versatility, high specific heat input, and flexibility [55, 108–111]. Although fusion welding resulted in severe vaporization of Mg alloy at typical welding temperatures, the selection of optimum welding conditions is crucial for successful joining. Therefore, to control the severe vaporization of Mg, considerable number of authors concentrated on laser welding brazing and laser hybrid techniques.

Laser welding brazing (LWB) techniques offered additional advantages such as increased flexibility and adoptability when welding dissimilar metals [28, 91, 92, 112, 113]. To this end, LWB technique is suitable for joining dissimilar materials having large differences in melting points such as Mg/Ti [18, 114–116], Al/Ti [117], and Al/steel [118, 119]. In particular, welding of magnesium alloys to steel by LWB process was achieved with addition of third material or mutual diffusion of alloying elements. Thin interlayers such as Al [28, 120, 121], Ni [90, 113], Sn [31], Zn [41, 42, 87, 88, 91], and Fe-Al [41], have been used to improve the interfacial bonding between the immiscible Mg and Fe. Miao demonstrated the feasibility of joining 2.4-mm AZ31B Mg alloy to 1.7 mm Q235 steel by laser welding brazing process using high power CO2 laser without welding wire [122–124]. The analysis of the interface characteristics revealed that transition layer consisted mainly of IMCs, and metal oxides were observed at the interface. The compounds were identified by TEM as Al-rich phases, such as $Mg_{17}Al_{12}$, Mg_2Al_3, FeAl, and Fe_4Al_{13} [124]. Although tensile strength of the butt joints could reach 182 MPa (81% of Mg alloy BM), but severe oxidation and vaporization of the magnesium alloy, coupled with immiscibility between magnesium and iron, led to weld defects such as spatters and porosity.

Therefore, to address the severe vaporization of the Mg, laser brazing or laser welding brazing with magnesium-based filler wire was proposed [28]. The benefits of using Al-12Si [28], Ni [113], Sn [31], and Zn [42] as transition material between Mg alloy to steel were explored by Nasiri and co-workers. It was found that the presence of the interlayer significantly improved the spreadability of the liquid magnesium-based filler on steel surface but deteriorated the joint performance because of the weak bonding of IMCs products formed along the interface.

However, several authors observed mutual diffusion of alloying elements when joining Mg alloys to different grades of stainless steel [92, 125, 126]. For instance, [92] compared the joint performance of lap welded 1.5 mm-AZ31B/1.5 mm-mild steel (MS) joints and 1.5 mm-AZ31B/1.5 mm 201 stainless steel (SS) joints produced by LWB. Mechanical bonding was observed for Mg alloy to MS joints, whereas for Mg alloy to SS joints, thin reaction layer identified as FeAl by TEM as shown in **Figure 1**. The Al diffusion from magnesium BM to the interface was accelerated by

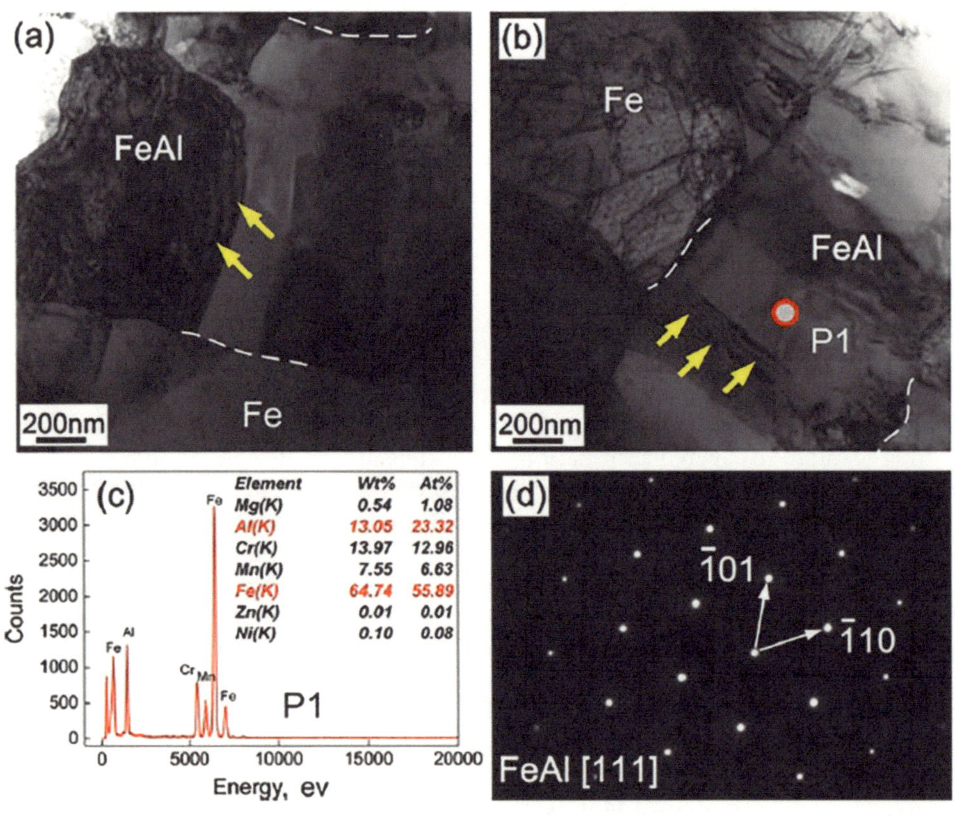

Figure 1.
Mg/201 stainless steel interface characteristics [73].

chemical potential induced by alloying elements of Cr and Ni was responsible for the interfacial reaction layer obtained.

A comparison of the Mg alloys/steel joints maximum tensile shear strength produced by LWB is shown in **Table 5**. The selection of suitable interlayer is crucial for successful bonding. For instance, using Ni and Sn interlayers, which led to formation of the Fe(Ni) and Al8Mn5-Fe(Al) reaction products along the α-Mg-Fe interface, respectively, resulted in formation of strong interfaces with low mismatch strain energy and strong bonds. Therefore, using both Ni and Sn interlayers is recommended for dissimilar joining of steel sheet to magnesium sheet. Furthermore, the diffusion of the alloying elements in BM and the bonding mechanism were extensively studied. A comparison of the joints properties reveals that relatively good static strength of magnesium alloy to steel joints has been obtained [126]. However, the behavior of the Mg alloys/steel joint under dynamic loading is yet to be explored. Among the LWB techniques, dual-beam mode with flux produced joint with an excellent mechanical resistance. To further enhance the reliability of the joint and improve its performance, the possibility of using different interlayers should be explored.

The excessive evaporation of magnesium during fusion welding can be overcome through the method of TIG-assisted laser process [33]. Generally, the addition of TIG could improve absorption of laser power and penetration of molten pool. During laser-TIG process, the TIG torch melts the magnesium, whereas, the laser is used to create deep penetration into the steel [36, 39]. Several authors reported that conventional fusion-welding technology such as TIG or laser welding alone could not be competent for joining of magnesium to steel [33, 89, 94]. For instance, [33, 89] demonstrated during TIG welding of Mg alloy to steel with higher heat input, either the magnesium alloys melt but the steel remained in solid state or the steel melt while blowholes formed in the magnesium alloys. In both cases, the bonding between the immiscible couple could not be obtained. Similarly, for conventional laser-welded magnesium alloys to steel joints, the immiscible materials could be joined with poor weld appearance and poor strength, coupled with large amount of electricity consumed by laser due to low absorptivity of magnesium alloys at room temperature [127]. For TIG-assisted laser process, the reflectivity of the magnesium to laser was reduced and thus the absorption of laser beam improved, creating a deeper penetration in the steel [33, 35, 128]. Consequently, laser-TIG process offers great potential for improving the joint performance for both similar and dissimilar metal welding [33, 39, 40, 89, 129, 130].

Like other joining techniques, interfacial bonding between magnesium alloy to steel could be achieved by insertion of third material or mutual diffusion of alloying elements from the BM [95]. Therefore, the benefits of using Cu [35, 39, 40], Ni [35, 36, 40, 93, 96], Sn [94], Zn [89], Cu-Zn [40, 97] to bond magnesium to steel had been extensively studied. These interlayers formed Mg_2X IMC (where X is the interlayer elements) and a solid solution of interlayer in iron along the joint interface, which consequently enhanced the joints mechanical strength [40].

Liu examined the weldability of AZ31B/304 SS by laser-GTA process. A transitional zone consisting of Mg-Fe and magnesium diffusion into the matrix of iron in the form of oxide was observed at the interface. The formation of the complex MgO, ZnO, Fe_2O_3, or Al_2O_3 oxides deteriorates the joint strength [39, 94]. Furthermore, the use of laser-TIG techniques to join AZ31B and Q235 steel with Ni, Cu, Sn, and Cu-Zn interlayers was also explored [35, 36, 38–40, 93, 94]. These interlayers were heated and melted to react with magnesium and steel, and formed a transitional layer in the FZ and solid solution along the steel side. The analysis of the strengthening

Techniques	Materials	Joint design	Transition material	Maximum strength	Failure mode	References
Laser welding	3 mm AZ31B/1.2 mm SP781 steel	Lap (Mg on top)	Zn Coating	6182 N	AZ31B FZ & Transition zone	[43]
Laser penetration brazing	2.4 mm AZ31B/1.7 mm Q235 steel	Butt (0.6 mm offset on Mg)	No interlayer	182 MPa	Interface	[122–124]
Laser brazing	2 mm AZ31B-H24/1 mm Al-12Si coated steel 2.4-mm- Dia Ti-Braze Mg 600 wire	Single flare bevel Lap (Mg on top)	Al-12 wt-% Si coating (20 ± 2 μm thick)	767 ± 138 N	Interface	[28]
Laser brazing	2 mm AZ31B-H24/1 mm Al-12Si coated steel 2.4-mm- Ti-Braze Mg 600 wire	Single flare bevel Lap (Mg on top)	Pure Ni coating(5 μm)	1506.3 ± 24.5 N	FZ	[113]
Laser brazing	2 mm AZ31B-H24/0.6 mm AISI 1008	Lap (Mg on top)	Pure Sn coating (3.7 ± 0.7 μm thick)	2064 ± 85 N	Steel BM	[31]
Laser brazing	2 mm AZ31B-H24/0.8 mm AISI 1008	Lap (Mg on top)	Pure Zn coating (2.6 ± 0.5 μm thick)	1086.4 ± 150.2 N	Interface	[42]
Laser welding brazing	1.5 mm AZ31B-H24/1.5 mm DP980	Lap (Mg on top)	uncoated	160 N/mm	Interface	[87, 88, 91]
			Zn coated (10 to 15 μm)	228 N/mm		
Laser welding brazing	1.5 mm AZ31B-H24/1.5 mm DP980 with 1.2 mm AZ31 filler	Lap (Mg on top)	uncoated	190 N/mm	Interface	[41]
			Zn + Fe-Al Phase	180 N/mm	Interface	
			Zn coating (10 μm)	160 N/mm	Interface	
			Fe-Al coating (0.9 μm)	240 N/mm	Weld Seam	
Laser welding brazing	1.5 mm AZ31/1.5 mm 22MnB5 with 1.2 mm AZ61 filler	Lap (Mg on top)	AlSi10Fe3-coating	3090 N	AZ61 filler	[121]
Laser welding brazing	1.5 mm AZ31B/1 mm Q235	Lap (Mg on top)	Al (0.3 mm)	133 N/mm	FZ	[120]
Laser welding brazing	1 mm AZ31B-H24/1 mm Q235A with 1.2 mm AZ92D filler	Lap (Mg on top)	Ni coating (4.5 ± 0.5 μm)	190 N/mm	FZ	[90]

Techniques	Materials	Joint design	Transition material	Maximum strength	Failure mode	References
Laser Welding brazing	AZ31B/mild Steel	Lap (Mg on top)	No interlayer	142 N/mm	Interface	[92]
	AZ31B/stainless steel			270 N/mm	Mg FZ	
Laser welding brazing	1.5 mm AZ31B-H24/1.5 mm 201 stainless steel and 2 mm AZ31 Filler	Lap (Mg on top)	No interlayer	274.5 N/mm	Mg HAZ	[125, 126]
Laser offset welding	3 mm AZ31/3 mm AISI 316	Butt (3 mm Offset on Steel)	No interlayer	100 MPa	AZ31 BM	[30]

Table 5.
Comparison of the Mg alloys/steel joints maximum tensile shear strength produced by laser welding brazing.

mechanism revealed that with the addition of suitable interlayers, the joint shear strength could reach a significantly high value or even surpasses Mg alloy BM.

Generally, the corrosion behavior of magnesium alloys is affected by the microstructural variation imposed by welding process [27, 43, 56, 63, 64, 125, 126, 131]. For similar Mg/Mg welds, the grain refinement imposed by welding process was reported to improve the corrosion resistance [132]. However, Liu et al. studied the corrosion behavior of the magnesium to steel joints produced by TIG-assisted laser process in NaCl solution. It was found that the grain refinement and iron splashes imposed by welding process were observed in the weld, which accelerated the corrosion of the magnesium alloy. Interestingly, the use of Al coating was reported to raise the lifespan of the dissimilar joints [47].

A comparison of the Mg alloys/steel joints maximum tensile shear strength produced by laser-TIG process is shown in **Table 6**. In hybrid laser-TIG welding, excessive vaporization of the magnesium was observed due to the penetration of the laser

Techniques	Materials	Joint design	Transition material	Maximum tensile shear strength (MPa)	Failure mode	References
Laser-GTA	1.7 mm AZ31B/1.2 mm 304 steel	Lap (Mg on top)	No interlayer	90	Interfacial	[33]
Laser-TIG	1.7 mm AZ31B/1.2 mm Q235	Lap (Mg on top)	No interlayer	120	Interfacial	[36, 39, 96]
			Pure Cu foil (0.1 mm thick)	170	FZ	[38]
			Pure Ni foil (0.1 mm thick)	166		
			Pure Sn foil (0.1 mm thick)	117		
			Cu-Zn (H80)-0.1 mm thick	161		
			Cu-Zn (H62)-0.1 mm thick	157		
Laser-TIG	1.5 mm AZ31B/1.5 mm DP980	Lap (Mg on top)	Zn coating (10 to 15 μm thick)	68	Interfacial	[89]
Laser-TIG	1.6 mm AZ31B/1.1 mm Q235	Butt (0.2 mm offset on steel)	Cu-Zn alloy (H62) 0.1 mm thick	203	Weld seam	[97]

Table 6.
Comparison of the magnesium alloys/steel joints maximum tensile shear strength produced by laser-TIG process.

Techniques	Materials	Joint design	Transition material	Maximum tensile strength	Failure mode	References
CMT Welding	1 mm AZ31B/ Mild steel	Lap (Mg on top)	HDG Zn coating (10 μm)	224 N/mm	Interfacial	[99, 101]
			Uncoated	258 N/mm	Mg HAZ	
Bypass current-MIG-welding	2.5 mm AZ31B/ 2 mm Q235, with 1.6 mm AZ31 filler	Lap (Mg on top)	Zn coating	133.02 MPa	weld metal	[135]
MIG spot welding	1.8 mm AZ31B/1.5 mm Q235 with 1.2 mm ER50-6 filler	Lap (Steel on top)	Pure Cu foil (0.1 mm)	3200 N	Interfacial	[72]
MIG spot welding	3 mm AZ31B/3 mm Q235, 1.6 mm Mg based filler	Butt (V groove)	Uncoated	160 MPa	Interfacial	[100]
			Pure Cu foil (0.1 mm)	185 MPa	weld metal	
CMT brazing	2 mm AZ31/1 mm steel with 1.2 mm AZ31 filler	Lap (Mg on top)	Zn coating (10 μm)	3100 N	Interfacial	[102]
			Zn-Fe (8 μm)	4100 N		
			Uncoated	7000 N	Weld metal	
			AlSi Coating (25–30 μm)	6000 N		

Table 7.
Comparison of the Mg alloys/steel joints maximum tensile shear strength produced by MIG welding brazing.

from upper magnesium plate and the interlayer into the bottom steel. Furthermore, the violent stirring of molten pool restricted metallurgical bonding adjacent to the steel interface, which limited the application of this welding technique. Despite the fast heating and cooling rate of the TIG-assisted laser process, the interfacial reaction was achieved through diffusion and combination of alloying elements from the BM. Based on the existing literature, the interfacial characteristics and the mechanism of wetting in Mg alloys to steel joints produced using TIG-assisted laser process with addition of interlayers were thoroughly investigated. The feasibility of using Cu, Ni, Sn, Zn, and Cu-Zn intermediate elements was also explored. The presence of interlayer was essential for successful joining Mg to steel. The selection of a suitable interlayer was essential for successful bonding. For instance, the Sn added joint shows comparatively lower value due to inhomogeneous compositions in the FZ. A comparison of the joints mechanical properties shows that excellent static strength has been achieved, even surpassing that of magnesium alloy base metal with insertion of Cu and Ni intermediate elements. The high joint shear strength obtained was associated with better wettability and deeper penetration in the weld. However, limited studies focused on corrosion behavior of the TIG-assisted laser magnesium to steel joints. Furthermore, no study has focused on the behavior of the joints under dynamic loading.

3.2.3 Arc welding

Arc welding involves joining the materials surface permanently using power supply to obtain an electric arc between the electrode mounted in a torch and a metal. Among the arc welding processes, so far only metal inert gas welding has been used for this material combination.

Metal inert gas (MIG) welding has been widely used in automobile industries due its inherent properties, such as high efficiency, lower cost, and excellent adoptability to material geometry [133, 134]. Therefore, obtaining a reliable Mg alloy to steel joint by MIG welding is essential. However, joining magnesium to steel by MIG is rarely reported, including AZ31B/Zn-coated steel [99, 101, 102, 135], AZ31B/Q235 with Cu interlayer [72, 100] and AZ31B/Aluminized steel [102]. Generally, the addition of suitable interlayer improves the spreadability and the nucleation of magnesium on the steels.

Cold metal transfer (CMT) modified gas metal arc welding of AZ31B Mg to galvanized and bare mild steel sheets showed that welded brazed joints formed in both joints [99, 101].

A comparison of the Mg alloys/steel joints maximum tensile shear strength produced by MIG welding brazing is shown in **Table 7**. Generally, the investigation on the Mg alloys/steel joints produced by MIG welding is still in its infancy. Therefore, more interlayers should be tested. Moreover, the behavior of the joints under dynamic loading should be studied.

4. Summary and outlook

As there is a desire in the aerospace, aircraft, and automotive industries to join magnesium alloys to steel in order to achieve lighter weight, versatile, and tailored properties in one composite part, and development of a welding technology for Mg alloys/steel with a strong metallurgical bond will expedite increased applications of magnesium alloys in these industries. The major challenge in welding of magnesium alloys to steel is the huge differences in physical properties and limited solubility that make the welding process difficult. Different joining processes have been used to join Mg alloys to steel sheets but metallurgical bonding can only be possible with insertion of intermediate interlayer elements or alloy or mutual diffusion of alloying elements from BM. The existence of intermetallic phases or solid solutions between Mg and the interlayer and also the interlayer and Fe is an indication that metallurgical bonding between Mg and Fe using the interlayer may be possible. Formation of thick, brittle intermetallic compounds along the interface between Mg and steel can cause significant deterioration of mechanical properties. Therefore, when choosing the interlayer and the joining process that will be used, minimization of the thickness of any brittle IMCs that might form at magnesium alloy-interlayer-steel joint interfaces and minimization of intermixing between the Mg and Fe in the liquid state are main factors that must be considered.

Solid-state bonding techniques based on FSW, USW, and diffusion and eutectic bonding have been used. Generally, the intimate contact between the dissimilar materials and the formation of IMC with insertion of the suitable interlayer played a significant role in controlling the joint performance of solid-state bonded magnesium alloys to steel.

Joining magnesium alloys to different grades of steels by fusion processes such as RSW, laser welding, and arc welding with and without insertion of interlayers were also reported. Despite the fast heating and cooling rate of the fusion process, the interfacial reaction was achieved through diffusion and combination of alloying elements from the BM. The phase formation in the magnesium-interlayer-steel alloy system and the mechanism of wetting in magnesium-interlayer-steel alloy system were extensibility investigated. The benefits of using Cu, Ni, Sn, Zn, and Cu-Zn interlayers were also explored. The addition of suitable interlayer improves the spreadability and the nucleation of magnesium on the steel substrate. Therefore, more interlayer materials that form the eutectic phase with Mg either in a pure form such as Ag and Al or as alloys such as Al-Cu and Ag-Sn should be investigated.

Currently, a great deal of research has been conducted on the interface characteristics and mechanical performance of Mg alloys to steel joints, particularly under static loading. Under optimized processing conditions, excellent static strength has been achieved, even surpassing that of Mg alloy base metal with insertion of Ag, Cu, and Ni intermediate elements. However, few experiments have been carried out on the corrosion behavior of the jointed parts and the joints performance under dynamic loading. Thus, cost-effective and reliable joining techniques for Mg/steel will still require further development. The need for the industries for more advanced materials to accommodate the huge demands for strong, rigid and light structures may be the powerful drivers for further development of the welding techniques of hybrid structures of Mg alloys to steel.

Author details

Shamsu Tukur Auwal[1*], Murtala Sule Dambatta[1], Singh Ramesh[2] and Tan Caiwang[3]

1 Faculty of Engineering, Department of Mechanical Engineering, Kano University of Science and Technology, Kano, Nigeria

2 Faculty of Engineering, Center of Advanced Manufacturing and Materials Processing (AMMP), Department of Mechanical Engineering, University of Malaya, Kuala Lumpur, Malaysia

3 Shandong Provincial Key Laboratory of Special Welding Technology, Harbin Institute of Technology at Weihai, Weihai, China

*Address all correspondence to: engrstauwal@kust.edu.ng

References

[1] Schiermeier Q. Increased flood risk linked to global warming: Likelihood of extreme rainfall may have been doubled by rising greenhouse-gas levels. Nature. 2011;**470**(7334):316-317

[2] Pall P et al. Anthropogenic greenhouse gas contribution to flood risk in England and Wales in autumn 2000. Nature. 2011;**470**(7334):382-385

[3] Auwal S et al. Recent developments and challenges in welding of magnesium to titanium alloys. Advances in materials Research. 2019;**8**(1):47-73

[4] Xu W et al. Microstructure and mechanical properties of weld-bonded and resistance spot welded magnesium-to-steel dissimilar joints. Materials Science and Engineering a-Structural Materials Properties Microstructure and Processing. 2012;**537**:11-24

[5] Hirose A. Trends of applications and joining technologies of aluminum alloys in automobiles. Journal-Society of Automotive Engineers of Japan. 2007;**61**(4):18

[6] Sasabe S. Dissimilar Metal Joining Technology of Aluminum Alloys to Steel. Journal-Society of Automotive Engineers of Japan. 2007;**61**(4):24

[7] Kumar DS et al. Magnesium and its alloys in automotive applications—A review. American Journal of Materials Science and Technology. 2015;**4**(1):12-30

[8] Nasiri AM. Laser Brazing of Magnesium to Steel Sheet. University of Waterloo; 2013

[9] Manabe A. Weight reduction of vehicles and light metals. Zairyo-to-Kankyo. 1999;**48**(8):463-468

[10] Auwal ST et al. A review on laser beam welding of titanium alloys. The International Journal of Advanced Manufacturing Technology. 2018;**97**(1):1071-1098

[11] Humpenöder F et al. Land-use and carbon cycle responses to moderate climate change: Implications for land-based mitigation? Environmental science & technology. 2015;**49**(11):6731-6739

[12] Kim HC, Wallington TJ. Life-cycle energy and greenhouse gas emission benefits of lightweighting in automobiles: Review and harmonization. Environmental Science & Technology. 2013;**47**(12):6089-6097

[13] Liu L, Ren D, Liu F. A review of dissimilar welding techniques for magnesium alloys to aluminum alloys. Materials. 2014;**7**(5):3735-3757

[14] Agnew SR. Wrought magnesium: A 21st century outlook. Jom. 2004;**56**(5):20-21

[15] Watarai H. Trend of Research and Development for Magnesium Alloys-Reducing the Weight of Structural Materials in Motor Vehicles. NISTEP Science & Technology Foresight Center; 2006;**18**:84

[16] Cole GS. Issues that influence magnesium's use in the automotive industry. In: Materials Science Forum. Washington, DC: Trans Tech Publications; 2003

[17] Manladan SM et al. A review on resistance spot welding of magnesium alloys. The International Journal of Advanced Manufacturing Technology. 2016;**86**(5):1805-1825

[18] Auwal ST et al. Influence of electrodeposited Cu-Ni layer on interfacial reaction and mechanical properties of laser welded-brazed Mg/Ti lap joints. Journal of Manufacturing Processes. 2019;**37**:251-265

[19] Emley EF. Principles of magnesium technology. Oxford: Pergamon Press; 1966

[20] Jana S, Hovanski Y. Fatigue behaviour of magnesium to steel dissimilar friction stir lap joints. Science and Technology of Welding and Joining. 2012;**17**(2):141-145

[21] Chen YC, Nakata K. Friction stir lap welding of magnesium alloy and zinc-coated steel. Materials Transactions. 2009;**50**(11):2598-2603

[22] Zhang CQ, Wang XJ, Li BQ. A technological study on friction stir blind rivet jointing of AZ31B magnesium alloys and high-strength DP600 steel. In: Advanced Materials Research. Washington, DC: Trans Tech Publications; 2011

[23] MacNew T. More emphasis on stainless steel for passenger car components. Autom Ind. 1956;**114**:52-53

[24] Ratte E et al. Energy absorption behaviour of austenitic and duplex stainless steels in a crash box geometry. Steel Research International. 2006;77(9-10):692-697

[25] Campbell FC Jr. Manufacturing Technology for Aerospace Structural Materials. Amsterdam, Netherlands: Elsevier; 2011

[26] Göken J et al. New development in magnesium technology for light weight structures in transportation industries. In: Materials Science Forum. Washington, DC: Trans Tech Publications; 2003

[27] Jana S, Hovanski Y, Grant GJ. Friction stir lap welding of magnesium alloy to steel: A preliminary investigation. Metallurgical and Materials Transactions a-Physical Metallurgy and Materials Science. 2010;**41a**(12):3173-3182

[28] Nasiri A et al. Microstructure and properties of laser brazed magnesium to coated steel. Welding journal. 2011;**90**(11):211-219

[29] Liyanage T et al. Joint formation in dissimilar Al alloy/steel and Mg alloy/steel friction stir spot welds. Science and Technology of Welding and Joining. 2009;**14**(6):500-508

[30] Casalino G et al. Laser offset welding of AZ31B magnesium alloy to 316 stainless steel. Journal of Materials Processing Technology. 2017;**242**:49-59

[31] Nasiri A, Weckman D, Zhou Y. Interfacial microstructure of laser brazed AZ31B magnesium to Sn plated steel sheet. Welding Journal. 2015;**94**:61-72

[32] Liu L et al. The mechanisms of resistance spot welding of magnesium to steel. Metallurgical and Materials Transactions A—Physical Metallurgy and Materials Science. 2010;**41a**(10):2651-2661

[33] Liu L, Zhao X. Study on the weld joint of Mg alloy and steel by laser-GTA hybrid welding. Materials Characterization. 2008;**59**(9):1279-1284

[34] Liu L et al. Bonding of immiscible Mg and Fe via a nanoscale Fe 2 Al 5 transition layer. Scripta Materialia. 2011;**65**(11):982-985

[35] Liu L, Qi X. Strengthening effect of nickel and copper interlayers on hybrid laser-TIG welded joints between magnesium alloy and mild steel. Materials & Design. 2010;**31**(8):3960-3963

[36] Qi X, Song G. Interfacial structure of the joints between magnesium alloy

and mild steel with nickel as interlayer by hybrid laser-TIG welding. Materials & Design. 2010;**31**(1):605-609

[37] Elthalabawy WM, Khan TI. Microstructural development of diffusion-brazed austenitic stainless steel to magnesium alloy using a nickel interlayer. Materials Characterization. 2010;**61**(7):703-712

[38] Qi X, Liu L. Investigation on welding mechanism and interlayer selection of magnesium/steel lap joints. Welding Journal. 2011;**90**(1):1S-7S

[39] Liu L, Qi X. Effects of copper addition on microstructure and strength of the hybrid laser-TIG welded joints between magnesium alloy and mild steel. Journal of Materials Science. 2009;**44**(21):5725-5731

[40] Qi X-d, Liu L-m. Comparative study on characteristics of hybrid laser-TIG welded AZ61/Q235 lap joints with and without interlayers. Journal of Materials Science. 2010;**45**(14):3912-3920

[41] Tan C et al. Comparative study of microstructure and mechanical properties of laser welded–brazed Mg/steel joints with four different coating surfaces. Science and Technology of Welding and Joining. 2013;**18**(6):466-472

[42] Nasiri A, Zhou Y. Effect of Zn interlayer on brazeability of AZ31B–Mg alloy to steel sheet. Science and Technology of Welding and Joining. 2015;**20**(2):155-163

[43] Wahba M, Katayama S. Laser welding of AZ31B magnesium alloy to Zn-coated steel. Materials & Design. 2012;**35**:701-706. http://textileinfo.com/

[44] Koba M et al. Bonding interface formation between Mg alloy and steel by liquid-phase bonding using the Ag interlayer. Metallurgical and Materials Transactions A. 2012;**43**(2):592-597

[45] Wielage B, Mücklich S, Grund T. Corrosion behaviour of soldered joints of magnesium alloys and dissimilar materials. Microchimica Acta. 2006;**156**(1-2):151-157

[46] Pan TY, Santella ML. Corrosion of ultrasonic spot welded joints of magnesium to steel. Magnesium Technology. 2012;**2012**:265-269

[47] Liu L, Xu R. Investigation of corrosion behavior of Mg-steel laser-TIG hybrid lap joints. Corrosion Science. 2012;**54**:212-218

[48] Shen Z et al. Comparison of fatigue behavior in Mg/Mg similar and Mg/steel dissimilar refill friction stir spot welds. International Journal of Fatigue. 2016;**92**:78-86

[49] Sakkinen D. Magnesium alloy metallurgy. Advanced Materials & Processes. 1995;**5**:26-27

[50] Zhang M-X, Kelly P. Edge-to-edge matching and its applications: Part I. Application to the simple HCP/BCC system. Acta Materialia. 2005;**53**(4):1073-1084

[51] Mishin Y, Asta M, Li J. Atomistic modeling of interfaces and their impact on microstructure and properties. Acta Materialia. 2010;**58**(4):1117-1151

[52] A.A. Nayeb-Hashemi, JBC., and L.J. Swartzendruber, Binary alloy phase diagrams. 1722, ASM International, : Materials Park, OH

[53] Mao H, Bell P. Equations of state of MgO and ε Fe under static pressure conditions. Journal of Geophysical Research: Solid Earth. 1979;**84**(B9):4533-4536

[54] Baqer YM et al. Challenges and advances in laser welding of dissimilar light alloys: Al/Mg, Al/Ti, and Mg/Ti alloys. The International Journal of Advanced Manufacturing Technology. 2018;**95**:4353-4369

[55] Kobayashi K, Nishimoto K, Ikeuchi K. Introduction to Joining Engineering of Materials. Tokyo: Sampo; 2002. pp. 182-188

[56] Chen Y, Nakata K. Effect of surface states of steel on microstructure and mechanical properties of lap joints of magnesium alloy and steel by friction stir welding. Science and Technology of Welding and Joining. 2010;**15**(4):293-298*Bacillus*

[57] Thomas W, et al. Great Britain Patent Application No. 9125978.8. 1991

[58] Li Y et al. A review: Effect of friction stir welding on microstructure and mechanical properties of magnesium alloys. Metals. 2017;7(12):1-14

[59] Chen Y, Nakata K. Friction stir lap joining aluminum and magnesium alloys. Scripta Materialia. 2008;**58**(6):433-436

[60] Abe Y et al. Dissimilar metal joining of magnesium alloy to steel by FSW. In: Advanced Materials Research. Washington, DC: Trans Tech Publications.; 2007

[61] Watanabe T, Kagiya K, Yanagisawa A. Solid state welding of steel and magnesium alloy using a rotating pin. Quarterly Journal of the Japan Welding Society. 2006;**24**:108-115

[62] Chen YC, Nakata K. Effect of tool geometry on microstructure and mechanical properties of friction stir lap welded magnesium alloy and steel. Materials & Design. 2009;**30**(9):3913-3919

[63] Schneider C et al. Characterisation of interface of steel/magnesium FSW. Science and Technology of Welding and Joining. 2011;**16**(1):100-107

[64] Wei Y et al. Microstructures and mechanical properties of magnesium alloy and stainless steel weld-joint made by friction stir lap welding. Materials & Design. 2012;**33**(Supplement C):111-114

[65] Joo S. Joining of dissimilar AZ31B magnesium alloy and SS400 mild steel by hybrid gas tungsten arc friction stir welding. Metals and Materials International. 2013;**19**(6):1251

[66] Chen Y et al. Microstructures and properties of Mg alloy/DP600 steel dissimilar refill friction stir spot welds. Science and Technology of Welding and Joining. 2015;**20**(6):494-501

[67] Zhang Z-k et al. Friction stir keyholeless spot welding of AZ31 Mg alloy-mild steel. Transactions of Nonferrous Metals Society of China. 2014;**24**(6):1709-1716

[68] Uematsu Y et al. Comparative study of fatigue behaviour in dissimilar Al alloy/steel and Mg alloy/steel friction stir spot welds fabricated by scroll grooved tool without probe. Science and Technology of Welding and Joining. 2012;**17**(5):348-356

[69] Schilling C, dos Santos J. Method and device for joining at least two adjoining work pieces by friction welding. Google Patents; 2004

[70] Venukumar S, Yalagi S, Muthukumaran S. Comparison of microstructure and mechanical properties of conventional and refilled friction stir spot welds in AA 6061-T6 using filler plate. Transactions of Nonferrous Metals Society of China. 2013;**23**(10):2833-2842

[71] Tier M et al. The influence of weld microstructure on mechanical properties of alclad AA2024-T3 friction spot welded. SAE Technical Paper. 2008

[72] Ren D, Liu L. Interface microstructure and mechanical properties of arc spot welding Mg–steel dissimilar joint with Cu interlayer. Materials & Design. 2014;**59**(Supplement C):369-376

[73] Patel V, Bhole S, Chen D. Formation of zinc interlayer texture during dissimilar ultrasonic spot welding of magnesium and high strength low alloy steel. Materials & Design. 2013;**45**:236-240

[74] Santella M et al. Ultrasonic spot welding of AZ31B to galvanized mild steel. SAE International Journal of Materials and Manufacturing. 2010;**3**:652-657

[75] Santella M et al. Details of Mg–Zn reactions in AZ31 to galvanised mild steel ultrasonic spot welds. Science and Technology of Welding and Joining. 2012;**17**(3):219-224

[76] Patel V, Bhole S, Chen D. Characterization of ultrasonic spot welded joints of Mg-to-galvanized and ungalvanized steel with a tin interlayer. Journal of Materials Processing Technology. 2014;**214**(4):811-817

[77] Pan DT-Y et al. Ultrasonic Spot Welding of AZ31B to Galvanized Mild Steel. Oak Ridge, TN: Oak Ridge National Laboratory (ORNL); 2010

[78] Yuan X-j et al. Microstructural characteristics of joint region during diffusion-brazing of magnesium alloy and stainless steel using pure copper interlayer. Transactions of Nonferrous Metals Society of China. 2013;**23**(3):599-604

[79] Elthalabawy WM, Khan TI. Eutectic bonding of austenitic stainless steel 316L to magnesium alloy AZ31 using copper interlayer. The International Journal of Advanced Manufacturing Technology. 2011;**55**(1):235-241for biomedical applications. Biodegradable Systems in Tissue Engineering and RegenerativeMedicine. CRC Press/Taylor & Francis Group/LLC, Boca Raton, FL. 301-324

[80] Elthalabawy W, Khan T. Liquid phase bonding of 316L stainless steel to AZ31 magnesium alloy. Journal of Materials Science & Technology. 2011;**27**(1):22-28Current Chemistry. 200. Springer-Verlag, Heidelberg, Germany, 95-126

[81] Araki T et al. Reactive transient liquid phase bonding between AZ31 magnesium alloy and low carbon steel. Materials Transactions. 2011;**52**(3):568-571

[82] Tachibana T et al. Effects of zinc insert and Al content in Mg alloy on the bondability in dissimilar joints of steel and magnesium alloys. Quarterly Journal of the Japan Welding Society. 2009;**27**(2):183s-186s

[83] Khan WEaTI. Diffusion bonding of austenitic stainless steel 316L to a magnesium alloy. Key Engineering Materials. 2010;**442**:26-33

[84] Liu L et al. Microstructure and fatigue properties of Mg-to-steel dissimilar resistance spot welds. Materials & Design. 2013;**45**:336-342

[85] Manladan S et al. Microstructure and mechanical properties of resistance spot welded in welding-brazing mode and resistance element welded magnesium alloy/austenitic stainless steel joints. Journal of Materials Processing Technology. 2017;**250**:45-54

[86] Feng Y et al. Resistance spot welding of Mg to electro-galvanized steel with

hot-dip galvanized steel interlayer. Journal of Materials Processing Technology. 2016;**236**:114-122

[87] Tan C et al. Interfacial microstructure and fracture behavior of laser welded–brazed Mg alloys to Zn-coated steel. The International Journal of Advanced Manufacturing Technology. 2013;**68**(5-8):1179-1188

[88] Li L et al. Influence of Zn coating on interfacial reactions and mechanical properties during laser welding-brazing of Mg to steel. Metallurgical and Materials Transactions A. 2012;**43**(12):4740-4754

[89] Tan C et al. Laser-tungsten inert gas hybrid welding of dissimilar metals AZ31B Mg alloys to Zn coated steel. Materials & Design. 2013;**49**(Supplement C):766-773

[90] Tan C et al. Influence of laser power on the microstructure and mechanical properties of a laser welded-brazed Mg Alloy/Ni-coated steel dissimilar joint. Journal of Materials Engineering and Performance. 2017;**26**:2983-2997

[91] Li L et al. CO2 laser welding–brazing characteristics of dissimilar metals AZ31B Mg alloy to Zn coated dual phase steel with Mg based filler. Journal of Materials Processing Technology. 2013;**213**(3):361-375

[92] Li L et al. Comparative study on microstructure and mechanical properties of laser welded–brazed Mg/ mild steel and Mg/stainless steel joints. Materials & Design. 2013;**43**:59-65J of Mole. Cata. B: Enzymatic

[93] Liu L, Shan C. Study on laser-tungsten inert gas hybrid welding of dissimilar Mg alloy and steel with Ni as interlayer. Materialwissenschaft und Werkstofftechnik. 2009;**40**(10):780-783

[94] Liu L, Qi X, Wu Z. Microstructural characteristics of lap joint between magnesium alloy and mild steel with and without the addition of Sn element. Materials Letters. 2010;**64**(1):89-92

[95] Liu LM, Qi XD, Zhang ZD. The effect of alloying elements on the shear strength of the lap joint of AZ31B magnesium alloy to Q235 steel by hybrid laser-TIG welding technique. Metallurgical and Materials Transactions A—Physical Metallurgy and Materials Science. 2012;**43a**(6):1976-1988

[96] Song G, Shan C. Microstructures of laser–tungsten inert gas hybrid welding of dissimilar AZ31 Mg alloy and Q235 steel with Ni as interlayer. Materials Research Innovations. 2011;**15**(2):145-149

[97] Song G, An G, Liu L. Effect of gradient thermal distribution on butt joining of magnesium alloy to steel with Cu–Zn alloy interlayer by hybrid laser–tungsten inert gas welding. Materials & Design. 2012;**35**(Supplement C):323-329

[98] Zeng Z et al. Numerical and experiment analysis of residual stress on magnesium alloy and steel butt joint by hybrid laser-TIG welding. Computational Materials Science. 2011;**50**(5):1763-1769

[99] Cao R et al. Feasibility of cold-metal-transfer welding magnesium AZ31 to galvanized mild steel. Welding Journal. 2013;**92**(9):274s-282s

[100] Sun Y. Effects of Cu addition on microstructure characteristics and tensile behaviors of metal Inert-gas arc welded Mg-Steel dissimilar joints. Materials Transactions. 2015;**56**(11):1868-1874

[101] Cao R et al. Effects of zinc coating on magnesium alloy–steel joints produced by cold metal transfer method. Materials Science and Technology. 2016;**32**(18):1805-1817

[102] Kang M, Kim C, Kim Y-M. Joining of AZ31 magnesium alloy and steel sheet under four different coating conditions based on gas metal arc weld-brazing. Materials Transactions. 2017;**58**(1):95-102

[103] Lang B et al. Effects of welding parameters on microstructure and mechanical properties of resistance spot welded magnesium alloy joints. Science and Technology of Welding and Joining. 2008;**13**(8):698-704

[104] Min D, Yong Z, Jie L. Dissimilar spot welding joints of AZ31-443 ferritic stainless steel with cover plate. The International Journal of Advanced Manufacturing Technology. 2016;**85**(5-8):1539-1545

[105] Senkara J, Zhang H. Resistance Welding Fundamentals and Applications. London: CRC Press Taylor & Francis Group; 2006

[106] Chang B, Shi Y, Dong S. Studies on a computational model and the stress field characteristics of weld-bonded joints for a car body steel sheet. Journal of Materials Processing Technology. 2000;**100**(1):171-178

[107] Santos I et al. Weld bonding of stainless steel. International Journal of Machine Tools and Manufacture. 2004;**44**(14):1431-1439

[108] Auwal S et al. A review on laser beam welding of copper alloys. The International Journal of Advanced Manufacturing Technology. 2018;**96**(1-4):475-490

[109] Duley W. Heat Transfer and Modelling in Laser Welding. New York: Wiley-Interscience, Inc; 1999. pp. 67-113

[110] Steen W. Laser material processing—An overview. Journal of Optics A: Pure and Applied Optics. 2003;**5**(4):S3

[111] Peyre P et al. Generation of aluminium–steel joints with laser-induced reactive wetting. Materials Science and Engineering: A. 2007;**444**(1):327-338

[112] Kreimeyer M, Wagner F, Vollertsen F. Laser processing of aluminum–titanium-tailored blanks. Optics and Lasers in Engineering. 2005;**43**(9):1021-1035

[113] Nasiri A, Weckman D, Zhou Y. Interfacial microstructure of diode laser brazed AZ31B magnesium to steel sheet using a nickel interlayer. Welding Journal. 2013;**92**(1):1-10

[114] Tan C et al. Microstructure and mechanical properties of laser welded-brazed Mg/Ti joints with AZ91 Mg based filler. Materials & Design. 2016;**99**:127-134

[115] Auwal ST et al. Comparative study on characteristics of laser welded-brazed AZ31/Ti-6Al-4V lap joints with and without coatings. The International Journal of Advanced Manufacturing Technology. 2019;**101**:1023-1040

[116] Auwal S et al. Effect of copper-nickel interlayer thickness on laser welding-brazing of Mg/Ti alloy. Optics & Laser Technology. 2019;**115**:149-159

[117] Chen Y, Chen S, Li L. Effects of heat input on microstructure and mechanical property of Al/Ti joints by rectangular spot laser welding-brazing method. The International Journal of Advanced Manufacturing Technology. 2009;**44**(3):265-272

[118] Dharmendra C et al. Study on laser welding–brazing of zinc coated steel to aluminum alloy with a zinc based filler. Materials Science and Engineering: A. 2011;**528**(3):1497-1503

[119] Li L et al. Effect of groove shape on laser welding-brazing Al to steel. Journal

of Materials Processing Technology. 2018;**252**:573-581

[120] Tan C et al. Influence of Al Interlayer Thickness on Laser Welding of Mg/Steel. Welding Journal. 2016;**95**:384s-394s

[121] Windmann M et al. Laser beam welding of magnesium to coated high-strength steel 22MnB5. International Journal of Advanced Manufacturing Technology. 2016;**87**(9-12):3149-3156

[122] Miao YG et al. Microstructure and interface characteristics of laser penetration brazed magnesium alloy and steel. Science and Technology of Welding and Joining. 2010;**15**(2):97-103

[123] Miao Y et al. Effect of laser offsets on joint performance of laser penetration brazing for magnesium alloy and steel. Materials & Design. 2010;**31**(6):3121-3126

[124] Miao Y et al. Phase constitution in the interfacial region of laser penetration brazed magnesium–steel joints. Materials Characterization. 2014;**93**:87-93

[125] Tan C et al. Microstructural characteristics and mechanical properties of fiber laser welded-brazed Mg alloy-stainless steel joint. Welding Journal. 2014;**93**(10):399-409

[126] Tan C et al. Laser welding-brazing of Mg to stainless steel: Joining characteristics, interfacial microstructure, and mechanical properties. The International Journal of Advanced Manufacturing Technology. 2016;**86**(1-4):203-213

[127] Cao X et al. A review of laser welding techniques for magnesium alloys. Journal of Materials Processing Technology. 2006;**171**(2):188-204

[128] Zhang Y et al. A novel "sandwich" method for observation of the keyhole in deep penetration laser welding. Optics and Lasers in Engineering. 2008;**46**(2):133-139

[129] Gao M et al. High power fiber laser arc hybrid welding of AZ31B magnesium alloy. Materials & Design. 2012;**42**:46-54

[130] Gao M et al. Study of laser MIG hybrid welded AZ31 magnesium alloy. Science and Technology of Welding and Joining. 2009;**14**(4):274-281

[131] Yao ZX et al. Analysis about the jointing status for dissimilar metals of steel with magnesium. In: Applied Mechanics and Materials. Washington, DC: Trans Tech Publications; 2012

[132] Liu L, Xu R. Investigation of the corrosion behaviour of laser-TIG hybrid welded Mg alloys. Corrosion Science. 2010;**52**(9):3078-3085

[133] Wang XY, Sun DQ, Sun Y. Influence of Cu-interlayer thickness on microstructures and mechanical properties of MIG-welded Mg-steel joints. Journal of Materials Engineering and Performance. 2016;**25**(3):910-920

[134] Zhang Y, Jiang M, Lu W. Double electrodes improve GMAW heat input control. Welding Journal. 2004;**83**(11):39-41

[135] Miao YG et al. Effect of heat input on microstructure and mechanical properties of joints made by bypass-current MIG welding-brazing of magnesium alloy to galvanized steel. Acta Metallurgica Sinica-English Letters. 2014;**27**(6):1038-1045

Chapter 16

Quality and Fatigue Assessment of Welded Railway Bridge Components by Testing

Janusz Hołowaty and Bernard Wichtowski

Abstract

During a decades-long program from 1953 to 1990, the quality of welded joints in railway bridges in Poland was assessed and quantified. It was discovered that many welded joints have technological cracks, and their quality is poor, especially in old constructions. Nearly, 200 bridges were tested using X-ray examination. The number of joints tested was over 15,000; cracks were discovered in 400 welded joints in the 34 bridges tested. To solve the problem, repeated examinations on welded joints with imperfections were undertaken and laboratory fatigue tests were performed. The tests and numerical analysis allowed fatigue behavior and tensile stresses in welded butt splices with cover plates to be recognized and excluded such a structural solution in bridges. The existing discontinuities and imperfections in welded joints following many years in service show no growths or forming of new cracks, as the applied stresses are below the threshold fatigue strength. As a result of decades of service, steel bridges undergo functional aging, and their structural steels undergo structural aging. There is a need to both harmonize differentiated procedures and create national recommendations to assess their safe endurance. Therefore, of use may be the findings presented in the chapter.

Keywords: welded bridges, reliability, welded joints, fatigue, NDT, imperfections

1. Introduction

Bridges are engineering structures which are subjected to dynamic actions, variable in time—repeated millions of times, and inconsistently. The endurance of steel bridges in service is determined mainly by fatigue, which usually causes catastrophic failures, and corrosion, which ordinarily results in degradation failures [1–3]. Many early welded bridges in Europe were fractured over the period 1925–1936 and later [4]. Many failures and catastrophic events happened due to fatigue and fracture despite the pioneering works of August Wöhler and other scientists [5, 6]. As a result, the limit state analysis of structures with a probabilistic approach became necessary.

IntechOpen

Fractures led to a prohibition on using welding for early high-strength steels [7, 8]. As a result, welding was allowed only on mild steels. It was not until the late 1950s that some fatigue requirements were introduced into the design of steel railway bridges. It took another decade for these to be considered for road bridges; this started the modern approach to fatigue.

Each welded structure possesses discontinuity, or cracks and other imperfections resulting from manufacture or welding. These do not show any growth or instability while the loading rate is relatively low, below threshold fatigue strength. This means that such a stress range does not cause crack propagation. The endurance of steel structures or fatigue details with imperfections may be determined by testing small specimens or numerical analysis.

There are still many welded steel bridges in service which were manufactured in periods of poor quality of both materials and welded joints as well as few requirements for fatigue. After many years in service, each steel structure also undergoes what is known as functional aging, and its structural steel is subjected to structural aging. The mechanism of the aging process may be described by the classical separation theory or the newer dislocation blocking theory.

The quantitative results of welded butt splice quality obtained over 37 years of the testing of welded railway bridges in Poland are given. The radiographic examinations were conducted by the Steel Construction Chair at the West Pomeranian University of Technology in Szczecin [3]. The radiographic tests on the welded joints were supplemented by laboratory fatigue tests and more recently by Finite Element Method (FEM) analysis.

2. Fatigue-induced structural changes in steels

It is assumed that the basic factor describing the properties of a material is the changes in internal material structure resulting from the structural degradation processes described among other things in [3, 9–11]. This is mainly concerned with the decreasing value of the impact strength, sometimes even by several times. The comparison level for such a phenomenon is the difference in material properties of actual steel versus normalized, as the other has material properties from the time of the structure's construction. The simulation of these properties is carried out by thermal annealing. For this purpose, specimens are annealed at a temperature of 930°C (steels of $C \leq 0.26\%$) for an hour and then cooled in air. This way the minimal possible grain size in the steel is achieved.

This process increases yield strength and at the same time lowers the ductile-brittle transition temperature, i.e. significantly increases mechanical properties (**Figure 1**). Sometimes, astonishing results are obtained. For example, from the railway bridge over the Warta River in Gorzów Wielkopolski (western Poland), two types of steel specimens were tested for Charpy impact energy.

The bridge was constructed for the German Railways in 1938, using German normalized mild steel St37-12 (**Figure 1**). The tests refer to specimens which were

- naturally aged S without any measures;
- normalized N, i.e. annealed at 930°C for an hour and then cooled in air.

A significant aging effect was found in the structural steel after 77 years in service. At −20°C, the impact energy was 19.2 times higher.

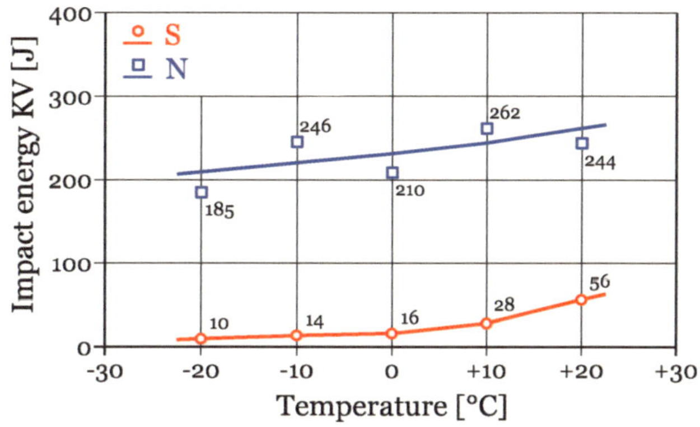

Figure 1.
Impact energy KV(T) for naturally aged (S) and normalized (N) specimens from a plate girder railway bridge constructed in 1938.

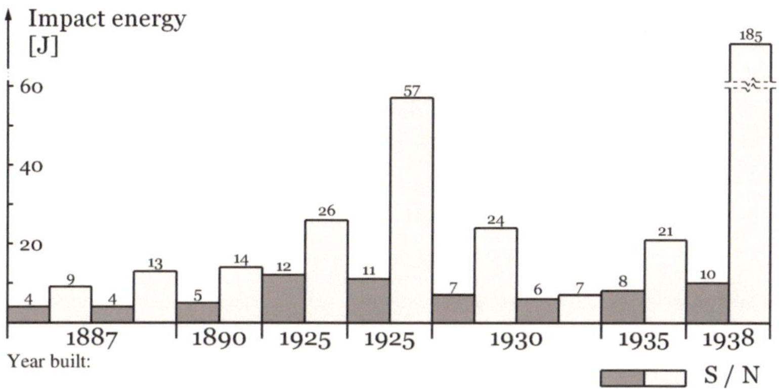

Figure 2.
Notch toughness of tested bridge steels at temperature −20°C for naturally aged (S) and normalized (N) specimens.

The results of impact energy tests at temperature −20°C for nine steel grades from eight bridges constructed in the years 1887–1938 are shown in **Figure 2**. Two types of specimens were tested: naturally aged and normalized. The steel in post-service conditions showed a very small KV impact energy value.

The actual ascertained KV values are only from 4 to 12 J. This dependence concerns all the steels tested independently of carbon content from 0.016% to 0.258%.

Such a condition shows brittleness in the material; this is a particular danger when it is located in areas of stress concentrations, for example around welding imperfections (WIs) in a weld – **Figure 3**. Welding imperfections (WIs) are crack initiators when the loads reach a prescribed critical value. The largest concentration of normal stresses σ_x is caused by ellipsoidal welding imperfections and longitudinal ones with elliptical cross sections. For these two groups of welding imperfections, the maximum stress gradient increases as the curvature radius value of the sharpest shape of welding imperfections lowers.

For example, for welding imperfections with shape characteristic $t/\rho = 100$, the shape coefficient values $u = y/t = 1$ are 13.63 and 21.00 **Figure 3**. In the case of globular welding imperfections of a small stress concentration—class III with a sharp shape—it

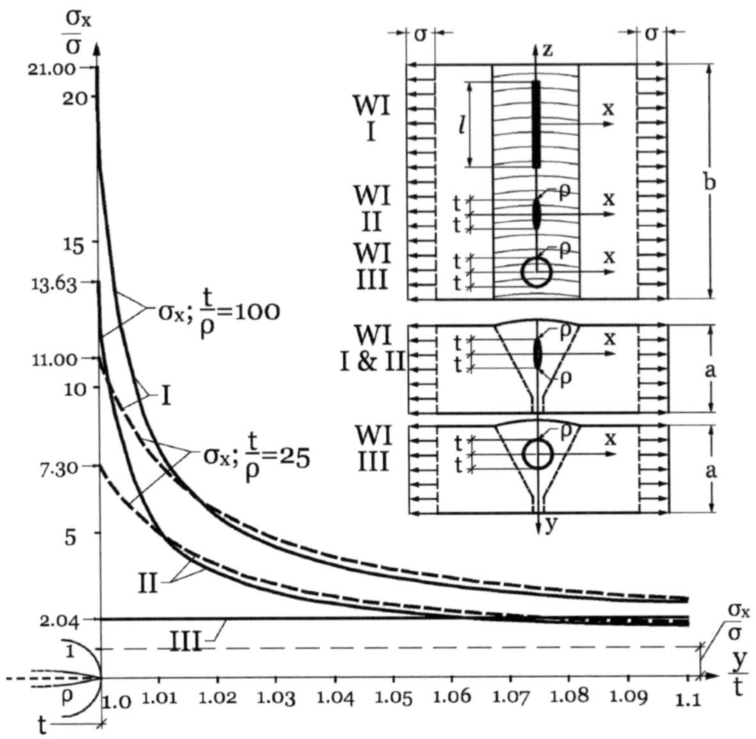

Figure 3.
Stress distribution σ_x near welding imperfections: longitudinal – I, ellipsoidal – II, and globular – III [3, 12].

is independent from the imperfection size and is $\sigma_x = 2.04{\cdot}\sigma$ [3, 12], where σ is the design stress.

Tensile stresses near a notch may be the effect of external loading, residual stresses, or both simultaneously [13]. It is not possible to exclude brittle fracture even if there is no external loading.

According to the literature [14], the fatigue strength at 10^5 cycles to strength at $2{\cdot}10^6$ cycles ratio is in the range 1.44–2.45, with the average value 1.85. Similar values of 1.75 and 1.73 were obtained for Polish mild steel (St3SX: C = 0.160%, Mn = 0.498%) [15] – **Figure 4**. Tested were as follows:

- naturally aged specimens S, for 35 years (f_y = 260 MPa, f_u = 405 MPa),
- artificially overaged specimens NN (f_y = 495 MPa, f_u = 515 MPa).

Specimens NN were cold-deformed up to 10% relative elongation and then heated up to temperature 250°C, kept at this temperature for an hour, and then cooled in air.

The value of the infinitive fatigue strength Z_{rj} = 145.3 MPa of naturally aged steel for 35 years is 55.9% of its yield strength f_y = 260 MPa and for the overaged steel this value is Z_{rj} = 157.3 MPa which constitutes only 31.8% of its yield strength. As a result of additional aging (specimens NN), there was an unexpected very large increase in steel yield strength f_y of 90.4%, and ultimate strength f_u increased by 27.1%.

The specific character of bridge loadings and the structural changes in steels due to aging are the main reasons why a general hypothesis for their fatigue estimation has yet

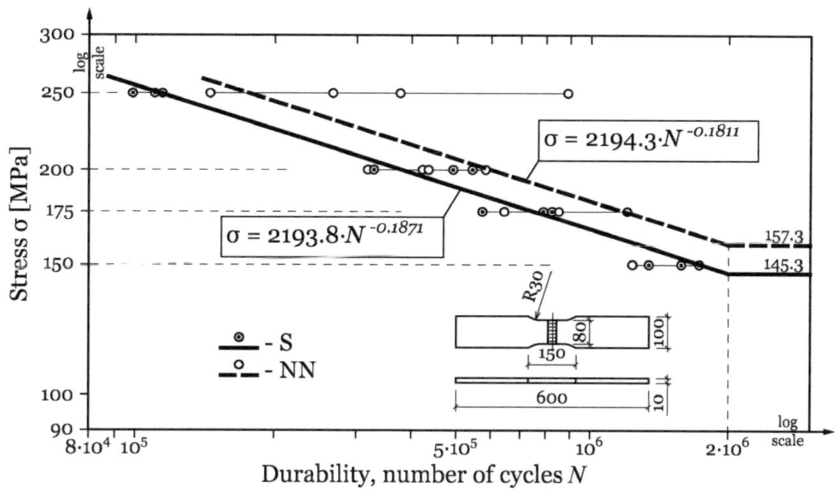

Figure 4.
Regression straights obtained from fatigue tests on: naturally aged specimens (S) and additionally aged specimens (NN) [15].

to be proposed [16]. Phenomenological models are still used despite the significant number of studies undertaken, especially for riveted structures, and more excellent research tools for testing have appeared as well as the possibility of numerical analysis. A problem has arisen as to how to adapt the information from laboratory fatigue tests to the design for the durability of structures in service. Knowledge of both the loading spectrum which a structure will be carrying and problems with butt splices have become necessary.

Welded structures under high stresses are damaged mainly by fatigue crack growth or brittleness. Hence, fracture mechanics has recognized the most important issues:

- discovering and locating flat welding imperfections in a welded joint,
- determining flat welding imperfection dimensions precisely, especially in the joint depth.

Over the years 1970–1980, the International Welding Institute introduced the "fitness for purpose" criterion, which relies on the formulation of fracture mechanics calculations for determining the permissible size of welding imperfections, thereby confirming the required quality and durability of structures according to standards and technical requirements. The determination of specific quality levels and example calculations as well as the determination of safety coefficients are given in [17, 18].

3. Quality of welds in railway bridges according to Non-Destructive Testing (NDT)

The aging of metallic materials favors the formation of brittle cracks as the ductile-brittle transition temperature clearly approaches higher values. Strains associated with aging depend on the location in the structure. They are particularly dangerous in stress concentration areas, e.g. around welding imperfections (WIs) and in heat-affected zones (HAZs). According to Neuber [3, 12], the largest concentration of stresses is associated with ellipsoidal and longitudinal welding imperfections with an elliptical cross section (**Figure 3**).These types of imperfections include cracks and lack of fusion, as well as band slag intrusion and incomplete side fusion. This issue is becoming particularly significant for bridges constructed after 1936, considering the poor quality of connecting welds. This has been confirmed by radiographic tests performed on bridges in service on Polish railway lines [3].

The Steel Structure Chair at the Technical University in Szczecin (now WPUT) carried out radiological tests on butt splices in the steel girders of around 200 railway bridges on the Polish railway network. For 154 bridges, including 124 plate girder bridges and 30 truss bridges, the exact time of construction was also established.

The range and results of in situ nondestructive radiological tests on the railway bridges are given in two histograms (**Figures** 5 and **6**). The tests consist of the sum of 5-year intervals which take into account the bridge construction period from 1936 to 1975. **Figure** 5, in the upper part of the histogram, shows the number of bridges tested and the number of X-rays taken over a particular 5-year interval. At the same time, the number of internal structural cracks in connecting welds is given. These cracks were discovered on 437 X-rays. They constitute 2.8% of the total number of welds tested, equaling 15,875 units. This number includes 10,507 X-rays on butt joints in tensile components and the remaining 5368 X-rays on compress components. Simultaneously, the lower part of the histogram gives the number of welded joints tested over given 5-year intervals.

Figure 6 shows the proportional and numerical specification of hot cracks detected in bridge structures for given 5-year intervals of construction. Apart from one crack from 1974, the remaining cracks were ascertained in bridges constructed before 1960. According to Eurocode 1993-1-9 for the design of steel structures, such cracks do not

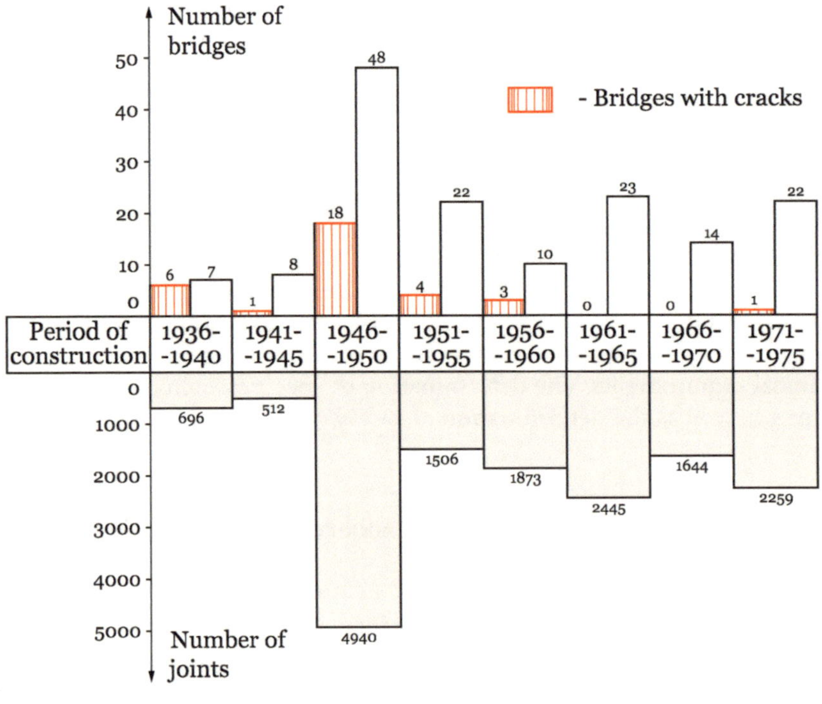

Figure 5.
Numerical specification of the bridges and welded joints tested.

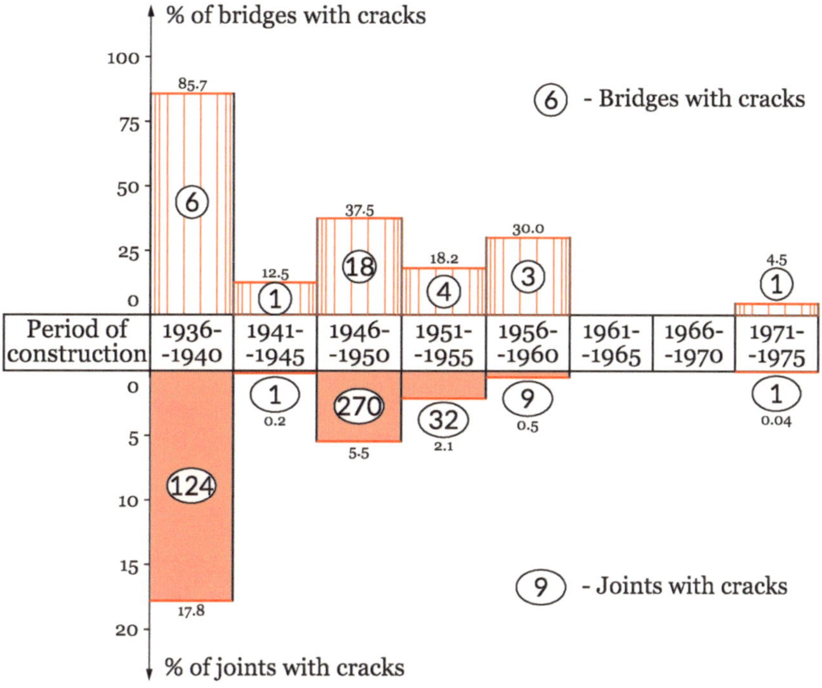

Figure 6.
Specification for bridges and joints with hot cracks.

exclude structures from service. The upper parts of the histogram (**Figure 6**) show the proportional and numerical specification of bridges with internal cracks in welded butt splices. Of note is that the majority of cracks were found in the oldest historical bridges from the years 1936–1940. For the seven bridges from that period tested, cracks were discovered in six structures, which constitute 85.7%. They were discovered on 124 welded butt splices from the general lot of 696 pieces tested.

Similar unfavorable test results were obtained during examination of the bridges constructed over the years 1946–1950. Cracks were found in 18 structures (37.5%) on 270 X-rays (5.5%).

4. Fatigue tests on butt welds

All the cracks in the butt splices of the bridge structures are internal hot cracks. They arise in the weld metal and HAZ of a joint during the crystallization process in the liquidus–solidus temperature range (**Figure 3**).

Service fatigue cracks appear in stress concentration zones caused by structural details and increase their value by a concentration factor caused by nonmetallic inclusions. Therefore, the development of a fatigue crack depends on many factors: the shape and dimensions of a structural component as well as the way and magnitude of loading. The larger the structure, the smaller the critical crack length necessary to cause the final brittle fracture [3, 18]. Development of fatigue cracking appears in stages under fluctuating loads, and their increase is caused by the weakening of a structure's strength. This is why, in the literature, the stress is put down to the influence of the imperfection's geometry and its location in the weld.

The literature gives different models for the initiation and proliferation of fatigue cracks in a nondeterministic approach. Despite many attempts to describe the fatigue mechanism, given in 64 hypotheses [19], no general hypothesis has yet to be formulated [18, 20, 21]. We are still stuck in the phenological description, despite having more and better research tools and numerical calculations. It appears that when assessing the endurance of bridges with cracks in their welded joints, it is highly useful to analyze their service behavior and the results of laboratory tests.

In Poland, the pioneer of the in situ field testing of welded butt splices on railway bridges was Professor Andrzej Fabiszewski from the Technical University of Szczecin. The procedure understood the principle that a weld is the weakest point in welded structures. The results of these tests were a great surprise to the organizers. In 34 bridges, internal technological cracks and hot cracks were ascertained in 437 welded butt splices (**Figure 6**).

To answer the question, "What do we do with theses bridges?", laboratory fatigue strength tests were carried out on three typical structural solutions which reflected the details in the early welded bridges. Specimens U, C, and P and the test results are given in **Figure 7**. The tests were carried out on 60 specimens, each time loaded at 5 loading levels. The tests are presented more precisely in [3]. They allowed, using the least-square method, fatigue class values according to EN ISO 5817: 2014 to be estimated [22, 23]. The following fatigue classes ($\Delta\sigma_C$) were obtained for individual specimens from different constructions:

- specimens U with butt welds of an acceptable quality level, fatigue class $\Delta\sigma_C$ = 125 MPa.
- specimens C with an internal crack in the butt welds, fatigue class $\Delta\sigma_C$ = 90 MPa.
- specimens P with butt welds covered by one-sided rhombic cover plates, fatigue class $\Delta\sigma_C$ = 79 MPa.

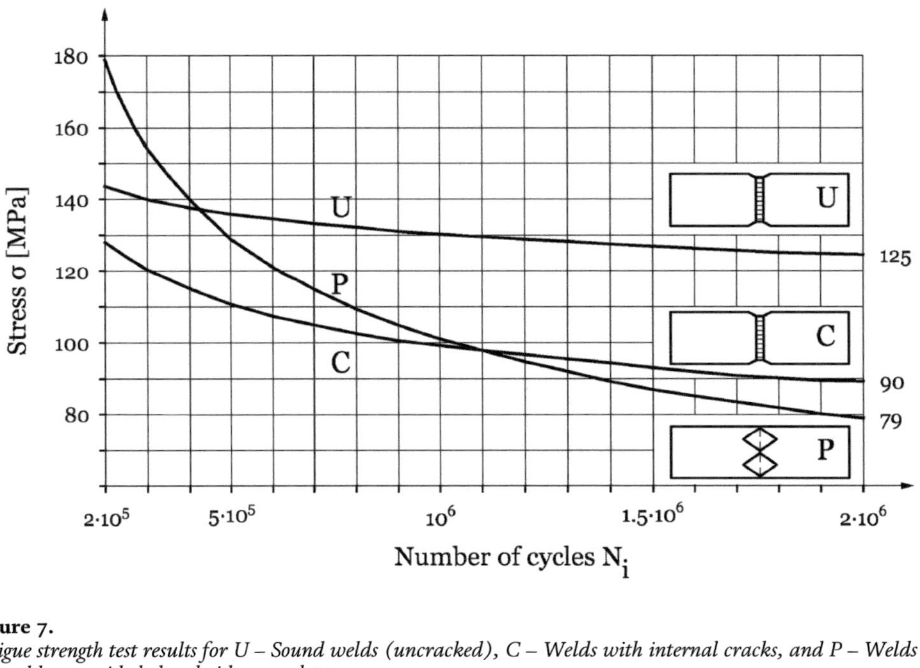

Figure 7.
Fatigue strength test results for U – Sound welds (uncracked), C – Welds with internal cracks, and P – Welds covered by one-sided rhomboid cover plates.

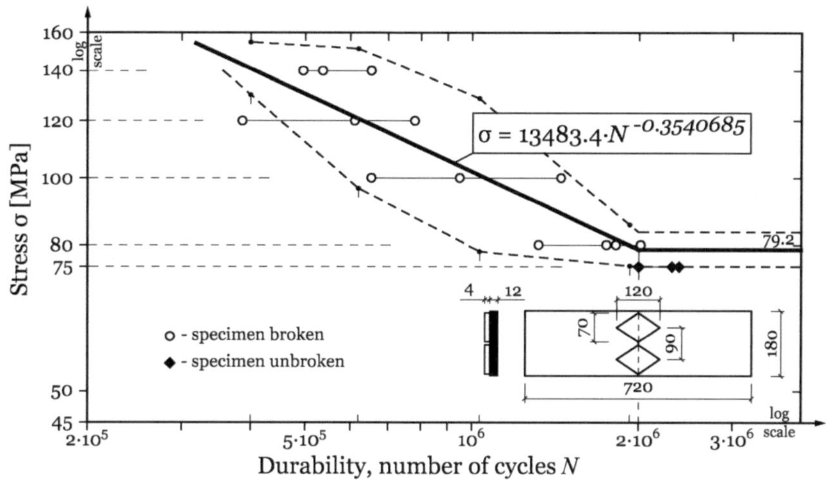

Figure 8.
Fatigue test results on 16 specimens and regression analysis.

The tests were carried out using a pulsator of frequency 5 Hz and stress ratio R = 0.1. Of note is the low fatigue class $\Delta\sigma_C$ = 79 MPa for specimens "strengthened" with rhombic cover plates. The rhombic cover plates had been intended to secure welded butt splices in early welded bridges, but the fatigue effects appeared to be quite the opposite. The results of the tests (**Figure** 7) clearly show that for the number of load cycles N_i larger than $1.1{\cdot}10^6$, the fatigue strength of the specimens with cover plates is lower than the fatigue strength of the specimens with cracked butt welds (type C).

The results of fatigue tests on 16 specimens with rhomboid cover plates give cause for reflection (**Figure 8**). Specimens with dimensions 180×12×720 mm were manufactured from Polish mild steel St3M for bridges (C = 0.19%, Mn = 0.66%) of f_y = 312 MPa and f_u = 452 MPa. The tests were carried out according to the Polish standard on fatigue tests on metals using five stress levels: 75, 80, 100, 120, and 140 MPa. The tests were performed on a pulsator with 5-Hz frequency. The first cracks appeared near the welded end of the cover plate and spread toward the specimen edges. In three specimens with stress levels 80, 100, and 140 MPa, the cracks appeared at $99{\cdot}10^3$, $168.9{\cdot}10^3$, and $20{\cdot}10^3$ cycles before total fracture. However, two specimens at stress level 80 MPa were not damaged, despite being loaded by $1819.8{\cdot}10^3$ and $836.8{\cdot}10^3$ cycles after the first cracks appeared (**Figure 9**). The test results for 13 damaged specimens allow us to work out the logarithmic regression equation.

The tests show stress concentrations by rhomboid cover plates mainly at their ends [14]. The fatigue strength value resulting from using cover plates depends on their shapes, as well as their length (**Table 1**). The lowest value is reached when the additional element is shortened up to 300 mm.

5. Stresses in welds with cracks

The fatigue behavior of metals is determined precisely in [20, 24–26]. Fatigue hypotheses, dislocation structures, fatigue cracking, and their fractures are given there. The issue of no crack growth in existing structures in service is also discussed there. Such cracks appear when the initiation crack stress is lower than the value of

Specimen No.	Cover plate ending	Fatigue strength, MPa			Literature
		10^5 cycles	$2{\cdot}10^5$ cycles	%	
1	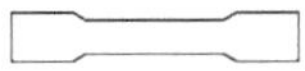	—	280	100	[24]
2		—	~90	32	
3	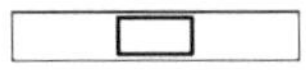	—	90–100	34	
4		182	79	28	[14]
5	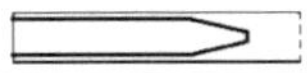	234			
6		202			
7	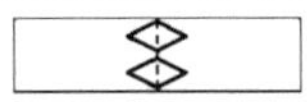	228	79	28	[3]

Table 1.
Effect of cover plates on the fatigue strength.

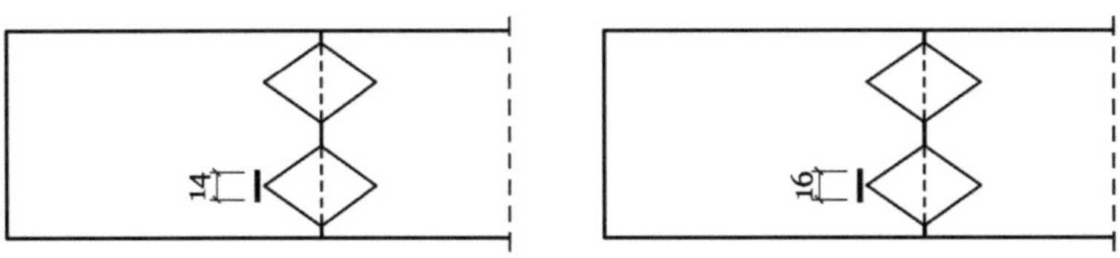

Figure 9.
Undamaged specimens after $3637{\cdot}10^3$ and $2827{\cdot}10^3$ load cycles (cracks after $1817.2{\cdot}10^3$ and $1990.2{\cdot}10^3$ load cycles).

stress necessary for crack propagation. The undamaged service of these structures and the lack of propagation cracks is because no situation has arisen during service which would lead to their appearance. Many hypotheses resulting from laboratory tests have been put forward, including the oldest tests on wagon axle models (Ø 50 mm), carried out by T.V. Buchwalter as early as 1938 [27, 28]. Generally, there is no one solution for the three-dimensional problem of fatigue fractures. However, a material experiment review laid out the directions for further research to find a more precise solution to the problem. There is the optimistic fact that as early as 1965, Kudriawcew [29] stated "structural sections in which non-propagating cracks develop may be stronger than sections constructed with notches."

The prediction of eventual fatigue cracking in welded butt splices in the railway bridges tested was assessed on the basis of strength analysis for three selected bridges. These are plate girder bridges constructed in 1938, 1938, and 1947 on different railway lines. Their technical characteristics are given in **Table 2**. This is a compilation of requirements collated in two papers [7, 16] relating to the structures.

The analysis was undertaken on the three bridges. They were constructed either side of WWII. In total, there were six examinations of the welded butt splices, and three additional tests limited to the testing of splices with cracks. The additional examinations were executed in 5–8-year time intervals. Overall, 632 X-rays were taken and 49 technological hot cracks were discovered. The calculated stress values at the butt splice locations in the lower chords of the plate girders are given in columns 5–7 in **Table 2**. Column 8 shows the results of service stresses measured on the welded joints of bridge III located on the Katowice–Tczew coal railway line. The permanent load and the electric locomotive ST-21 (live load) are taken into account. The great similarity of stresses calculated theoretically (column 7) and stresses measured "in situ" on bridge III (column 8) are of note.

Bridge No. (year built)	Number		Span length l, m	Stress σ, MPa			
	Girder (tested)	X-rays (cracks)		Dead load	Live actions		Dead + measured
					LM 71	Service	
1	2	3	4	5	6	7	8
I (1938)	6 (2)	128 (8)	22.40	a = 10/14	81/108	47/63	
				b = 12/14	94/111	54/64	
				c = 13/15	99/114	57/65	
II (1939)	6 (1)	144 (39)	20.20	a = 14/17	116/137	68/80	
				b = 15/17	122/138	74/84	
				c = 15/17	124/137	72/80	
III (1947)	12 (3)	360 (5)	21.20	a = 9/10	80/93	47/55	50/59
				b = 10/11	89/101	52/59	52/61

Table 2.
Bridges, their technical data, and normal stresses in butt welds with cracks as well as behind cover plates.

Locations of welded butt splices:

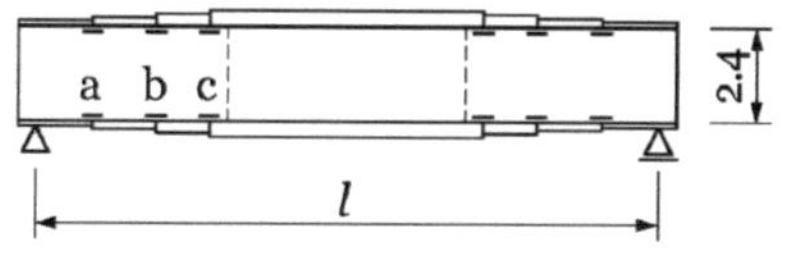

The calculations assume the creation of a national set of standards, i.e. Polish bridge standards: PN-85/S-10030 and PN-82/S-10052 for actions and steel bridges, respectively. The characteristic values of the live action effects with dynamic factor Φ are taken into account. The standard load model in the form of the contemporary Eurocode railway traffic model LM 71 for loading class k = +2 is under consideration. The stresses were determined on the butt weld and cover plate axis as well as in the flange plates just behind the ends of the rhombic plates (values in denominators). It is easy to see that the service values of stresses in column 7 are from 60 to 70% of the stress values for the standard loading (LM 71) in column 6. This means that they are also lower than the values of $Z_{rj} = \Delta\sigma_C = 79$ MPa determined according to **Figure 7**, i.e. the service stresses are lower than the limit value for such joints determined in [3].

Considering the load spectra recommended by the JRC (Joint Research Centre) for railway bridges [1], the above value of loading will be smaller. The authors conducted such analyses and described them in [2, 30].

The load spectrum given in **Figure 10** according to the old British standard BS 153 was established in structural calculations [29–31]. The method allows the service life of bridge structures to be prolonged by as much as three times. The authors recommend this method for the endurance assessment of historical bridges.

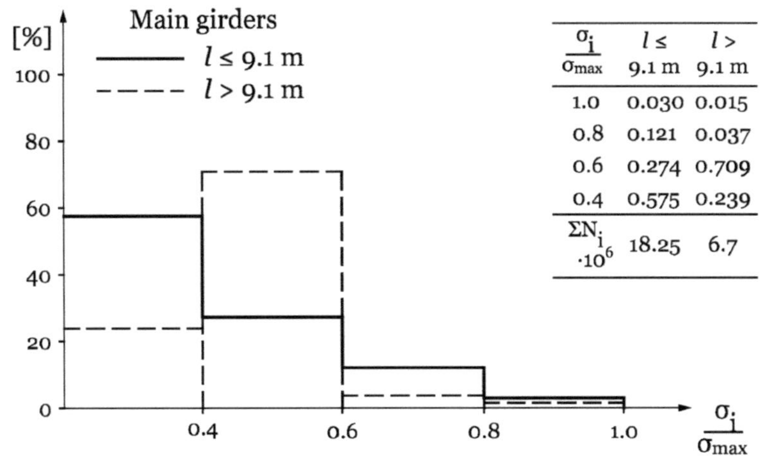

σ_i/σ_{max}	$l \le$ 9.1 m	$l >$ 9.1 m
1.0	0.030	0.015
0.8	0.121	0.037
0.6	0.274	0.709
0.4	0.575	0.239
$\Sigma N_i \cdot 10^6$	18.25	6.7

Figure 10.
Stress ratio frequency.

In some countries, the results of traffic load measurements have been published, giving the load spectra for analyzing existing bridges [32, 33]. New standards and guides for the testing and assessment of existing bridges have appeared [19, 34–37].

6. Numerical analysis of welded butt splices with cover plates in bridges

The first welded bridge designers were aware that "a weld is the weakest place in the structure." Because of welding imperfections, their resistance is lower than that of the welded material. The simplest and the most economical way to eliminate these differences seemed to be enlargement of the welded joint section by adding cover plates which compensated for the weakened section. In the welded plate girders of railway bridges constructed up to 1939 and in the period from 1945 to 1953, the butt splices of webs and flanges were covered with one- or two-sided cover plates [3].

In the 154 railway bridges which were checked radiographically, internal cracking was discovered in 438 welded joints. In this group, there were 28 plate girder structures; the constructions of their lower flanges are shown in **Figure 11**. In 18 structures, their butt splices are covered with one-sided rhomboid cover plates from the side of the girder longitudinal axis. The rhomboid cover plates are from 90 to 200 mm in width and from 160 to 340 mm long.

To assess the endurance of such types of joints, fatigue strength tests were undertaken, which were discussed in Section 4. The results of the tests and the regression line are given in **Figure 8**. The determined infinitive fatigue strength value Z_{rj} = 79 MPa at $N_i = 2{\cdot}10^6$ load cycles constitutes only 26% of yield strength f_y = 302 MPa for the steel of the specimens tested. It is worth mentioning that for three stress levels σ = 80, 100, and 140 MPa, on five specimens seven cracks appeared, as shown in **Figure 12**.

The results of the fatigue tests show a very low fatigue limit value for the welded butt splices covered with rhomboid plates. The problem was solved numerically using an FEM model as shown in **Figure 13**. More details of the numerical analysis are given in [38, 39].

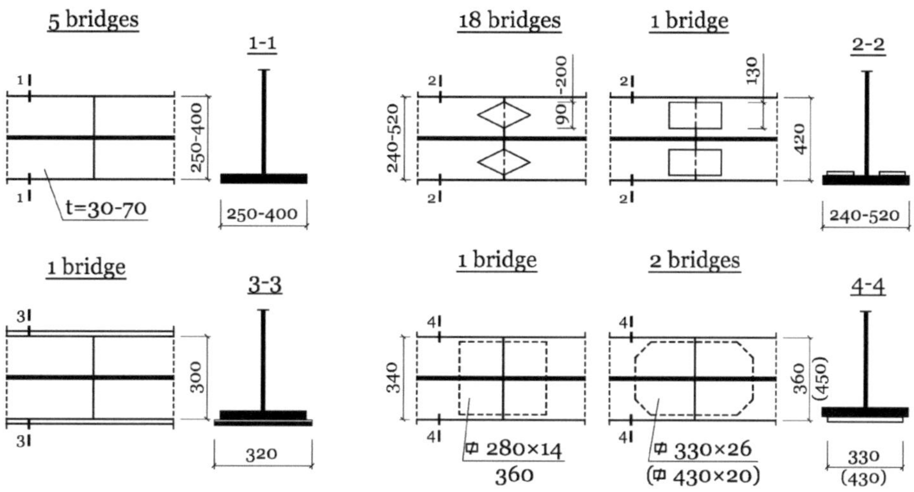

Figure 11.
Details of welded butt splices with cracks in 28 plate girder bridges.

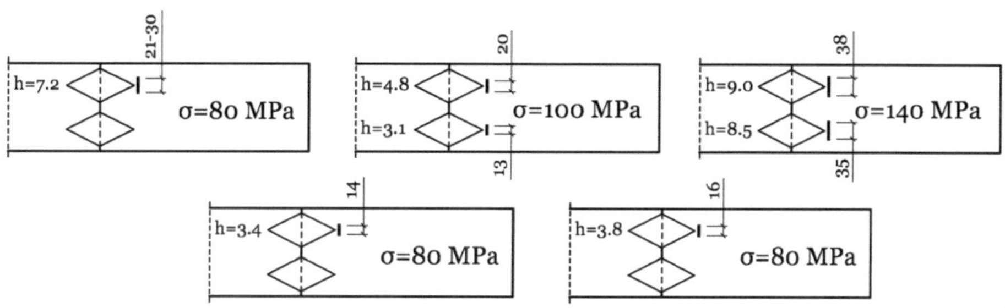

Figure 12.
Cracks in flanges with rhomboid cover plates after fatigue tests: The top three specimens – Damaged and the bottom two specimens – Undamaged.

For the numerical analysis, the welded splice was modeled using the FEM method (**Figure 14**) with Inventor Nastran software. Material parameters for structural steel are f_y = 249 MPa and f_u = 360 MPa. The stresses were calculated in four cross sections and on nine points for each section. Loading was modeled as 162, 173, 216, 260, and 303 kN tensile forces with 75, 80, 100, 120, and 140 MPa course tensile stresses in the flange. The same stress levels were formulated as for the laboratory fatigue tests.

Analysis of the tensile stresses in the welded joint allowed us to formulate some remarks:

- rhomboid cover plates do not lower the stresses in a butt weld as was initially assumed (see **Table 1**); it was expected that the resistance of the joint would be increased by 25.9%;
- cover plates appeared to be unnecessary components, causing some additional fatigue problems;
- at the ends of the cover plates, the concentration of stresses appear within a range of 1.47–1.69;

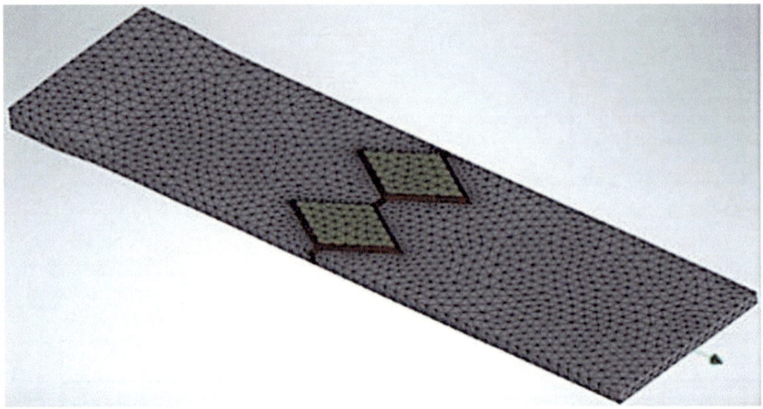

Figure 13.
Numerical model for analysis of a welded butt splice with cover plates.

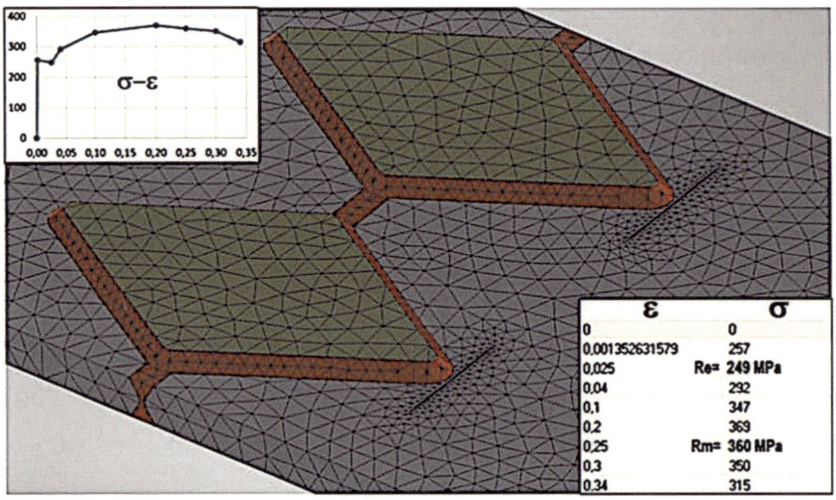

Figure 14.
Details of the numerical model with cracks and structural steel material data.

- the stress concentration together with the smallest concentration factor for globular (spherical) nonmetallic inclusions of 2.04 are the reasons for the formation of one-sided stochastic cracking already at the 80-MPa stress level (**Figure 13**); thus $\sigma = 80.1.69.2.04 = 275$ MPa which is greater than the steel yield strength $f_y = 249$ MPa [39].

Cracks appeared at three stress levels, $\sigma = 80$, 100, and 140 MPa, with a varied number of load cycles from 535,000 to 990,200.

These are fatigue cracks developing in stages, as opposed to the rapidly developing cracks in the fatigue tests of welded joints on specimens U, C, and P (**Figure 8**). All the cracks had a similar fracture as shown in **Figure 15**, with three developing trajectories: I – crack initiations, II – growth, and III – final fracture. The scheme of fatigue crack zones is shown in **Figure 16**.

The stress distribution on the circumference of the cracks is similar, with the smallest values in the upper zone. The values are equal to the upper values of yield

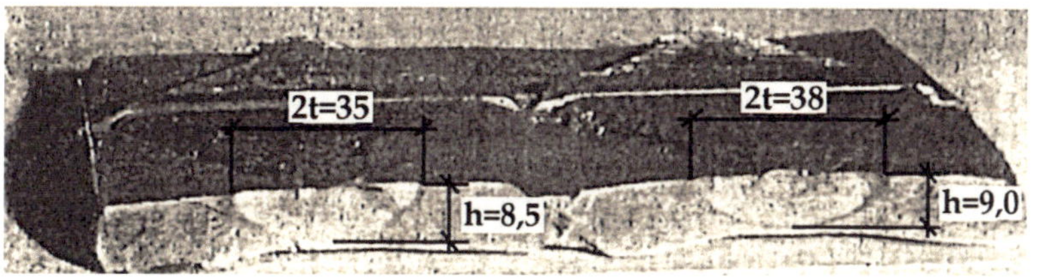

Figure 15.
*Fracture surface of a broken specimen after the fatigue testing (see **Figure 13**).*

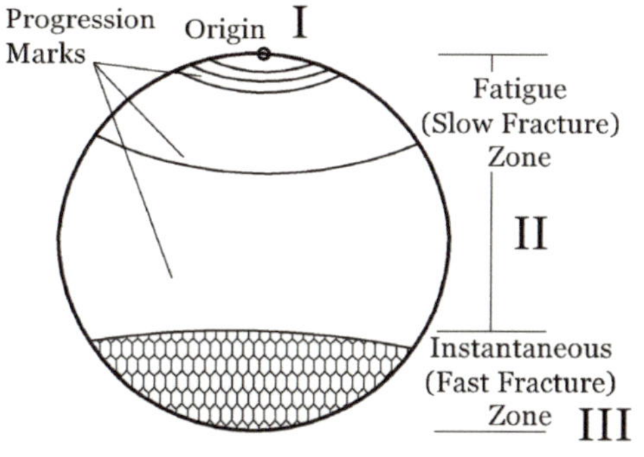

Figure 16.
Scheme of a fatigue crack: Zone I – Origin, zone II – Fatigue zone and zone III – Final fracture.

strength f_y = 280 MPa, while the maximum stress $\sigma \sim 306$ MPa appeared at the crack tip.

A study of the literature shows that no direct criterion has been established for precise cracking in zones I and II, e.g. the zones of settled crack growth, and zone III (unstable crack growth). This has not been achieved since 1913 (C. E. Inglis) despite the development of 64 growing hypotheses at the microstructure level and thousands of publications [27, 28, 40, 41]. For example, after the chapter, "Fatigue crack growth" in [27], there is a list of 469 supplementary readings. Crack growth is described there probabilistically in a way that is comprehensible only for specialists.

Considering the results of the numerical calculations of stresses in cracked joints in **Figure 17**, a new way for describing ductile fracture growth (zone II) may be suggested. The analysis takes into account two laws of physics:

- ductile fracture growth may be restrained at any moment by lowering stresses below the material's yield strength;
- the moving peaks of the upper size of the fracture with length 2 t on the top surface will first reach the elastic stress area σ = 99–100 and 100–104 MPa at stress level 100 MPa as well as σ = 172–180 and 176–180 MPa at stress level 140 MPa – see **Figure 17**. This is guaranteed by the geometry of the fracture $t/h \sim 2$ and is assured by fracture surface analysis which shows that the fracture growth is along the top surface rather than into the material.

There is a reduction in edge surface stresses on the top surface to the measured values 274–310 MPa, i.e. to the upper yield strength of the material f_{yH} = 280 MPa. The

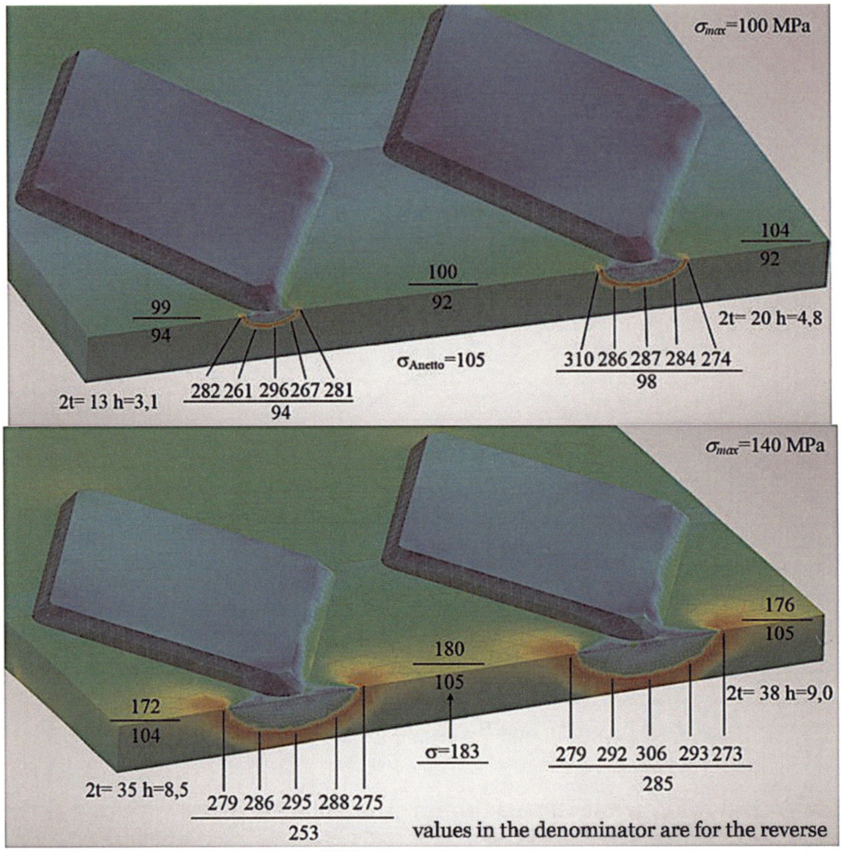

Figure 17.
Concentration of stresses at the ends of cover plates for stress levels: 100 and 140 MPa.

growth of ductile fracture disappears at the edge points on the top surface. This phenomenon evolves in the nearby "deep" points of the fracture and according to the stress equalizing rule, it gradually restrains a two-sided fracture from proceeding to the tip of the fracture. The cracking growth in zone II disappears totally.

Generally, it should be stated that no comprehensive model for a general description of fatigue fractures has yet been devised. All models described in the literature relate only to growth zones I and II. The only known model for transition from fracture zone to final fracture, zone III, was devised by A.H. Cottrel and N.J. Petch [40, 42]. The Cottrel–Petch theory describes the ductile-brittle transition properties of steel. A basis for the transition is assumed yield strength σ_{pl}.

When the yield strength is larger than the fracture growth stress, then the material is brittle and vice versa. "Brittle fracture will occur when the work of applied stress σ during fracture growth reaches the effective energy of newly formed surfaces." This means that brittle fracture will occur under stress $\sigma = \sigma_{pl}$ (**Figure 17**).

An explanation for this phenomenon in relation to the five cracks in the three damaged joints (**Figure 12**) is given in [38]. In the analysis, the results obtained during fatigue strength tests for three types of joints were used (**Figure 7**). Practically, this applies to the infinitive fatigue strength values Z_{rj} given as a function of load cycles N_i.

7. Conclusions

During their service life, bridges, like other structures, undergo functional and material degradation. In many cases, tests show that despite non-strict conformity with actual design standards, they are able to carry actual service loadings. According to [1], calculations of existing steel bridges resistance should be carried out more precisely and in a more readable manner following a three-phase assessment: a preliminary evaluation, a detailed investigation, and an expert investigation. Some other studies have recommended using five assessment levels: a bridge rating, a preliminary evaluation, a detailed investigation, an expert investigation, and advance testing [16, 35–37]. Application of the assessment procedure led to improvements in knowledge and a calibration of confidence factors.

At the same time, the project "Sustainable bridges – assessment for future traffic demands and longer lives" with nine packages, deals with the increasing capacity and service life of existing railway bridges [35]. New methods for the resistance assessment of existing bridges were developed in standards and guidelines using calibrated partial safety coefficients.

The Steel Structure Chair at the Technical University in Szczecin (now WPUT) contributed to these issues by carrying out a multiyear program for the quality-level assessment of welded butt splices in existing railway bridges in service in Poland. Radiological tests on butt splices in the steel girders of around 200 railway bridges were carried out. Nearly, 200 bridges were tested using X-ray examination over a 37-year period. The number of joints tested was over 15,000; cracks were discovered in 400 welded joints in 34 bridges tested. Repeated tests on welded joints with cracks were undertaken as well as laboratory investigations for their complex assessment. Partial results of the laboratory fatigue testing are given in the paper.

Numerical analysis of tensile stresses in welded splices was undertaken to support and confirm the results of laboratory fatigue tests. The results of the FEM analysis of tensile stresses in welded joints with rhomboid cover plates are given. The FEM method allowed us to determine tensile stresses at each service stage. The analysis allowed for:

- negative assessment of the welded joint type with cover plates – total disqualification of cover plates;
- suggesting a new hypothesis for the initiation and growth of the surface fatigue fractures;
- conformity assessment for the Cottrell–Petch theory for the transition from ductile fracture to brittle state at the final fracture.

Author details

Janusz Hołowaty* and Bernard Wichtowski
West Pomeranian University of Technology in Szczecin, Poland

*Address all correspondence to: jah@wp.pl

References

[1] Kühn B, Lukic M, Nussbaumer A, Günther H, Helmerich R, Herion S, et al. . In: Sedlacek G, Bijlaard F, Geradin M, Pinto Vieira A, Dimova S, editors. Assessment of Existing Steel Structures —Recommendations for Estimation of the Remaining Fatigue Life. EUR 23252 EN. Luxembourg (Luxembourg): Office for Official Publications of the European Communities; 2008 89 p. JRC43401

[2] Scheer J. Failed Bridges. Case Studies, Causes and Consequences. Berlin: Ernst &Sohn; 2010 307 p

[3] Wichtowski B. Fatigue Strength of Welded Butt Joints in Steel Railway Bridges. Szczecin: PS Scientific Works No. 572; 2002 203 p

[4] Barsom JM, Rolfe ST. Fracture and Fatigue Control in Structures. Applications of Fracture Mechanics. 3rd ed. Philadelphia: ASTM; 1999. 516 p. DOI: 10.1520/MNL41-3RD-EB

[5] Zenner H, Hinkelmann K. August Wöhler – Founder of fatigue strength research. On the 200th anniversary of August Wohler birth. Steel Construction. 2019;**2**:156-162. DOI: 10.1002/ stco.201900011

[6] Hołowaty J, Wichtowski B. Properties of structural steels in bridges on a railway line No. 275—The line of August Wöhler fatigue research. Inżynieriai Budownictwo. 2020;**1-2**:31-35

[7] Hołowaty J. Properties of high tensile steels in historical railway bridges. Proceedings of the Institution of Civil Engineers—Construction Materials. 2018;**171**(6):234-245. DOI: 10.1680/ jcoma.17.00012

[8] Hołowaty J. Structural steels of historical welded bridges. Welding Technology Review. 2018;**90**(5):99-102. DOI: 10.26628/ps.v90i5.907

[9] Wichtowski B, Jasiński W. Microstructural degradation processes in normalized mild steel from a railway bridge. Welding Technology Review. 2015;**87**:94-99

[10] Hołowaty J. Toughness tests on steels from old railway bridges. Procedia Structural Integrity. 2017;**5**:1043-1050. DOI: 10.1016/j.prostr.2017.07.067

[11] Wichtowski B, Hołowaty J. Degradation processes in normalized mild- and low-alloy steel building structures in service. In: F. Wald and M. Jandera, editors. Stability and Ductility of Steel Structures 2019: Proceedings of the International Colloquium on Stability and Ductility of Steel Structures (SDSS 2019). September 11–13, 2019; Prague, Czech Republic; CRC Press; 2019: 1275–1282

[12] Neuber H. Theory of Notch Stresses—Principles for Exact Stress Calculation. Ann Arbor, Michigan: Edwards Inc.; 1946 181 p

[13] Sonsino CM, Fricke W, de Bruyne F, Hoppe A, Ahmadi A, Zhangd G. Notch stress concepts for the fatigue assessment of welded joints—Background and applications. International Journal of Fatigue. 2012;**34**:2-16

[14] Gurnay TR. Fatigue of Welded Structures. Warszawa: PWN; 1976 356 p

[15] Wichtowski B. Fatigue life of the oldest welded bridges in Poland in the light of research. Welded Technology Review. 2017;**89**:75-79

[16] Pipinado A, editor. Innovative Bridge Design Handbook. Construction,

Rehabilitation and Maintenance. Amsterdam, Boston, Heidelberg, London, New York, Tokyo: Elsevier; 2016 844 p

[17] Klimpel A. Control and Quality Assurance in Welding. Gliwice: WPŚ; 1998 266 p

[18] Rykaluk K. Cracks in Steel Structures. Wrocław: DWE; 1999 232 p

[19] Bakhat B, Humar J, Jalali J, Mufti A, Newhook J, Rahman S, et al. Guidelines for Structural Health Monitoring. Design Manual No. 2. ISIS Canada. 2001. 125 p

[20] Fisher JW. Fatigue and Fracture in Steel Bridges. Case Studies. New York: John Wiley & Sons; 1984 315 p

[21] Fisher JS, Mertz DR. Hundreds of bridges—Thousands of cracks. Civil Engineering (ASCE). 1985;**55**:64-67

[22] Di Battista JD, Adamson DEJ, Kulak GL. Evaluation of remaining fatigue life for riveted truss bridges. Canadian Journal of Civil Engineering. 1998;**15**(4): 678-691

[23] Wichtowski B, Wichtowski M. Fatigue strength of butt welds with cover plates according Eurocode 3. Welding Technology Review. 2011;**83** (3):36-43

[24] Stüssi F. Entwurf und Berechnung von Stahlbauten. Berlin–Heidelberg: Springer–Verlag; 1958 577 p

[25] Caiza P, Ummenhofer T, Correia J, De Jesus A. Applying the Weibull and Stüssi methods that derive reliable Wöhler curves to historical German bridges. Practice Periodical on Structural Design. 2020;**25**(4) 9 p

[26] Kik T, Slovaček M, Moravec J. Hot cracking of welded joints—Tests and numerical simulations. Welded Technology Review. 2012;**23**(4):125-130

[27] Kacańda S. Fatigue Cracking of Metals. Warszawa: WNT; 1985 491 p

[28] Kacanda S. Fatigue Failure of Metals. Netherlands: Springer; 2011 382 p

[29] Czudek H, Pietraszek T. Endurance of Steel Bridge Structures under Variable Loads. Warszawa: WKŁ; 1980 140 p

[30] Wiśniewski DF, Casas JR, Ghosn. Codes for safety assessment of existing bridges—Current state and further development. Structural Engineering International. 2012;**4**:552-561. DOI: 10.2749/101686612X13363929517857

[31] British Standard BS 153: Specification for Steel Girder Bridges. Part 3A. Loads. BSI; 1972 39 p

[32] Herzog M. Realistischer Betriebsfestigkeitsnachweisfürstählerne. Eisenbahnbrücken Der Stahlbau. 1974; **10**:316-318

[33] Pelikan W. Eine Betrachtungüber die Größe der Betriebslasten von Eisenbahn – und Straßenbrücken und ihre Auswirkung auf die Bemessungdieser Bauwerke. Die Bauingenieur. 1968;**6**:207-214

[34] Helmerich R, Kühn B, Nussbaumer A. Assessment of existing steel structures. A guideline for estimation of the remaining fatigue life. Structure and Infrastructure Engineering. 2007;**1**: 245-255

[35] Guideline for load and resistance assessment of existing European railway bridges: Advices on the use of advanced methods. Sustainable Bridges– Assessment for Future Traffic Demands and Longer Lives. SB-LRA. Technical

Report. Skanska Sverige Ab, COWI A/S. 2007, 428 p

[36] Al-Amrani M, Aygül M. Fatigue design of steel and composite bridges. In: Raport. Chalmers University of Technology; 2014 156 p

[37] Lukić M, Al-Amrani M, Aygül M, Bokesjö M, Urushadze S, Frýba L, et al. Bridge fatigue guidance—Meeting sustainable design and assessment (BRiFaG). In: Final Report EUR 25866 EN. Luxembourg: Publication Office of the European Union; 2013 158 p

[38] Wichtowski B, Konecki K. Fatigue strength of steel plate girder railway bridges with butt joints reinforced with one-sided rhomb-shaped cover plates. Welded Technology Review. 2021;**93**(3): 31-40. DOI: 10.26628/wtr.v93i3.1139

[39] Wichtowski B, Konecki K. The finite element analysis of fatigue cracks in a welded butt splice with one-sided rhomboid cover plates. Inżynieriai Budownictwo. 2021;**3**:107-113

[40] Wyrzykowski JW, Pleszakow E, Sieniawski J. Deformations and Cracking of Metals. Warszawa: WNT; 1999 491 p

[41] Neville W, Sachs PE. Understanding the surface features of fatigue fractures: How they describe the failure cause and the failure history. Journal of Failure Analysis and Prevention. 2005;**5**(4): 11-15

[42] Amstrong A. 60 years of hall-Petch: Past to present nano-scale connections. Materials Transactions. 2014;**55**:2-12